BIOTECHNOLOGICAL ADVANCES IN PROCESSING MUNICIPAL WASTES FOR FUELS AND CHEMICALS

Biotechnological Advances

in Processing

Municipal Wastes

for Fuels and Chemicals

Edited by

A.A. Antonopoulos

Energy and Environmental Systems Division
Argonne National Laboratory
Argonne, Illinois

NOYES DATA CORPORATION
Park Ridge, New Jersey, U.S.A.
1987

Copyright © 1987 by Noyes Data Corporation
Library of Congress Catalog Card Number 86-31144
ISBN: 0-8155-1122-1
Printed in the United States

Published in the United States of America by
Noyes Data Corporation
Mill Road, Park Ridge, New Jersey 07656

10 9 8 7 6 5 4 3 2 1

Library of Congress Cataloging-in-Publication Data

Biotechnological advances in processing municipal
    wastes for fuels and chemicals.

    Proceedings of the Symposium on Biotechnological
Advances in Processing Municipal Wastes for Fuels and
Chemicals held in Minneapolis, Minn., on Aug. 15–17,
1984, and sponsored by the U.S. Department of Energy's
Division of Energy from Municipal Waste and by Argonne
National Laboratory's Energy and Environmental Systems
Division (Energy from Municipal Waste Program).
    Includes bibliographies and index.
    1. Refuse as fuel--Congresses. 2. Refuse and refuse
disposal--Congresses. 3. Biotechnology--Congresses.
I. Antonopoulos, Antonios A. II. Symposium on Biotechno-
logical Advances in Processing Municipal Wastes for
Fuels and Chemicals (1st : 1984 : Minneapolis, Minn.)
III. United States. Dept. of Energy. Energy from
Municipal Waste Division. IV. Argonne National Lab-
oratory. Energy and Environmental Systems Division.
TD796.2.B56 1987   628.4'458   86-31144
ISBN 0-8155-1122-1

Foreword

Research and development efforts in the area of municipal-waste bioconversion to fuels and other valuable chemicals has generated a wealth of information. This book presents an amalgam of this knowledge in a form which should be helpful to researchers, policy makers, and entrepreneurs.

The book is based on a symposium held in Minneapolis, Minnesota in late 1984. The symposium presentations were prepared by a forum of experts in various related fields and covered many research areas, with primary emphasis on understanding the physiology, biochemistry, and genetics of microbes involved in anaerobic digestion; the various microecological interactions; the biological production of fuels and chemicals; and the generation and extraction of landfill gas. The authors appraised the current status of their fields and offered some research perspectives. The information presented, it is hoped, will be used to stimulate further improvement of municipal-waste bioconversion technology. The use of biotechnology in processing waste could be an important tool for environmental engineers.

The information in the book is from *Biotechnological Advances in Processing Municipal Wastes for Fuels and Chemicals,* edited by A.A. Antonopoulos of Argonne National Laboratory for the U.S. Department of Energy, issued December 1985.

The table of contents is organized in such a way as to serve as a subject index and provides easy access to the information contained in the book.

Preface

The Symposium on Biotechnological Advances in Processing Municipal Wastes for Fuels and Chemicals—sponsored by the U.S. Department of Energy's Division of Energy from Municipal Waste and by Argonne National Laboratory's Energy and Environmental Systems Division (Energy from Municipal Waste Program)—was held in Minneapolis, Minnesota, on August 15-17, 1984. At that time, recent research and development efforts in the area of municipal-waste bioconversion to fuels and other valuable chemicals had generated a wealth of information, and a forum of experts was needed to amalgamate this knowledge and present it in a form that could be helpful to researchers, policy makers, and entrepreneuers.

The symposium presentations covered many research areas, with primary emphasis on understanding the physiology, biochemistry, and genetics of microbes involved in anaerobic digestion; the various microecological interactions; the biological production of fuels and chemicals; and the generation and extraction of landfill gas. The authors appraised the current status of their fields and offered some research perspectives. It is hoped that the information presented will be used by those concerned to stimulate further improvement of municipal-waste bioconversion technology.

The symposium committee and its chairman offer their sincere thanks to the symposium speakers, poster-presenters, and all participants. Special appreciation is extended to Donald Walter of the U.S. Department of Energy (DOE) and Harvey Drucker of Argonne National Laboratory (ANL) for their support and encouragement, and to the keynote speaker Robert Hungate (University of California at Davis) for an inspiring presentation. Special recognition is extended to session chairmen: J. Gregory Zeikus (Michigan State University), John Pfeffer (University of Illinois, Urbana), Robert Mah (University of California, Los Angeles), John Rees (BioTechnica Ltd., Cardiff, U.K.), Nicholas Lailas (DOE), and James Lazar (ANL), as well as to session cochairmen: Jean Bogner (ANL), Karel Grohmann (Solar Energy Research Institute), Richard Speece (Drexel University), Ross McKinney (University of Kansas), Edward Wene (ANL), and Michael Torpy (ANL). Thanks are also extended to Kathy Macal and Floyd Bennett for their editorial assistance.

A.A. Antonopoulos
Symposium Chairman and Editor of the Proceedings

Contributors

G. Albagnac
Station de Technologie Alimentaire
Institut National de la Recherche
  Agronomique
BP 39-59651 Villeneuve d'Ascq
  Cedex, France

Jean-Paul Aubert
Unité de Physiologie Cellulaire,
Département de Biochimie et
  Génétique Moléculaire
Institut Pasteur
28 rue du Dr. Roux
75724 Paris Cedex 15, France

P.S. Beaty
Dept. of Botany and Microbiology
University of Oklahoma
Norman, OK 73019

Negash Belay
Dept. of Microbiology
University of Iowa
Iowa City, IA 52242

Lakshmi Bhatnagar
Unité de Physiologie Cellulaire
Département de Biochimie et
  Génétique Moléculaire
Institut Pasteur
28 rue du Dr. Roux
75724 Paris Cedex 15, France

David R. Boone
Environmental and Occupational
  Health Sciences
School of Public Health
University of California
Los Angeles, CA 90024

F.M. Bordeaux
Dept. of Microbiology and Cell Science
University of Florida
Gainesville, FL 32611

E.C. Clausen
Dept. of Chemical Engineering
University of Arkansas
Fayetteville, AR 72701

M. Coccia
Rosenstiel Basic Medical Sciences
  Research Center
Brandeis University
Waltham, MA 02254

Daniel R. Coleman
Southern Research Institute
Birmingham, AL 35255

Lacy Daniels
Dept. of Microbiology
University of Iowa
Iowa City, IA 52242

Kenneth E. Daugherty
Dept. of Chemistry
North Texas State University
Denton, TX 76203

D.V. Dervartanian
Dept. of Biochemistry
University of Georgia
Athens, GA 30602

L.F. Diaz
Cal Recovery Systems, Inc.
Richmond, CA 94804

H.C. Dubourguier
Station de Technologie Alimentaire
Institut National de la Recherche
  Agronomique
BP 39-59651 Villeneuve d'Ascq
  Cedex, France

Michael H. Eley
University of Alabama in Hunstsville
Huntsville, AL 35807

G. Fauque
Association pour la Recherche en
  Bioenergie Solaire
E.C.E., CEN-Cadarache, B.P.N° 1
13115-Saint Paul Lez Durance, France

Daniel Feldman
Laboratory of Renewable Resources
  Engineering
Purdue University
West Lafayette, IN 47907

J.L. Gaddy
Dept. of Chemical Engineering
University of Arkansas
Fayetteville, AR 72701

S. Ghosh
Institute of Gas Technology
Chicago, IL 60616

C.G. Golueke
Cal Recovery Systems, Inc.
Richmond, CA 94804

K. Grohmann
Biotechnology Branch
Solar Fuels Division
Solar Energy Research Institute
Golden, CO 80401

H.O. Halvorson
Rosentiel Basic Medical Sciences
  Research Center
Brandeis University
Waltham, MA 02254

Thomas D. Hayes
Gas Research Institute
Chicago, IL 60631

Marc Henriquet
Unité de Physiologie Cellulaire
Départment de Biochimie et
  Génétique Moléculaire
Institut Pasteur
28 rue du Dr. Roux
75724 Paris Cedex 15, France

J. Michael Henson
Dept. of Microbiology and Cell Science
University of Florida
Gainesville, FL 32611

M. Himmel
Biotechnology Branch
Solar Fuels Division
Solar Energy Research Institute
Golden, CO 80401

Nancy W.Y. Ho
Laboratory of Renewable Resources
  Engineering
Purdue University
West Lafayette, IN 47907

Robert E. Hungate
Dept. of Bacteriology
University of California
Davis, CA 95616

H. Ronald Isaacson
Gas Research Institute
Chicago, IL 60631

M.K. Jain
Dept. of Bacteriology
University of Wisconsin-Madison
Madison, WI 53706

Richard L. Jenkins
Getty Synthetic Fuels, Inc.
P.O. Box 1900
Long Beach, CA 90801

Melvin V. Kilgore, Jr.
Southern Research Institute
Birmingham, AL 35255

J.A. Krzycki
Dept. of Bacteriology
University of Wisconsin
Madison, WI 53705

Thomas J. Laughlin
Southern Research Institute
Birmingham, AL 35255

J. Legall
Association pour la Recherche en
  Bioenergie Solaire
E.C.E., CEN-Cadarache, B.P.N° 1
13155-Saint Paul Lez Durance, France

P.A. Lespinat
Association pour la Recherche en
  Bioenergie Solaire
E.C.E., CEN-Cadarache, B.P.N° 1
13115-Saint Paul Lez Durance, France

A.R. Lino
Centro de Quimica Estrutural
Complexo I
1000 Lisbon, Portugal

Chris L. Lishawa
Southern Research Institute
Birmingham, AL 35255

Rocco L. Mancinelli
Washburn University
Topeka, KS 66621

NASA-Ames Research Center
Moffett Field, CA 94035

Perry L. McCarty
Dept. of Civil Engineering
Stanford University
Stanford, CA 94305

M.J. McInerney
Dept. of Botany and Microbiology
University of Oklahoma
Norman, OK 73019

Christopher P. McKay
NASA-Ames Research Center
Moffett Field, CA 94035

Ross E. McKinney
Dept. of Chemical Engineering
University of Kansas
Lawrence, KS 66045

William E. Meyers
Southern Research Institute
Birmingham, AL 35255

R. Moletta
Station de Technologie Alimentaire
Institut National de la Recherche
  Agronomique
BP 39-59651 Villeneuve d'Ascq
  Cedex, France

Christina J. Morris
Dept. of Microbiology
Ohio State University
Columbus, OH 43210

I. Moura
Centro de Quimica Estrutural
Complexo I
1000 Lisbon, Portugal

J.J.G. Moura
Centro de Quimica Estrutural
Complexo I
1000 Lisbon, Portugal

Dominique Pariot
Unité de Physiologie Cellulaire
Département de Biochimie et
  Génétique Moléculaire
Institut Pasteur
28 rue du Dr. Roux
75724 Paris Cedex 15, France

Gene F. Parkin
Drexel University
Philadelphia, PA 19104

H.D. Peck, Jr.
Dept. of Biochemistry
University of Georgia
Athens, GA 30602

John A. Pettus
Getty Synthetic Fuels, Inc.
P.O. Box 1900
Long Beach, CA 90801

John T. Pfeffer
Dept. of Civil Engineering
University of Illinois
Urbana, IL 61801

J.F. Rees
BioTechnica Limited
Cardiff CF4 5DL, United Kingdom

John N. Reeve
Dept. of Microbiology
Ohio State University
Columbus, OH 43210

C. Rivard
Biotechnology Branch
Solar Fuels Division
Solar Energy Research Institute
Golden, CO 80401

M.F. Roberts
Dept. of Chemistry
Massachusetts Institute of Technology
Cambridge, MA 02139

E. Samain
Station de Technologie Alimentaire
Institut National de la Recherche
  Agronomique
BP 39-59651 Villeneuve d'Ascq
  Cedex, France

G.M. Savage
Cal Recovery Systems, Inc.
Richmond, CA 94804

Peter Schönheit
Fachbereich Biologie, Mikrobiologie
Phillipps Universtität
D-3550 Marburg, Federal Republic
  of Germany

Lionel Sibold
Unité de Physiologie Cellulaire
Département de Biochimie et
  Génétique Moléculaire
Institut Pasteur
28 rue du Dr. Roux
75724 Paris Cedex 15, France

Reza Shamskorzani
Dept. of Chemical Engineering
University of Kansas
Lawrence, KS 66045

Daniel Smith
Dept. of Civil Engineering
Stanford University
Stanford, CA 94305

Paul H. Smith
Dept. of Microbiology and Cell Science
University of Florida
Gainesville, FL 32611

Richard Sparling
Dept. of Microbiology
University of Iowa
Iowa City, IA 52242

Richard E. Speece
Drexel University
Philadelphia, PA 19104

N. Suresh
Rosentiel Basic Medical Sciences
  Research Center
Brandeis University
Waltham, MA 02254

M. Teixeira
Centro de Quimica Estrutural
Complexo I
1000 Lisbon, Portugal

Rudolf K. Thauer
Fachbereich Biologie, Mikrobiologie
Phillipps Universität
D-3550 Marburg, Federal Republic
  of Germany

M. Timmerman
Air Products and Chemicals, Inc.
Allentown, PA 18105

J.P. Touzel
Station de Technologie Alimentaire
Institut National de la Recherche
  Agronomique
BP 39, 59651 Villeneuve d'Ascq
  Cedex, France

George T. Tsao
Laboratory of Renewable Resources
  Engineering
Purdue University
West Lafayette, IN 47907

Robert E. Van Heuit
EMCON Associates
San Jose, CA 95112

Barney J. Venables
TRAC Laboratories, Inc.
113 Cedar Street
Denton, TX 76201

R. Warburg
Rosentiel Basic Medical Sciences
  Research Center
Brandeis University
Waltham, MA 02254

J. Wells
Rosentiel Basic Medical Sciences
  Research Center
Brandeis University
Waltham, MA 02254

N.G. Wofford
Department of Botany and Microbiology
University of Oklahoma
Norman, OK 73019

V. Kenneth Wright
Cyclic Energies, Inc.
Denton, TX 76201

A.V. Xavier
Centro de Quimica Estrutural
Complexo I
1000 Lisbon, Portugal

J.G. Zeikus
Dept. of Bacteriology
University of Wisconsin-Madison
Madison, WI 53706

Michigan Biotechnology Institute
Michigan State University
East Lansing, MI 48824

NOTICE

This report was prepared as an account of work sponsored by the U.S. Department of Energy. Neither the United States Government nor any agency thereof, nor any of their employees, nor the Publisher, makes any warranty, express or implied, or assumes any legal liability or responsibility for the accuracy, completeness, or usefulness of any information, apparatus, product, or process disclosed, or represents that its use would not infringe privately owned rights. Reference herein to any specific commercial product, process, or service by trade name, trademark, manufacturer, or otherwise, does not necessarily constitute or imply its endorsement, recommendation, or favoring by the United States Government, or any agency thereof, or the Publisher. The views and opinions of authors expressed herein do not necessarily state or reflect those of the United States Government, or any agency thereof, or the Publisher.

Final determination of the suitability of any information, procedure, or product for use contemplated by any user, and the manner of that use, is the sole responsibility of the user. The book is intended for informational purposes only. Expert advice should be obtained at all times when implementation is being considered.

Contents

DEVELOPMENT OF IDEAS ON THE NATURE AND AGENTS OF BIOMETHANOGENESIS

Robert E. Hungate

Department of Bacteriology
University of California
Davis, California

My task this evening is to review for you some of the early developments leading to this First Symposium on Bio-technological Advances in Processing Municipal Wastes for Fuels and Chemicals. This review must be an anecdotal account because what are ultimately grand insights into broad biological truths do not come suddenly, full-fledged from a super mind. Rather, they arise in fragmentary fashion, with each investigator voicing his interpretation of his fragment, with the same fragment discovered repeatedly in other contexts by other investigators, and with correct interpretations some-times discovered early but not appreciated and accepted until much later. This review is anecdotal also because it contains many personal recollections.

So far as we know the first random activity dealing with methane production in the western world was that of Father Charles Joseph Campi who reported to Alexander Volta that the gas arising as a stream of bubbles in a pond near St. Colomban, Italy, was combustible. On 14 November 1767, Volta wrote a letter to the Journal of Milan, reporting in part as follows *translated from Italian and published in French at Strasbourg University in 1778, the French into English 15 August 1984): "When you wrote me the first time about the source of the combustible air you discovered at the beginning of the last autumn and which subsequently we have discussed together, you will recall how many explanations and conjectures we have indulged in on a matter so interesting and marvelous as the different kinds of air and above all the air you have found near the beautiful hills of St. Colomban. .... and if it is a great advantage for your source that the air arises spontaneously in great abundance and continuously, in the other sources the gas bubbles equally when one stirs up the bottom." The excitement and interest engendered in these first discoverers has persisted, and almost all scientists studying the problem have been imbued with the idea of exploiting biomethanogenesis to human benefit.

It spurred Magendie in 1816 to garner the bodies of just executed Parisien criminals and prove that the gas collected from their colon contained methane whereas that from the stomach and intestine did not. Although modern studies show that only dihydrogen accumulates in the colon of some individuals, there is no evidence that colonic methane correlates with criminal tendencies.

In 1863 Reiset demonstrated that methane was an important component of the gas in the ruminant stomach, and Tappeiner (1882) found methane in the hindgut of most domestic animals. It is also found in termites (Breznak, 1982). The methane in traces in the surface waters of the oceans is presumably produced in the alimentary tracts of marine animals. Methane in coal mines has caused many deaths, and its abundance as natural gas has made it a prime source of energy in industrial and other nations.

This abundance of methane is explained by the fact that under anaerobic conditions the carbon compounds methane and carbon dioxide have a lower potential energy state than any other forms of carbon (Symons and Buswell, 1933). These two molecules accumulate in habitats accessible to bacteria but devoid of oxygen, nitrate and sulfate. Since Earth was originally anaerobic, methanogenesis has been assumed to be a very primitive metabolic activity (Hungate, 1955), with carbohydrate selected as an important metabolite because the exactly intermediate oxidation state of its carbon permits a maximum amount of anerobic metabolic work through conversion to methane and carbon dioxide. The classic studies of Woese and collaborators supply impressive evidence of the primitiveness of the methanogenic bacteria, and it is not surprising that numerous species, fitted to the diversity of modern anaerobic Earth habitats, have evolved. The numbers in which new methanogens are being described support the idea that one of the avenues of research to improve the processing of wastes will be the detection of additional strains in common (Paynter and Hungate, 1968) as well as bizarre and exotic habitats.

But let us now turn to the problems encountered during domestication of methanogenic bacteria. Bechamp (1868) was apparently the first to ascribe methanogenesis to microorganisms, but Pasteur explained the putrefaction in the bodies of dead animals as due to bacteria and was presumably aware of the findings of Magendie and Reiset.

The first experimental work on methanogenesis was that of Leo Popoff in 1873, showing that cellulose added to sewage sludge gave rise to methane. This was followed by similar experiments of Hoppe-Seyler (1876) and Tappeiner (1882). In most of these early studies the bodies of plants were used as substrate. It must have been inferred that since plants were the materials undergoing decomposition in marshes and pond sediments, they or components such as cellulose should be the substrate in enrichment attempts. This assumption was fortunate; the rate of breakdown of the plant cell walls is often sufficiently slow that the methanogens can keep pace with the bacteria attacking the wall components. However, Hoppe-Seyler in 1887 demonstrated methane production also from acetate. He even postulated that the methane arose from the methyl group of acetate but thought that the energy yield was not sufficient to account for the microbial growth.

The first serious attempt to identify and culture the bacteria producing methane were by Omeliansky (1904), the only student of the great microbiologist, Winogradsky. Omelianski started work on the assumption that the bacteria attacking cellulose were the ones producing methane, but showed also that subcultures in which the inoculum was pasteurized produced hydrogen instead of methane. He carried stocks producing these respective products, and felt that he could detect microscopic differences in the kinds of cells in the two cultures, but his photomicrographs are not convincing.

Mazé (1903) was much more successful. Roux had given him a methane-producing culture of leaves, and in subcultures he could detect what he termed a pseudosarcina; the planes of division were not at the 90-degree angles characteristic of a true sarcina. Subcultures into acetate and butyrate enrichments containing a little beef extract showed even greater preponderance of the pseudosarcina, and on inoculating into an agar dilution series he obtained among others discrete sarcina-like colonies. These alone were unable to grow when subcultured, but if two accompanying sporeforming non-methanogenic strains were included, methane was formed. Also with acetone as substrate Mazé obtained methane production in crude cultures, recovering one CO_2 and two methanes, but concluded that other reactions provided the energy for growth because, according to his information the combustion of the two methanes yielded more energy than did combustion of the acetone used. Actually there is a free energy decrement of about 20 kilojoules.

Hoppe-Seyler and Mazé may have tested acetate and butyrate because already by 1832 (Sprengel) acetate and butyrate were known to be produced in the rumen, and since methane also was a known product, it was logical to inquire whether it came from the fermentation acids.

Over the years the two-stage hypothesis for the methane fermentation has been proposed many times. Mazé (1903) stated (broad translation from the French): "It is in old cultures that the pseudosarcina is best seen. The bacteria which accompany it are there of course, but they ferment the initial substrate, producing as gas only carbonic acid and hydrogen, whereas in the cultures producing methane hydrogen is always absent. Analysis of the H_2 cultures discloses acetate and butyrate. Thus the methane fermentation is at the expense of these products, including also H_2."

At this point we should mention the first of a series of Dutch investigators of methanogenesis. Holland (Low Countries, Nederland, Pay-Bas) had a great interest in methane. The flooded nature of most of it makes anaerobic conditions almost universal. Methane continuously bubbles to the surface of canals, and in some parts of Holland has been exploited to provide energy for light and heat (Schnellen, 1947). Thus it is understandable, first, that a doctoral thesis on the microbiology of methanogenesis should have been carried out in Holland; second, that it should have been carried out at Delft, the city producing Leeuwenhoek, the discoverer of bacteria; third, that it should have been the first doctoral thesis (1906) to be defended at Delft; and, fourth, that it should have been at the Technische Hoogeschool where Beijerinck was at the time a world renowned microbiologist, Söhngen was the candidate and on completion of his doctorate he became professor at Wageningen.

Söhngen thoroughly reviewed the literature, and taking up in the direction pursued by Mazé he proved that hydrogen was used in the methane fermentation and that actually a methanogenic fermentation could be supported with only dihydrogen and carbon dioxide as substrates. But in addition he obtained methane from enrichments of fatty acids with even numbers of carbon atoms up to capric acid, and also obtained methane from sugar, starch and cellulose. The acids from these and also from pectin he regarded as the actual substrate for the bacteria producing methane, in accord with the 2-stage concept. At Wageningen Söhngen continued his interest in methane, and one of his students (Coolhaas, 1927) demonstrated that thermophilic cultures formed methane from acetate, also from formate, isobutyrate, oxalate, lactate and gluconate, and

after a longer latent period fermented propionate and butyrate.
Additionally, Groenewege (1920) found that mixtures of fatty
acids and primary alcohols were fermented to methane.

But not all of the new knowledge came from detailed
laboratory experiments. Buswell at Illinois examined an
extremely large number of substrates (Tarvin and Buswell, 1934),
including benzoate and other ring compounds, using inoculum
from an anerobic digestor at Urbana, Illinois. Buswell's
activities spanned more than 30 years, and I suspect that his
studies on anaerobic digestion may have been in part responsible
for the supplementation in Palo Alto in the 1930s of the municipal gas supply with methane from the municipal sewage plant.

During this period Stephenson and Stickland (1933)
obtained methane in broth dilutions of a crude culture given
hexamethylenetetramine as a source of formate, but could not
obtain growth in agar media. It was believed that a pure culture
had been obtained by picking a single cell to start a broth
culture which could use H_2/CO_2, formate, formaldehyde and
methanol but which also reduced sulfate to H_2S, a reaction not
substantiated thus far in a proven pure culture. Addition of
H_2S to their medium allowed them to dispense with the
sterilized exhausted culture fluid at first required to obtain
growth in subcultures.

In the meantime the mechanisms of the biological oxidation
of organic matter had been of great interest. Warburg with
his superb knowledge and techniques was able to elucidate
some of the catalysts and carriers concerned with the reactions
involving dioxygen, but Wieland envisioned most oxidations as
consisting of dehydrogenations followed by hydrations. The
Wieland scheme for oxidation was applied to microbial fermentations by Kluyver and Konker (1926) in their classic paper
on "The Unity of Biochemistry," and these ideas came to
Stanford with C.B. Van Niel. He had worked with Kluyver at
Delft and as conservator of the microbial culture collection
while working for his doctorate had acquired first hand
acquaintance with a great variety of microbes and with their
metabolic reactions. He was brought to the Hopkins Marine
Station by Baas-Becking, and joined the faculty in 1929.

Lewis Thayer, working at the Hopkins Marine Station on
the biogenesis of petroleum (1931), advised me to take van
Niel's microbiology course which I did in the summer of 1931
as his first student. H. A. Barker, a classmate and friend
at Stanford, took the course the following year, and we
constituted a nucleus for a small spontaneous group meeting
with van Niel when he came to the campus from Pacific Grove.

I well remember reporting on the Kluyver and Konker paper showing the universality of the dehydrogenations. Van Niel
postulated that methane was formed by reduction of CO_2 with
hydrogen from H_2 or organic sources, thus accounting for the
fact that methane was the only reduced carbon compound formed.
After Barker completed his Ph.D. in chemistry, working with
the physical chemist McBain, he obtained a National Research
Council Fellowship to take postdoctoral work at Hopkins
Marine Station with van Niel, and undertook to obtain evidence
bearing on the reduction of CO_2 as the mechanism for methane
formation. The problem was approached by postulating that
ethanol, known to be fermentable to methane (V.L. Omeliansky,
1916; Groenwege, 1920), was converted to acetate, with the
thereby derived hydrogen reducing CO_2 to methane. Cultures
were set up and an enrichment obtained in which methane formed
and acetate was produced. Barker obtained a fellowship at
Delft where he extended and refined his cultures and identified
*Methanosarcina methanica, Methanosarcina mazéi, Methano-
bacterium Söhngenii* and *Methanobacterium omelianskii.*
Methanobacterium omelianskii converted ethanol and CO_2
quantitatively into methane and acetic acid. As we are now
aware this subsequently proved to be a mixture of two
organisms, but the belief in the existence of a pure culture,
together with the extensive information of the other cultures,
constituted a powerful stimulus to study the bacteria-producing
methane.

Ch. G.T.P. Schnellen started work with Kluyver to attempt
isolation of additional pure cultures of methanogenic bacteria.
Already in 1938 he succeeded, using specially narrowed shake
culture tubes with paraffin above the restricted neck. His
work was a masterpiece. He obtained what is now accepted as
the first indisputably pure culture of a methanogen, *Methanosarcina barkeri,* and also *Methanobacterium formicicum.* The
former was isolated on methanol, and fermented also acetate and
H_2/CO_2. *Methanobacterium formicicum* fermented only formate and
H_2/CO_2. Schnellen showed also that carbon monoxide could be
converted to methane, the first step being the addition of water
to CO to form dihydrogen and carbon dioxide. Schnellen's chief
concern was to test rigorously the Van Neil hypothesis that CO_2
was the source of carbon in methane. He concluded that his
results fully supported this view.

Schnellen's work was interrupted when he was conscripted
into the army but he returned and completed his thesis after
the war (1947). It did not have as great or immediate an
effect on many investigators as might have been expected from
the quality of the work, but it undoubtedly influenced
Theresa Stadtman (Stadtman and Barker, 1951) to undertake the

isolation of a formate-fermenting strain,*Methanococcus vannielii,* the first methanogenic organism to be isolated within the U.S. It was a rapidly growing prototroph fermenting formate to methane.

But methanogenesis depends not only on methanogenic bacteria but also on bacteria that decompose plant materials. The nature of the methanogens and also of the cellulolytic bacteria had been a mystery to microbiologists because of the lack of easily obtainable and conserved pure cultures. Thus the experiences with cultivation of cellulolytic bacteria are also pertinent to our topic. Perhaps the chief discrete element in the evolution of reliable methods for cultivation of the cellulolytics was the belief that there might be degrees of anaerobiosis (Hungate, 1947, 1950), and that it might not be enough just to remove dioxygen and prevent its access, but that it might be essential also to reduce the oxidation-reduction potential to a point at which essential enzymes could be maintained in an effective state.

But success in identifying cellulolytics was due also to observation of ecological factors simulating the natural habitat and to direct dilution of habitat material into agar dilution series so that preliminary liquid enrichment did not change the character of the natural population. There had been several reports of successful production in agar media of cellulolytic colonies producing clear spots in the agar due to digestion of the cellulose (Kellerman and McBeth, 1912), but these had been variously discredited by the current "authorities" in bacteriology, including the Delft Laboratory, as being solubilization of the $CaCO_3$ included in the media, or use of a substrate that really wasn't the "true" cellulose. At any rate the isolation of these bacteria was a major undertaking. Thus it was a great stimulus to my own investigations when in an old shake tube inoculated with fluid from an in vitro culture of rumen protozoa I observed an unmistakable clearing of the cellulose around a central bacterial colony, apparently the causal agent of the cellulose digestion. Obviously a thinner layer of cellulose agar would make it easier to detect digestion, hence the roll tube. Fortunately the biotin required by *Clostridium cellobioparum* (Hungate, 1944) was contained in the agar, and the circular clearings of the cellulose in subcultures strikingly demonstrated the advantages of the method. For the rumen cellulolytics rumen fluid was added as part of the medium, and bacteria more stringently anaerobic than most of those previously described were isolated (1947).

Marvin Bryant enrolled at Pullman after the war and as an undergraduate laboratory assistant took to the anaerobic methods

and principles that enabled him soon to publish his thesis
on the isolation of the rumen spirochete. He was supported
by Washington State while he went with me to work with Bob
Dougherty at Cornell in the summer of 1950 on acute indigestion
in sheep. During that time a position for someone in rumen
microbiology opened up at Beltsville, and it was arranged that
Marvin would complete his doctorate at Maryland, working with
Ray Doetsch. Marvin now claims in fun that I "kicked him out,"
but I felt that he had been such an apt student and had already
acquired most of what I could impart that I believed he could
develop admirably on his own, a feeling fully justified by his
subsequent accomplishments. At any rate Marvin went to
Beltsville, developed (Bryand and Burkey, 1953) the procedure
for using a non-selective (habitat-simulating) medium to
obtain pure strains of the most abundant rumen bacteria, and
embarked on his illustrious career including personal instr-
uction in the roll-tube technique.

The spread of the anaerobic methods was also by other
students in my laboratory, though Paul Smith at Blacksburg and
Albrecht Kistner at Pretoria successfully mastered it from the
description published in 1950.

Richard McBee (1948) successfully applied the anaerobic
method to the cultivation of *Clostridium thermocellum*. Leroy
Maki (1953) as a masters student isolated cellulolytic bacteria
from sewage sludge and carried the procedure to Mike Foster's
lab at Wisconsin. Betty Hall isolated *Ruminococcus* from the
rabbit gut (1952), and Kaars Sijpesteijn published on the rumen
cocci at Delft (1948).

The isolation of methanogens was accomplished by the roll-
tube technique when Bob Mylroie, who had worked with Buswell at
Illinois, came to Pullman and undertook to obtain a pure
culture using H_2/CO_2 in a completely inorganic medium containing finely divided platinum as catalyst. He demonstrated
(Mylroie and Hungate, 1954) that simple provision of a redox
potential as low as that of H_2 allowed methane production by a
pure culture. Apparently the methanogens have no metabolism
capable of generating low potentials, but must depend on
accompanying forms.

The general idea that methanogens are killed by dioxygen
may not be true. Thorough washing under aerobic conditions of
many methanogenic cells with subsequent centrifugation and
inoculation of myriads of cells practically devoid of oxidized
medium into a thoroughly reduced subculture should be examined
to see if survivors can be detected. Reduction of an oxidized

culture can fail because of the buffering action of the oxidized materials in the system. Zehnder (Zehnder and Wuhrmann, 1977) has shown that at least some dioxyduric methanogens occur.

Mylroie's work was, I believe, the first isolation from a methanogenic habitat by direct dilution without previous enrichment. An important step in the adaptation of methods for culturing methanogens was the demonstration (Smith and Hungate, 1958) that also complex media, in this case rumen fluid, could be adapted to direct isolation of a methanogen from the habitat. In the course of this work Paul Smith introduced the syringe transfer method in which the culture is never opened during transfer. Subsequently at Florida he has not only isolated a number of strains, but has also served as a bridge between fundamental microbiology and practical applications of methanogenesis. The isolation of *Methanomicrobium mobile* (Paynter and Hungate, 1968) showed variability in the methanogens occupying a given habitat, a finding indicating the desirability of continually isolating new strains.

With the improvement of anaerobic chambers to the point that methanogens can be grown in them on plates, the cultivation and direct isolation of methanogenic bacteria in pure culture has become frequent. The Woese postulate of the antiquity of methanogens led to studies showing that their antibiotic susceptibility differed so much from that of eubacteria that suitable antibiotics could effectively suppress the eubacteria in inocula taken directly from habitats, and with fluorescence detection based on knowledge of key inter- mediates and cofactors in the metabolism of methanogens, the ready detection and isolation of new methanogenic species is greatly facilitated.

This history of the microbiology of methanogenesis is a good example of the difficulty in tracing in logical or orderly fashion the development and acceptance of ideas. Anaerobic methods were available during the 1950s, pure cultures had been obtained, the biochemistry was being studied, sparks were needed to excite the interest of additional workers and cause increased interest in methanobacteriology. One of these was probably the demonstration by Bryant that when he cultured *Methanobacterium omelianskii* in dilution series containing H_2/CO_2 the isolated methanogen was no longer able to attack ethanol. The coculture nature of the inoculum was suspected and the accompanying organism attacking the ethanol was revealed. It is of interest that chiefly the ethanolotrophic organism must have been cultured also by Johns and Barker (1960)

when they grew *"Methanobacterium omelianskii"* in a medium
containing no carbon dioxide, and showed the accumulation of
acetate and dihydrogen.

Elucidation of the coculture nature of *M. omelianskii*
reduced to five compounds (H_2/CO_2, methanol, CO, formate and
acetate, the substrates known to be used by methanogens) as an
additional methanogenic substrate. The special importance of
acetate in methanogenesis needs to be stressed. It had been
postulated as a substrate by Hoppe-Seyler as early as 1887, but
he though its demethylation might provide too little energy to
support growth. This idea persisted in spite of Mazé's and
Söhngen's evidence of its use by a pure culture. It is of
interest that the final acceptance of acetate as an unquestion-
ably useful methanogenic substrate was marked by the invitation
to Bob Mah to address the Royal Society of London on this
topic three hundred and eight years after the first Leeuwenhoek
letter to that society, and if we count Leeuwenhoek, Beijerinck,
Kluyver and van Niel as a scientific lineage the acetate address
to the Royal Society was appropriately by the Leeuwenhoek's
sixth generation scientific offspring.

The finding of methylamines (Hippe et al., 1979) recalls
to my mind the question posed some years ago in Norway by
Helge Larsen, at Trondheim, namely, how could one account for
the great quantities of methane accumulating in buildings
storing large quantities of fish at low temperatures? The
answer is clear: methylamine oxide is an abundant nitrogenous
waste in fish, rapidly reduced to methylamine by euryoxic
bacteria on the fish surfaces, and converted to methane by
anaerobes.

Realization of the coculture nature of *M. omelianskii*,
and of H_2 as the intermediate between the two species, coupled
with the endergonic nature under standard conditions of most
of the dehydrogenations required to account for anaerobic
oxidations led Bryant to realize that the most practical method
to demonstrate this interaction was to couple in a single
culture the organism capable of dehydrogenating the substrate
to be oxidized with an organism having an affinity for
dihydrogen making the concentration so continuously low that
the dehydrogenation reaction remained endergonic. Bryand and
his coworkers have shown in a beautiful series of papers how
propionate (Boone and Bryant, 1980), butyrate and even
benzoate can be oxidized anaerobically by dehydrogenation,
with the hydrogen oxidized by CO_2 to form methane, or by
sulfate to form H_2S or even by fumarate to form succinate.

The concentrations of H_2 are so small that the absolute amount transferred per unit time is not large, and in consequence the growth of these proton-reducing bacteria and of the accompanying methanogen is very slow. But the life span of the individual cells is long. Under favorable conditions they should remain viable for years, a factor of probable value in the practical processes of waste disposal. There are indications of rather extensive anaerobic oxidations by single bacterial strains (Grbic-Galic, Barik and Bryant, personal communications). Further exploration should disclose additional examples.

It should be noted that Buswell, McCarty and others had early proved that the crude populations developing in sewage disposal plants completely oxidized complex organic matter with carbon dioxide, but the details of the reactions and proof that dihydrogen is the reductant awaited thorough documentation of the syntrophy needed to maintain continuously the low concentrations of dihydrogen that mark these anaerobic oxidations.

We cannot leave these aspects of our subject without at least mentioning also Peter Mitchell's new insights into accomplishment of biological work in membranes. These open up new possibilities for many reactions by which methane can be biologically generated. Bryant's recent success in using the membrane phosphorylation by *Wolinella succinogenes* to oxidize benzoate suggests that additional anaerobic membrane phosphorylations useful in speeding up anaerobic oxidations will be found.

An overview of biomethanogenesis is incomplete without at least mention of the recent beautiful elucidations of its biochemistry. The fascinating accomplishments of Ralph Wolfe and his associates, and the increasing perceptions and contributions of other laboratories to knowledge of the molecular mechanisms by which methane is formed have been a powerful stimulus to research. We look forward expectantly to knowing more about the reactions, catalysts, and expediters involved in the conversion not only of H_2/CO_2 but also of acetate, methanol, formate, carbon monoxide, methylamines, and probably additional methyl groups.

It is impossible to predict which of the advances in our understanding of the detailed mechanisms of methanogenesis and their bacterial agents will lead to better exploitation of this process. But for those concerned with the practical

applications of methanogenesis the added insights they provide
are bound to suggest ideas, testable in the normal processes
of waste disposal, with high probability that some of them
will be of value.

It is important to point out that practical methanogenesis
depends not only on the knowledge of the organisms and the
processes they effect, but also on the nature of the materials,
chiefly plant cell walls, they must use. A very large proportion of plant material is not readily digestible by
presently known enzymes or microbes to products that can
serve as substrates for methanogenesis. It should be realized
that plant materials readily digestible and fermentable to
methane may constitute also a favorable source of food for
ruminants, and ruminants, producing high quality protein, also
represent conserved energy, in this case in the form of food
rather than power. Wise social, economic and political
judgments are needed for a judicious allocation of resources
to these various demands. The wastes with which this symposium
is concerned are unsuitable for ruminants and thus ideal for
production of energy and feedstocks.

I cannot refrain also from emphasizing the importance of
preserving on land the inorganic nutrients essential for life.
As humans and their domesticated animals become an increasingly
large element of the animal population on Earth it becomes
increasinly important that the inorganic elements in their
excreta be retained on the land mass rather than accumulated
in the ocean where their dispersion in solution greatly
increases the energy cost of recovery in useful concentrations.

I realize that his rather lengthy presentation does not
contribute much of direct value in the solution of the problems
we face, but I hope that review of your predecessors efforts
may encourage your own endeavors to advantageously process
municipal wastes for fuel and chemicals.

REFERENCES

Bechamp, A.,*Ann. Chim. Phys. 13,* 103 (1868)(cited by Zehnder,
    1982).
Boone, D.R. and M. P. Bryant, *Appl. Environm. Microbiol. 40,*
    626, (1980).
Breznak, J. A., *Annual Rev. Microbiol. 36,* 323 (1982).
Bryant, M. P. and L. A. Burkey, *J. Dairy Sci. 36,* 205 (1953).
Bryant, M. P., E. A. Wolin, M. J. Wolin, and R. S. Wolfe, *Arch.
    Microbiol. 59,* 20 (1967).
Coolhaas, C., "Bijdrage tot de kennis der dissimilatie van
    vetzure zouten en koolhydraten door thermophile bacterien.
    Dissertation, Wageningen. (1927).

Groenewege, J., "Mededeelingen v.d. Burgerlichen Geneeskundigen Dienst. in Nederlandsch-Indie," (1920)(Cited by Schnellen, 1947).

Hall, E. R., *J. Gen. Microbiol.* 7, 350 (1952).

Hippe, H., D. Caspari, K. Fiebig, and G. Gottschalk, *Proc. Nat. Acad. Sci., U.S.A.*, 76, 494 (1979).

Hoppe-Seyler, F., *Pflüger's Arch. ges. Physiol:* 12, 1 (1876).

Hoppe-Seyler, F., *Z. Physiol. Chem.* 11, 561 (1887).

Hungate, R.E., *J. Bacteriol.* 48, 499 (1944).

Hungate, R.E., *J. Bacteriol.* 53, 631 (1947).

Hungate, R.E., Bacteriological Rev., 14, 1 (1950).

Hungate, R.E. in: "Biochemistry and Physiology of Protozoa," Vol. II,]95-197, Appendix II, Why Carbohydrates?, (S.H. Hutner and A. Lwoff, eds.) Academic Press, New York (1955).

Johns, A. T. and H. A. Barker, *J. Bacteriol.* 80, 837 (1960).

Kellerman, K. F. and I. G. McBeth, *Zentralbl. Bakt, Parasitenk, Abt. II, Orig.*, 34, 485 (1912).

Kluyver, A. J. and H. L. Donker, *Chem. d. Zelle u. Gewebe 13*, 134 (1926).

McBee, R. H., *J. Bacteriol.* 56, 653 (1948).

Magendie, F., *Ann. Chim. Physique, Ser. 2,2*, 292 (1816).

Maki, L. R., *Antonie van Leeuwenhoek 20*, 185 (1953).

Mazé, P., *Compt. Rend. Acad. Sci., Paris 137*, 887 (1903).

Mylorie, R. L. and R. E. Hungate, *Canadian J. Microbiol. 1*, 55 (1954).

Omelianski, W. L., *Centralbl. f. Bakteriol., Abt. II, Orig. 11*, 369 (1904).

Omelianski, V. L., *Ann. Inst. Past. 30*, 56 (1916).

Paynter, M.J.B. and R.E. Hungate, *J. Bacteriol.* 95, 1943 (1968).

Popoff, L., *Pflüger's Arch. ges. Physiolog. 10*, 123 (1873).

Reiset, J., *Ann. Chim. Physique, Ser. 3*, 69, 129 (1863).

Schnellen , Ch.G.T.P., "Onderzoekingen Over de Methaangisting" Dissertation, Technische Hoogeschool, Delft (1947).

Sijpesteijn, A.K., "Cellulose-Decomposing Bacteria from the Rumen of Cattle," Dissertation, Technische Hoogeschool, Delft (1948).

Smith, P. H. and R. E. Hungate, *J. Bacteriol.* 75, 713 (1958).

Söhngen, N. L., "Het Onstaan en Verdivijnen van Watersstof an Methaan Onder den Invloed van het Organische Leven," Dissertation, Technische Hoogeschool, Delft (1906).

Sprengel, K., "Chemie für Forstmänner, Landwirte und Cameralisten," Vol. I, pp. 793, Vandenhouk and Ruprecht, Göttingen (1832).

Stadtman, T. C. and H. A. Barker, *J. Bacteriol.* 62, 269 (1951).

Stephenson, M. and L. H. Stickland, *Biochem. J.* 27, 1517 (1933).

Symons, G. E. and A. M. Buswell, *J. Amer. Chem. Soc.* 55, 2028 (1933).

Tappeiner, H., *Ber. d. deutsch, Chem. Gesellschaft 15*, 999 (1882).

Tappeiner, H., *Ber. d. deutsch. Chem. Gesellschaft 16,* 1740 (1883).

Tarvin, D. and A. M. Buswell, *J. Amer. Chem. Soc. 56,* 1751 (1934).

Thayer, L. A., *Bull. Amer. Assoc. Petrol. Geologists 15,* 441 (1931).

Volta, A., Lettres sur L'Air Inflammable des Marais auquel on a ajouté trois Lettres du meme Auteur tirées du Journal de Milan. Traduites de L'Italien avec permission. Strasbourg. J.H. Hertz, Imprimeur de l'University (1778). [The first letter, to Father Charles Joseph Campi, was written at Come on the 14 Nov. 1767.]

Warburg, O., *Biochem. Zeitschr. 152,* 479 (1924).

Wieland, H., *Ber. d. deutsch. chem. Gesellschaft 46,* 3327 (1913).

Woese, C. R. and G. E. Fox., *Proc. Nat. Acad. Sci., U.S.A., 74,* 5088 (1977).

Zehnder, A.J.B., in: "Anaerobic Digestion," D.E. Hughes et al., eds., Elsevier Biomedical Press, New York, pp. 45-68 (1982)

Zehnder, A.J.B. and K. Wuhrmann, *Arch. Microbiol. 111,* 199 (1977

2
ACETATE METABOLISM BY *METHANOSARCINA*

J.A. Krzycki

J.G. Zeikus

Department of Bacteriology
University of Wisconsin
Madison, Wisconsin

Michigan Biotechnology Institute
Michigan State University
East Lansing, Michigan

ABSTRACT

Acetate metabolism is a key process in methanogens. *Methanosarcina barkeri* forms methane by the following reactions during growth on acetate as energy source: $CH_3COOH \dashrightarrow CH_4 + CO_2$. In addition, acetate or an acetyl derivative is synthesized as an early cell carbon precursor by *M. barkeri* growing on one carbon compounds. Both biodegradation and synthesis of acetate appear to share common biochemical features such as the involvement of methyl-CoM, methyl-B_{12}, and the enzyme carbon monoxide dehydrogenase. Methanogenesis from acetate occurs via two mechanisms; a primary route where methane is derived from an intact methyl group, and a secondary pathway where methane is formed by coupling methyl group oxidation to the reduction of the carboxyl group. A unifying hypothesis for ATP synthesis during methanogenesis from acetate by the primary route is proposed and is based on information obtained from the substrate dependent regulation of carbon monoxide dehydrogenase activity, characterization of the purified enzyme, and analysis of methane production from acetate in cell-free extracts. The hypothesis suggests that an acetyl derivative is metabolized to methyl-CoM and a carbonyl intermediate. The latter is oxidized to CO_2 by carbon monoxide dehydrogenase. The electrons produced are transferred along a membrane bound electron transport chain and used to reduce methyl-CoM to methane. ATP synthesis is then coupled to the free energy of the resultant proton motive force.

INTRODUCTION

Acetate is a key metabolite of the methanogenic bacteria. Many methanogens readily assimilate exogenous acetate into cell carbon (3,5,16,54,56). An acetate derivative is also formed as an early intermediate in the synthesis of cellular material from single carbon substrates (16,22,23, 51). Furthermore, certain methanogenic species can dissimilate acetate into CH_4 and CO_2 and gain energy by this process. Acetate degradation is of considerable importance in waste digestion because it is the main precursor to methane.

This paper will review the physiological and biochemical aspects of acetate metabolism by methanogens. Studies on *Methanosarcina barkeri* strain MS, which is capable of both acetate synthesis and degradation, will be highlighted. This organism grows on acetate as the sole carbon and energy source for metabolism. Also, during growth on one carbon compounds, such as methanol or hydrogen-carbon dioxide, acetate is synthesized by *M. barkeri* as an anabolic precursor.

PHYSIOLOGY AND BIOCHEMISTRY OF ACETATE ASSIMILATION

Acetate is an easily assimilated source of cell carbon for many methanogenic bacteria, but as illustrated in Table 1 the amounts incorporated vary considerably. Some methanogens, such as *Methanobacterium thermoautotrophicum* strain ΔH, will synthesize only small amounts of cell carbon from acetate present in the growth medium (3). In contrast, *Methanobrevibacter ruminantium* will derive up to 60% of total cellular carbon from exogenous acetate. This is by necessity since the organism obligately requires acetate as a growth factor, and is not able to synthesize it (5).

Table 1. Acetate Assimilation by Methanogenic Bacteria Grown on H_2-CO_2

Species	Obligate Acetate Requirement	% Cell Carbon From Acetate	Carbon Monoxide Dehydrogenase	References
Methanobrevibacter ruminantium	+	63	- -	5, R. Thauer, Personal Communication
Methanobacterium thermoautotrophicum				
Strain ΔH	- - -	1.3	+	3,8
Strain Marburg	- - -	65	+	16,51
Methanobacterium ivanovii	- - -	21	+	1
Methanosarcina barkeri strain MS	- - -	40	+	8,55
Methanococcus voltae	+	20	?	57

Although *M. barkeri* incorporates exogenous acetate into cell carbon, it does not require acetate for growth (54). A study of the incorporation of acetate radioactively labelled in the methyl or carboxyl group indicated that acetate was incorporated as a unit (54). These results led to the hypothesis that in the absence of exogenous acetate, acetate, or an acetyl derivative, is synthesized as an early step in the synthesis of cell carbon from single carbon substrates. Analysis of enzymes capable of activating acetate and transforming it into other precursors of cell carbon indicated that *M. barkeri* utilized enzymes associated with the tricarboxylic acid cycle to assimilate acetate (56). Short term labelling studies performed with cell suspensions (56) demonstrated that radioactive acetate was rapidly transformed into three amino acids, alanine, aspartate and glutamate. These results suggested that cell carbon was synthesized in *M. barkeri* via reactions common to the TCA cycle. This pathway of acetate assimilation differs from the cell carbon assimilation pathway of *M. thermoautotrophicum* because *M. barkeri* forms α-ketoglutarate via isocitrate dehydrogenase, whereas α-ketoglutarate dehydrogenase is used by *M. thermoautotrophicum* (16,65; see Thauer, of this volume).

The path of acetyl synthesis from single carbon substrates in *M. barkeri* is suggested to occur via a direct condensation of a methyl and carboxyl precursor (23). The nature of the methyl donor was first indicated by the inhibition of growth by iodopropane, an inhibitor of corrinoid coenzymes (22). It was demonstrated that at low iodopropane concentrations, the addition of acetate to media relieved this growth inhibition. This result suggested that a corrinoid, presumably at the methyl level (22), was involved in two carbon synthesis. In addition, methyl-B_{12} dependent acetate synthesis was demonstrated in cell extract studies with carbon monoxide as the acetate carboxyl precursor (23). Acetate formation was independent of the presence of ATP, and neither methyl-CoM nor methyltetrahydrofolate replaced methyl-B_{12} as the methyl donor. The synthesis of acetate using carbon monoxide also suggested involvement of carbon monoxide dehydrogenase in acetate synthesis, presumably for generation of a carbonyl intermediate from carbon dioxide. A similar mechanism for the catabolic synthesis of acetate had been previously suggested for the acetogenic bacteria (9).

Figure 1 summarizes the current perception of cell
carbon synthesis from CO_2 plus H_2 in *M. barkeri*. The generation of the methyl precursor of an acetyl moiety or
acetate from CO_2 is thought to involve common intermediates
in the methanogenic pathway for CO_2 reduction (23). After
reduction to an unidentified methyl intermediate the methyl
group is transferred to a corrinoid. The identity of methyl
X is unknown, but a likely candidate is methyltetrahydro-
methanopterin, a coenzyme recently identified as the first
methyl level in the reduction of CO_2 to methane (13). Methyl
donor formation from methanol may occur in a more direct
manner since corrinoid proteins are involved in the initial
steps of methanol catabolism (60). The generation of the
carboxyl precursor can occur from methanol, CO, or CO_2.
The nature of this C_1 intermediate is unclear, but it is
presumed to be the in vivo substrate for CO dehydrogenase.
The synthesis of an acetyl moiety from a CH_3 corrinoid and
a C_1 unit appear common to both *M. barkeri* and *M. thermoautotrophicum* (51). It is interesting to note that all
the acetate assimilating methanogens in Table 1 possess CO
dehydrogenase, with the exception of the obligate acetate
assimilating species, *Methanobacterium ruminatium,* which
does not have a detectable CO dehydrogenase activity (R.
Thauer, personal communication).

THE CATABOLISM OF ACETATE

Significance of Methanogenesis from Acetate

Methanogenic bacteria occur in a wide variety of anaerobic environments, yet they produce methane from only
a limited number of simple compounds such as hydrogen-carbon
dioxide, methanol, formate, carbon monoxide, methylamines,
and acetic acid (52,64). Acetate remains as one of the
less studied of methanogenic substrates, yet in some ecosystems it is the primary methane precursor. In environments
such as rice paddies and lake sediments, from 60 to 95% of the
methane produced is generated directly from acetate. Table 2
lists the major sources of biologically produced methane and
the contribution acetate makes in each as direct methane precursor. Collectively, these environments produce from 50 to
80% of the methane present in the atmosphere (15,43). It is
obvious that the conversion of acetate to methane and carbon
dioxide is a process of some significance in the global carbon
and methane cycles. Acetotrophic methanogens function in the
important terminal step of degrading organic matter into
CH_4 and CO_2 (Figure 2). This assures that the majority
of carbon continues to flow through the environment and
does not accumulate. Aside from this direct role, they

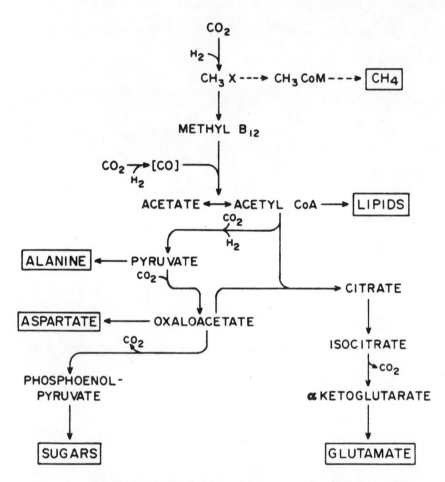

Figure 1.   Proposed biochemical model for autotrophic metabolism of *Methanosarcina barkeri*. CO_2 is reduced to the methyl level and either reduced to
CH_4 or carbonylated to an acetyl moiety which
serves as the precursor to further cell carbon
synthesis. The major C_3, C_4 and C_5 cell carbon
precursors are synthesized from the TCA cycle
related reactions indicated.

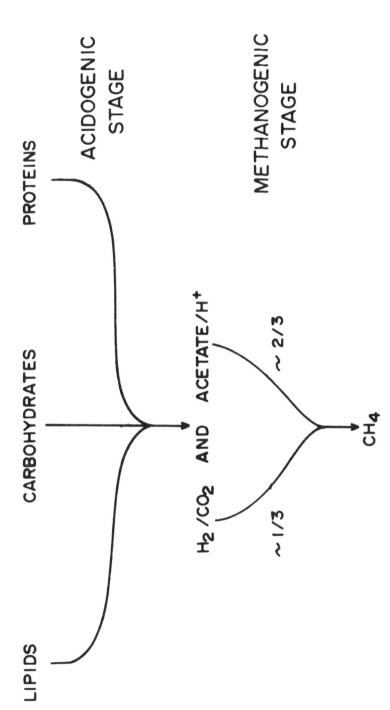

Figure 2. Significance of acetate metabolism in waste treatment. Acetic acid is the major precursor to methane, and its degradation prevents inhibition or organic destruction by proton toxicity.

function in maintenance of neutral pH by the consumption
of toxic protons produced by the fermentation of carbohydrates and amino acids (62,63). The importance of methanogenesis from acetic acid in this process is most evident
in anaerobic waste digestion where up to 90% of the methane
produced can be formed from acetate (33).

Table 2.   The Contribution of Acetate to Biogenic Methane
           Production

Source of Methane	$10^{12}g$ CH$_4$/ Year	Derived from Acetate	References
Ruminants	45-99	3-6	18,34,43
Paddy Soils	30-280	60	11,25,43
Freshwater Swamps	13-300	N.R.	11,43
Lake Sediments	1-25	60-95	7,10,37,58
Ocean Sediments	1-16	50	10,39
Sewage Digestors	N.R.	70-90	6,33,48

N.R. = Not Reported.

Physiological Characteristics of Acetotrophs

Acetotrophic methanogens represent those which are able
to derive energy and cell carbon from acetate metabolism.
It was not until 1978 that multiple investigators were able
to achieve repeatable growth of *Methanosarcina* at the expense of acetate (32,46,54).   These cultures were supplemented with yeast extract and this left some doubt as to
whether growth could occur solely on acetate.   Later it
was demonstrated that growth could occur in a defined medium
with acetate as the only energy source (26,47).   Since that
time many methanogens have been isolated using acetate as
a sole carbon and energy source.   However, only two genera,
Methanosarcina and *Methanothrix,* have been described.   Table
3 provides some of the physiological attributes of the
better characterized acetotrophic methanogens.   Several
common characteristics can be observed.   In general the K_s

Table 3. Physiological Comparison of Acetotrophic Methanogens

Species	Doubling Time (h)	K_s (mM)	Yield g cells/mol CH_4	% Oxidation of methyl Group	CO Dehydrogenase μmol/mg prot/min	References
Methanosarcina barkeri						
Strain MS	55	N.R.	3.9	5-14	5	26,59
Strain 227	30-38	5	2.0-3.4	1-3	-	32,46
Strain Fusaro	30	3	3-4.1	2%	N.R.	4,41
Methanosarcina						
Strain TM-1	12	4.5	1.8	1	10	35,68
Methanosarcina mazei	17	N.R.	N.R.	N.R.	N.R.	31
Methanosarcina acetivorans	24	N.R.	N.R.	N.R.	4	35,49
Methanothrix soehgenii	84-216	0.5-0.7	1.1-1.4	3.1	4	17,24,61

N.R. = Not Reported.

for acetate is high; 5 mM for the *Methanosarcina* strains, and about 0.7 mM for *Methanothrix*. The doubling time on acetate (or rather as is often reported, the time required for the doubling of methane) ranges from 10 to 60 hours. Generally, *Methanosarcina* grows faster on methanol or hydrogen-carbon dioxide than on acetate (52). The reported molar cell yields on acetate vary considerably, but in general these values are lower for acetate than for other substrates.

The majority of isolated acetotrophs were obtained from sewage sludge. *Methanosarcina acetivorans,* a marine isolate, is unusual in the possession of a protein cell wall (49), unlike the heteropolysaccharide wall found in the other *Methanosarcina* isolates (21). Most described acetotrophs are mesophilic, with the exception of the thermophilic *Methanosarcina* strain TM-1 and a *Methanothrix* sp. not presently in pure culture (36,66).

Origin of Methane During Acetotrophic Growth

It was first demonstrated by Buswell and Sollo in 1948 (6) that methane in an acetate enrichment culture was not produced from carbon dioxide. From this it was inferred that the carboxyl of acetate was not converted to methane and that the methyl group was directly reduced to methane. Further work on enrichments confirmed this hypothesis. Stadtman and Barker demonstrated that the methyl group was converted primarily to methane and only slightly to CO_2 (50). Pine and Barker showed that the hydrogen atoms on the methyl moiety were retained after conversion to methane (38). These results suggested that the primary route of methanogenesis was via a mechanism which has been called the "aceticlastic reaction":

$$\overset{*}{C}H_3 \, ^\circ COOH \dashrightarrow \overset{*}{C}H_4 + {}^\circ CO_2 .$$

Methane is formed from acetate by an apparent decarboxylation but the exact biochemical mechanism differs as shown below.

This appears to be the major route of methane formation from acetate in nature. A secondary route also exists, in which the methyl group is oxidized to carbon dioxide and the reducing equivalents generated are used to reduce carbon dioxide to methane (26). When methyl labeled acetate is introduced into a culture of *M. barkeri* strain MS both radioactive methane and carbon dioxide are produced.

This phenomenon occurs to varying extent in cultures of *M. barkeri* strain MS (26,59), and a recent *Methanosarcina* isolate from an acid lake (37). Other strains of *Methanosarcina* such as *M. barkeri* strain 227 or TM-1 oxidize the methyl group to a lesser extent, but do not produce methane from the carboxyl. The reducing equivalents generated are presumably used for biosynthesis. Recently, a co-culture was described in which methane was derived to an equal extent from both acetate carbons (67) with one organism oxidizing acetate to H_2-CO_2 and the other converting these into CH_4.

Mixotrophic Growth on Acetate and Other Substrates

The ability of acetotrophic methanogens to convert acetate to methane simultaneously with methanol or hydrogen-carbon dioxide is at present controversial. It was first reported that *M. barkeri* strain 227 utilized methanol, then acetate, as methanogenic substrates in a sequential fashion (46). This phenomenon could not be demonstrated in *M. barkeri* strain MS (19,26), or in strain Fusaro (39). For both strains it was observed that methanol and acetate were simultaneously removed from the medium (26). When radiotracers were employed to follow the course of methane production from either substrate, it was found that significant production of methane occurred from acetate while methanol was still present in the medium (Figure 2). Trimethylamine has also been reported to depress methanogenesis from acetate in strain Fusaro (4).

Although strain differences do exist with regard to simultaneous methanogenesis from acetate and methanol, all of the strains examined oxidize the methyl group of acetate to a greater extent in the presence of methanol (26,54,68). A similar effect is also observed in cells growing on trimethylamine and acetate (4). In both cases the oxidation of methanol or the methylamine is decreased. The electrons generated from acetate oxidation are then used to reduce the methanol or methylamine to methane. Little is known about the significance of preferential oxidation of substrates. Since all substrates appear to be converted to methane via methyl CoM (44), it appears that entry into oxidation pathways is controlled at another methyl level intermediate. The recent deuterium labelling study of Blaut and Gottschalk indicated that some intermediates of the oxidation pathway for acetate and for methanogenesis are linked (4).

Hydrogen has also been reported to inhibit methanogenesis from acetate, but again different effects have been found with different acetotrophs. *Methanosarcina* strains TM-1 and

M. barkeri strain 227 cease to catabolize acetate when incubated under hydrogen (46,68). *M. barkeri* strain 227 and *Methanosarcina mazei* have been reported to preferentially utilize hydrogen prior to acetate as a methanogenic substrate (14). A similar phenomenon was observed with *M. barkeri* strain MS, but inhibition by hydrogen could be relieved if carbon dioxide was added to the medium. Schink and Zeikus demon'strated that in a coculture of *Clostridium butyricum* and *M. barkeri* strain UBS, methane was simultaneously derived from carbon dioxide (with hydrogen), methanol and the methyl group of acetate (42). *Methanothrix,* which does not utilize hydrogen for methanogenesis, is not inhibited by hydrogen (17). Cell suspension studies with different strains have also yielded various results. *M. barkeri* strain MS transforms the methyl of acetate to methane at similar rates when incubated under one atmosphere of hydrogen or nitrogen (B. Morgan, unpublished results). On the other hand, *M. barkeri* strain Fusaro will cease forming methane from acetate in the presence of even small amounts of hydrogen (12).

The apparently conflicting results presented in the foregoing section emphasize the danger of ignoring possible differences in various isolates. That *Methanosarcina barkeri* was for a long period the only recognized genus and species of a coccoid, clustered methanogen has placed an undue emphasis on the supposed similarity of strains rather than their differences. As more is known of the metabolism of these strains, it is sure to become more evident that as much diversity exists among strains of *Methanosarcina barkeri* as among strains of *Escherichia coli.*

Biochemistry of Acetate Catabolism

There are several enzymes and coenzymes whose involvement in the pathway of acetate degradation to methane and carbon dioxide seem credible. The first of these is the central coenzyme of methanogenesis, coenzyme M. The methylated form of this coenzyme (or methyl-CoM) is acted on by the enzyme methylreductase as the terminal step in methane formation from hydrogen-carbon dioxide or methanol (44). Methylreductase is present in acetate grown cells of *M. barkeri* strain 227 (2), and *Methanosarcina* strain TM-1 (35). Bromoethanesulfonic acid, an analogue of methyl-CoM, inhibits methane formation in cultures of *M. barkeri* strain 227 grown on acetate (46). Finally, $^{14}CH_3CoM$ can be detected in cell suspensions and extracts actively converting ^{14}C-2 NaAcetate to methane (28, J. Krzycki and L. Lehman, unpublished data).

Methanosarcina strains contain large amounts of corrinoids (27,40). Some evidence points to their involvement in acetate catabolism. Iodopropane, an inhibitor of corrinoid enzymes, inhibits methanogenesis from acetate at much lower concentrations than required for inhibition of methane production from methanol or hydrogen-carbon dioxide (22). KCN is also a potent inhibitor of methane formation from acetate (12,17,45). One possible site of inhibition could be a corrinoid step in methane production from acetate but CO dehydrogenase is also inhibited by KCN.

Acetyl CoA has been suggested as an intermediate in methane production from acetate (53). This is based primarily on the high levels of acetate kinase and phosphotransacetylase present in acetate grown *M. barkeri* (23). Equally high activities of these enzymes are present when the organism is grown on hydrogen-carbon dioxide and, therefore, these activating enzymes may function in catabolism and/or anabolism.

One of the first clues to a possible mechanism of acetate dissimilation was the discovery of elevated activities of carbon monoxide dehydrogenase (i.e., COdh) in acetate grown cells of *M. barkeri* strains MS. During growth on acetate, specific activities of COdh from 5 to 10 μmole CO oxidized/min/mg protein are observed. This is about five times the activity seen in methanol grown cells (26). Similarly high activities have been recently observed for *Methanothrix* (24), *Methanosarcina* strain TM-1, and *M. acetivorans* (35). Purification of this nickel containing enzyme from *M. barkeri* grown on acetate revealed that as much as 5% of soluble cell protein was CO dehydrogenase (29). The enzyme also has a very poor affinity for CO, which is perhaps a reflection of the actual physiological function of the enzyme as something other than CO oxidation *per se*.

Understanding the biochemistry of methane formation from acetate was long impeded by the lack of a cell-free assay to study. Several labs attempted to develop such a system but were not successful. The first report of a non-cellular system came from Baresi (1), who was able to prepare a fraction from broken cells of *M. barkeri* strain 227 which could convert acetate to methane under a nitrogen atmosphere. The activity was entirely associated with a particular fraction following isolation on a sucrose gradient, which suggested that the methanogenic system required membranes in order to operate. The membrane requirement meant that fractionation and examination of the acetate cleaving enzyme array would be difficult. Recently it was discovered that the soluble protein fraction of *M. barkeri* strain MS converted acetate into methane at

five to ten percent of *in vivo* rates (30). Methane production
was dependent upon the addition of acetate, ATP, and hydrogen.
Despite the requirement for hydrogen, this *in vitro* system
appears to represent the actual catabolic pathway. The assay
is sensitive to *in vivo* inhibitors of methane production from
acetate. BES inhibited methane production at low concentrations, as expected if methyl-CoM is an intermediate in the
pathway. Cyanide also inhibited the reaction at low concentrations which almost completely inhibits carbon monoxide
dehydrogenase when assayed by methyl viologen reduction (29).
Radiotracer analysis demonstrated that most of the methane is
produced from the methyl group of acetate, while CO_2 is produced almost entirely from the carboxyl group. The hydrogen
requirement, which was not noted in the membranous preparations used by Baresi (1), implies that membrane bound proteins
may be required for electron transfer between the oxidative
and reductive portions of the pathway. Ultracentrifugation
has no effect on rates of methane production from acetate
which implies that the carbon transforming enzymes are not
integral membrane proteins.

The molar growth yields of acetotrophic methanogens, as
well as thermodynamic considerations (53), make it likely
that methanogens obtain only a fraction of an ATP for each
methane produced from acetate. At present only a chemiosmotic
model would allow such fractional conservation of energy.
Figure 3 illustrates our current biochemical hypothesis for
how *M. barkeri* strain MS dissimilates acetate with the conservation of ATP. This model provides a role for cytoplasmic
components (activating enzymes, CO dehydrogenase, and methylreductase) and membrane components (electron transport chain).
Acetate is first activated by transformation to an acetyl
moiety. The acetyl is then split to a methyl and a carbonyl
intermediate. The carbonyl intermediate is oxidized by CO
dehydrogenase to CO_2. The electron-proton carrier for CO
dehydrogenase then donates reducing equivalents to a membrane
bound electron transfer chain (e_2 and e_3) which effect a
vectorial proton translocation across the membrane to drive
ATP synthesis.

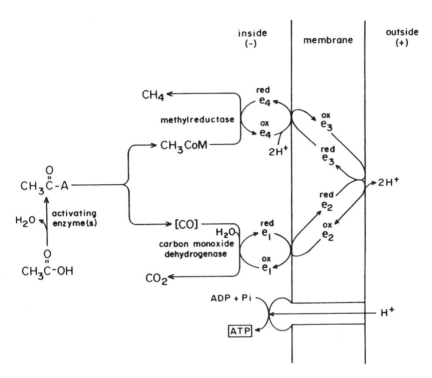

Figure 3.   Proposed biochemical model for acetate degradation and energy conservation by *Methanosarcina barkeri*. Acetate is activated to an acetyl moiety which is cleaved into methyl and carbonyl intermediates. The key redox process which generates a proton motive force is the coupling of soluble CO dehydrogenase and methyl reductase activities to vectorial membrane electron transport and methane production.

REFERENCES

1. Baresi, L., *Abstracts American So. Microbiol. Abstract 115,* (1983).

2. Baresi, L. and R.S. Wolfe, "Levels of Coenzyme F_{420}, Coenzyme M, Hydrogenase, and Methyl-CoM Reductase in Acetate Grown *Methanosarcina,*" *Appl. Environ. Microbiol., 41,* 388-391 (1981).

3. Bhatnagar, L., M.K. Jain, J.-P. Aubert, and J.G. Zeikus, "Comparison of Assimilatory Organic Nitrogen, Sulfur and Carbon Sources for Growth of *Methanobacterium* Species," *Appl. Environ. Microbiol. 48,* 785-790 (1984).

4. Blaut, M. and G. Gottschalk, "Effects of Trimethylamine on Acetate Utilization by *Methanosarcina barkeri,*" *Arch. Microbiol. 133,* 230 (1983).

5. Bryant, M.D., S.F. Tzeng, I.M. Robinson, A.E. Joyner, "Nutrient Requirements of Methanogenic Bacteria," *Advances in Chemistry Species, vol. 105* (F.G. Pohland, ed.), 23-40.   American Chemistry Society, Washington, DC.

6. Buswell, A.M., and F.W. Sailo, "The Mechanism of the Methane Fermentation," *J. Am. Chem. Soc., 70,* 1778-1780, (1948).

7. Cappenberg, Th.E., "Methanogenesis in the Bottom Deposits of a Small Stratifying Lake," *Microbial Production and Utilization of Gases (H_2,CH_4, CO_2)Symposium Proceedings,* (H.G. Schlegel, N. Pfenning, G. Gottschalk, eds.) E.Golte, Verlag, Gottingen.

8. Daniels, L., G. Fuchs, R.K. Thauer, and J.G. Zeikus, "Carbon Monoxide Oxidation by Methanogenic Bacteria," *J. Bacteriol. 132,*118-126 (1977).

9. Drake, H.L., S.-I. Hu, and H.G. Wood, "Purification of Five Components from *Clostridium thermoaceticum* which Catalysed Synthesis of Acetate from Pyruvate and Methyl-tetrahydrofolate: Properties of Phosphotransacetylase," *J. Biol. Chem. 256,* 11137-11144, (1977).

10. Ehhalt, D.H., "The Atmospheric Cycle of Methane", *Tellus 26,* 58-70 (1974).

11. Ehhalt, D.H. and O. Schmidt, "Sources and Sinks of Atmospheric Methane," *Pageoph 116,* 452-464, (1978).

12. Eikmanns, B. and R. Thauer, Personal Communication (1984).

13. Escalante-Semerena, J.C., K.L. Rinehart, Jr. and R.S. Wolfe, "Tetrahydromethanopterin, a Carbon Carrier in Methanogenesis," *J. Biol. Chem. 259,* 9447-9455 (1984).

14. Ferguson, T.J. and R.A. Mah, "Effect of H_2-CO_2 on Methanogenesis from Acetate or Methanol in *Methanosarcina* spp." *Appl. Environ. Microbiol., 46,* 348-355 (1983).
15. Freyer, H.-D., "Atmospheric Cycles of Trace Gases Containing Carbon," *The Global Carbon Cycle,* (B. Bolin, S. Kempe, P. Ketner, eds.), John Wiley and Sons, New York (1979).
16. Fuchs, G. and E. Stupperich, "Acetate Assimilation and the Synthesis of Alanine, Aspartate, and Glutamate in *Methanobacterium thermoautotrophicum,"* *Arch. Microbiol., 117,* 61-66.
17. Hauser, B.A., K. Wuhrmann, and A.J.B. Zehnder, "*Methanothrix soehgenii* gen. nov. sp. nov., A New Acetotrophic Non-hydrogen Oxidizing Methane Bacterium," *Arch. Microbiol., 132,* 1-9 (1982).
18. Hutchinson, G.E., "A Note on Two Aspects of the Geochemistry of Carbon," *Am. J. Sci., 247,* 27-32 (1949).
19. Hutten, T.J., H.C.M. Bongaerts, C. van der Drift, and G.D. Vogels, "Acetate, Methanol, and CO_2 as Substrates for Growth of *Methanosarcina barkeri,"* *Antonie van Leevwenhook, 46,* 601-610 (1981).
20. Jeris, J.S. and P.L. McCarty, "The Biochemistry of Methane Fermentation Using ^{14}C Tracers," *J. Water Pollution Control Fed., 37,* 178-192 (1965).
21. Kandler, O. and H. Konig, "Chemical Composition of the Peptidoglycan Free Cell Walls of Methanogenic Bacteria," *Arch. Microbiol., 118,* 141-152 (1978).
22. Kenealy, W. and J.G. Zeikus, "Influence of Corrinoid Antagonists on Methanogen Metabolism," *J. Bacteriol., 146,* 133-140 (1981).
23. Kenealy, W.R. and J.G. Zeikus, "One Carbon Metabolism in Methanogens: Evidence for Synthesis of a Two Carbon Cellular Intermediate and Unification of Catabolism and Anabolism in *Methanosarcina barkeri,"* *J. Bacteriol, 151,* 932-941 (1982).
24. Kohler, H.-P.E. and A.J.B. Zehnder, "Carbon Monoxide Dehydrogenase and Acetate Thiokinase in *Methanothrix soehgenii,"* *Fems Microbiol. Letts., 21,* 287 (1984).
25. Koyana, T., "Gaseous Metabolism in Lake Sediments and Paddy Soils and the Production of Atmospheric Methane and Hydrogen," *J. Geophys. Res., 68,* 3971-3973 (1963).
26. Krzycki, J.A., R.H. Wolkin, and J.G. Zeikus, "Comparison of Unitrophic and Mixotrophic Substrate Metabolism by an Acetate-Adapted Strain of *Methanosarcina barkeri,"* *J. Bacteriol., 149,* 247-254 (1982).
27. Krzycki, J. and J.G. Zeikus, "Quantification of Corrinoids in Methanogenic Bacteria," *Curr. Microbiol., 3,* 243-245 (1980).

28. Krzycki, J. and J.G. Zeikus, Abstracts of Annual Meeting of the American Soc. for Microbiology, Abstract Ill, (1983).
29. Krzycki, J. and J.G. Zeikus, "Characterization and Purification of Carbon Monoxide Dehydrogenase from *Methanosarcina barkeri*," *J. Bacteriol., 158,* 231-237 (1984).
30. Krzycki, J.A. and J.G. Zeikus, "Acetate Catabolism by Methanogenic Bacteria: Hydrogen Dependent Methanogenesis from Acetate by a Soluble Protein Fraction of *Methanosarcina,*" *Fems Micro. Letts.,* In press (1984).
31. Mah, R.A., "Isolation and Characterization of *Methanococcus mazei*," *Curr. Microbiol., 3,* 321-326 (1980).
32. Mah, R.A., M.R. Smith, and L. Baresi, "Studies on an Acetate Fermenting Strain of *Methanosarcina*," *Appl. Environ. Microbiol., 35,* 1174-1184 (1978).
33. Mountfort, D.O. and R.A. Asher, "Changes in Proportion of Acetate and Carbon Dioxide Used as Methane Precursors During the Anaerobic Digestion of Bovine Wastes," *Appl. Environ. Microbiol., 35,* 648-654 (1978).
34. Opperman, R.A., W.D. Nelson, and R.E. Bowman, "In vivo Studies of Methanogenesis in the bovine rumen: Dissimilation of Acetate," *J. Gen. Microbiol., 25,* 103-111 (1961).
35. Nelson, M.J.K. and K.R. Sowers, Abstracts of the Ann. Meet. Am. Soc. Microbiol., Abstract I25, (1984).
36. Nozhevnikova, A.N. and T.G. Yagodina, "A Thermophillic Acetate Methane-Producing Bacterium," *Microbiology, 51,* 534 (1983).
37. Phelps, T.J. and J.G. Zeikus, "Comparison of Proton Activity on the General Physiology of a Prevalent Acetogen and Methanogen Isolated from a Mildly Acid Lake Sediment," Manuscript in preparation (1984).
38. Pine, M.J. and H.A. Barker, "Studies on the Methane Fermentation. XII. The Pathway of Hydrogen in the Acetate Fermentation," *J. Bacteriol., 71,* 644-648 (1956).
39. Sansone, F.J. and C.S. Martens, "Methane Production From Acetate and Associated Methane Fluxes from Anoxic Costal Sediments," *Science, 211,* 709-767 (1981).
40. Scherer, P. and H. Sahm, "Growth of *Methanosarcina barkeri* on Methanol or Acetate in a Defined Medium," *Proc. 1st Inst. Sympos. on Anaerobic Digestion,* (D.A. Stafford and B.I. Wheatley, eds.), Scientific Press, Cordiff (1980).
41. Scherer, P. and H. Sahm, "Influence of Sulphur Containing Compounds on the Growth of *Methanosarcina barkeri* in a defined medium," *Europ. J. Appl. Microbiol. Biotechnol., 12,* 28-35 (1981).
42. Schink, B. and J.G. Zeikus, "Microbial Ecology of Pectin Decomposition in Anoxic Lake Sediments," *J. Gen. Microbiol., 128,* 393-404 (1982).

43. Seiler, W., "Contribution of Biological Processes to the Global Budget of Methane in the Atmosphere," *Current Perspectives in Microbial Ecology,* (M.J. Klug and C.A. Reddy, eds.), American Society for Microbiol., Washington, DC (1984).

44. Shapiro, S. and R.S. Wolfe, "Methyl Coenzyme M, An Intermediate in Methanogenic Dissimilation of C_1 Compounds by *Methanosarcina barkeri,*" *J. Bacteriol., 141,* 728-734 (1980).

45. Smith, M.R., Abstracts of Ann. Meet. Am. Soc. Microbiol., Abstract I7 (1983).

46. Smith, M.R. and R.A. Mah, "Growth and Methanogenesis by *Methanosarcina* Strain 227 on Acetate and Methanol," *Appl. Environ. Microbiol., 36,* 870-879 (1978).

47. Smith, M.R. and R.A. Mah, "Acetate as Sole Carbon and Energy Source for Growth of *Methanosarcina* Strain 227," *Appl. Environ. Microbiol., 39,* 993-999 (1980).

48. Smith, P.H. and R.A. Mah, "Kinetics of Acetate Metabolism During Sludge Digestion," *Appl. Microbiol., 14,* 368-371 (1966).

49. Sowers, K.R., S.F. Baron, and J.G. Ferry, "*Methanosarcina acetivorans* sp. nov., An Acetotrophic Methane Producing Bacterium Isolated from Marine Sediments," *Appl. Environ. Microbiol., 47,* 971-978 (1984).

50. Stadtman, T.C. and H.A. Barker, "Studies on the Methane Fermentation. IX. Origin of Methane in the Acetate and Methanol Fermentation by *Methanosarcina,*" *J. Bacteriol., 61,* 81 (1950).

51. Stupperich, E. and G. Fuchs, "Autotrophic Acetyl Coenzyme A Synthesis in vitro From Two CO2 in *Methanobacterium,*" *Febs. Letts., 156,* 345 (1983).

52. Taylor, G.T., "The Methanogenic Bacteria," *Progress in Industrial Microbiol., 16,* 231-330 (1982).

53. Thauer, R.K. and J.G. Morris, "Metabolism of Chemotrophic Anaerobes, Old Views and New Aspects," *The Microbe 1984: Part II. Prokaryotes and Eukaryotes,* (D.P. Kelly and N.G. Carr, eds.), *SGM Sympos., 36,* Cambridge Univ. Press, (1984).

54. Weimer, P.J. and J.G. Zeikus, "Acetate Metabolism in *Methanosarcina barkeri,*" *Arch. Microbiol., 119,* 175-182 (1978).

55. Weimer, P.J. and J.G. Zeikus, "One Carbon Metabolism in Methanogenic Bacteria: Cellular Characterization and Growth of *Methanosarcina barkeri,*" *Arch. Microbiol., 119,* 49-57 (1978).

56. Weimer, P.J. and J.G. Zeikus, "Acetate Assimilation Pathway of *Methanosarcina barkeri,*" *J. Bacteriol., 137,* 332-339 (1979).

57. Whitman, W.B., E. Ankwanela, and R.S. Wolfe, "Nutrition and Carbon Metabolism of *Methanococcus voltae*," *J. Bacteriol.*, *149*, 852-863 (1982).
58. Winfrey, M.R. and J.G. Zeikus, "Microbial Methanogenesis and Acetate Metabolism in a Meromictic Lake," *Appl. Environ. Microbiol.*, *37*, 213-221 (1979).
59. Winter, J. and R.S. Wolfe, "Complete Degradation of Carbohydrate to Carbon Dioxide and Methane by Syntrophic Cultures of *Acetobacterium woodii* and *Methanosarcina barkeri*," *Arch. Microbiol.*, *121*, 97-102 (1979).
60. Vogels, G.D. and C.M. Visser, "Interconnection of Methanogenic and Acetogenic pathways," *Fems Microbiol. Letts.*, *20*, 291-297 (1983).
61. Zehnder, A.J.B., B.A. Huser, T.D. Brock, and K. Wuhrmann, "Characterization of an Acetate Decarboxylating, Non-hydrogen Oxidizing Methane Bacterium," *Arch. Microbiol.*, *124*, 1-11 (1980).
62. Zeikus, J.G., "Microbial Population in Digesters," *First International Symposium on Anaerobic Digestion,* (D.A. Stafford, ed.), A.D. Scientific Press, Cordiff (1981).
63. Zeikus, J.G., "Chemical and Fuel Production by Anaerobic Bacteria," *Ann. Rev. Microbiol.*, *34*, 423-464 (1980).
64. Zeikus, J.G., "Metabolism of One Carbon Compounds by Chemotrophic Anaerobes," *Advances in Microbial Physiology*, *Vol. 24*, Academic Press, Inc., New York (1983).
65. Zeikus, J.G., G. Fuchs, W. Kenealy, and R.K. Thauer, "Oxidoreductase Involved in Cell Carbon Synthesis of *Methanobacterium thermoautotrophicum*," *J. Bacteriol.*, *132*, 604-613 (1977).
66. Zinder, S.H., S.C. Cardwell, T. Anguish, M. Lee, and M. Koch, "Methanogenesis in a Thermophillic (58°C) Anaerobic Digestor: *Methanothrix* as an important aceticlastic Methanogen," *Appl. Environ. Microbiol.*, *47*, 796-807 (1984).
67. Zinder, S.H. and M. Koch, "Non-aceticlastic Methanogenesis from Acetate: Acetate Oxidation by a Thermophilic Syntrophic Culture," *Arch. Microbiol.*, *138*, 263 (1984).
68. Zinder, S.H. and R.A. Mah, "Isolation and Characterization of a Thermophilic Strain of *Methanosarcina* unable to use H_2-CO_2 for Methanogenesis," *Appl. Environ. Microbiol.*, *38*, 996-1008 (1979).

3
ACETOCLASTIC METHANOGENS IN ANAEROBIC DIGESTERS

J.P. Touzel and G. Albagnac

Station de Technologie Alimentaire
Institut National de la Recherche Agronomique
Villeneuve D'Ascq, France

ABSTRACT

Four different strains of methane bacteria converting acetate into methane have been isolated from anaerobic digesters. Some of their main characteristics are presented.

INTRODUCTION

Methane bacteria able to convert acetate into methane (acetoclastic reaction) play a major role in anaerobic digesters where acetate accounts for more than 2/3 of the methane produced. The first acetoclastic methanogen ever obtained in pure culture was *Methanosarcina barkeri*. Since sarcina-like organisms were seldom observed in such ecosystems the purification and the isolation of new strains was of interest.

MATERIALS AND METHODS

The starting materials were the sludges of bench-scale anaerobic digesters treating wastewaters of the vegetable canning industry either at 35 and 55°C. The techniques designed for work with extremely oxygen-sensitive bacteria were used throughout these studies. The culture medium BCYT (Samain et al., 1982) was a carbonate buffered medium modified from Zeikus and Wolfe's (1972) CBBM. At least six successive enrichments were made with sodium acetate (50 mM) as the sole energy and carbon source and under a N_2+CO_2 (85+15) gas phase. Vanomycin (200 mg/l) was added in the last enrichments for eliminating non-methanogenic bacteria. Kinetic data were obtained from 200 to 400 ml cultures. Methane was analyzed gas-chromatographically using thermal conductivity detector.

Gas was sampled with a 1 ml gas-tight syringe fitted with a gas valve. Nitrogen served as internal standard. Growth rates were computed from the exponential part of the methane production kinetics with different concentrations of acetate. Estimation of the affinity constant and maximum growth rate was made by the distribution-free method of Cornish-Bowden (1979).

RESULTS

The main features of the four isolated strains are presented in Table 1.

Morphology

The post-typical sarcina is the MST-Al strain. Clumps are finely divided and they never form large aggregates like in MC3 and CHTI 55. These large aggregates are several mm in size and their aspects differ between MC3 and CHTI 55. But these two strains share the property of releasing coccoid units when for example pressure is applied on the microscope slide. Formation of coccoid units seems to depend on growth conditions as well. Strain MC3 grows entirely in the coccoid form on DSM medium nr. 120. Coccoid units are still made of numerous cells held together by a loose outer cell wall. The overall shape of the aggregate is spherical and number of cells devoid of cell wall are detected by electron microscopy.

Ultrastructure

Core-like structures extending through the cytoplasm were detected in the thermophilic strains only. In strain MST-Al these tubes are approximately 2µ in length and 0.15µ in diameter. Each cell contains 1 tube, sometimes 2. Tubes are open at each end and their contents appear similar to that of cytoplasm. Their function is unclear. Granular structures were seen commonly in the cytoplasm of the sarcina. Their typical "rosette" shape recalled the glycogen inclusions of mammalian tissues. The similarity of these two types of inclusions was further confirmed by performing the Thiery (1967) staining on thin sections of heart muscle and strain MST-Al.

KINETIC PARAMETERS

Kinetic parameters have been determined for the four isolated strains. Their values are compared with those given in the literature for the type strains or related species (Table 2).

Table 1. *Main Features of the Isolated Strains*

Strain nr.	FE	MC3	CHTI 55	MST-Al
Name	Methanothrix soehngenii	Methanosarcina mazei	Methanosarc. sp.	Methanosarc. sp.
Opt. temp.	35°C	35°C	57°C	55°C
Opt. pH	7	7	7	6.5-7
Substrates	acetate	acetate, methanol	Methanol + H₂	methylamines

Table 2. Kinetic Parameters of the Different Strains

Strain	Growth tempera- ture (°C)	Affinity constant for acetate K_s (mmol.l^{-1})	Specific growth rate		Min. gen. tim.	
			Theor. (h^{-1})	observ. (h^{-1})	theor. (h)	observ. (h)
Methanothrix soehngenii FE	35	1.04	.00996	.00924	70.8	75
(" " str. opfikon [1])	(35)	(.7)		(.00854)		(81)
Methanosarcina mazei MC3	35	1.27	.01965	.019378	35.3	35.7
(" " str. S6 [2])	(35)			(.04175)		(16.6)
Methanosarcina sp. CHTI 55	57	7.48	.0914	.0858	7.6	8.1
(" " str. TM1 [3])	(50)	(4.5)		(.0577)		(12)
Methanosarcina sp. MST-Al	55	20.98	.0745	.063388	9.3	10.9

[1] = Huser et al. (1982); [2] + Mah (1980); [3] = Zinder and Mah (1979)

CONCLUSION

With the exception of *M. soehngenii* which is commonly seen in anaerobic digester sludges, the occurrence of *Methanosarcina sp.* as major acetoclastic methanogens in such systems awaits further investigations. However, the morphological changes induced by the culture medium in some *Methanosarcina* strains could have been misleading for observers looking for sarcina-like organisms, as already noticed by Mah (1980). All the isolates are characterized by their inability to produce methane from $H_2 + CO_2$. Among other features, the occurrence of granular cytoplasmic inclusions possibly of polyosidic nature has been evidenced in all the *Methanosarcina* isolates together with tubular (core-like) structures in the thermophilic strains only. From a technological point of view, the kinetic characteristics clearly speak in favor of the mesophilic processes when an efficient acetate conversion into methane has to be reached as it is the case in wastewater treatment.

REFERENCES

Samain, E., G. Albagnac, M.C. Dubourguier, and J.P. Touzel, "Characterization of a New Propionic Acid Bacterium that Ferments Ethanol and Displays a Growth Factor Dependent Association with a Gram Negative Homoacetogen," *FEMS Microbiol. Lett., 15,* 69-74 (1982).

Zeikus, J.G. and R.S. Wolfe, *"Methanobacterium thermoautotrophicus* sp. nov., An Anerobic Autotrophic Extreme Thermophile," *J. Bact., 109,* 707-715 (1972).

Cornish-Bowden, A., "Fundamentals of Enzyme Kinetics, *Butterworths,* pp. 230 (1979).

Huser, B.A., K. Wuhrmann, and A.J.B. Zehnder, *"Methanothrix soehngenii* gen. nov. sp. nov., A New Acetotrophic Non-Hydrogen-Oxidizing Methane Bacterium," *Arch. Microbiol., 132*(1), 1-9 (1982).

Mah, R.A., "Isolation and Characterization of *Methanococcus mazei," Curr. Microbiol., 3,* 321-326 (1980).

Thiery, J.P., "Mise en Evidence des Polysaccharides sur Soupes Fines en Microscopie Electronique," *J. Microscopie, 6,* 978-1018 (1967).

Zinder, S.H. and R.A. Mah, "Isolation and Characterization of a Thermophilic Strain of *Methanosarcina* Unable to Use H_2-CO_2 for methanogenesis," *Appl. Environ. Microbiol., 38*(5), 996-1008 (1979).

4

METABOLISM OF H_2 AND CO_2 BY *METHANOBACTERIUM*

Peter Schönheit and Rudolf K. Thauer

Fachbereich Biologie, Mikrobiologie
Philipps-Universität
Marburg, Federal Republic of Germany

ABSTRACT

Methanobacterium species grow on H_2 and CO_2 as sole energy source: $4 H_2 + CO_2 \longrightarrow CH_4 + 2 H_2$. Per mol of methane formed from CO_2 about 1 mol of ATP is generated. Coupling is probably via the chemiosmotic mechanism. The reduction of CO_2 to methane proceeds via carrier bound intermediates involving methanofuran, tetrahydromethanopterin, CoM, component B, and a nickelporphinoid as coenzymes.

Most methanobacteria can grow on CO_2 as sole carbon source; i.e., they are autotrophs. Acetyl-CoA rather than 3-phosphoglycerate has been identified as an early CO_2 fixation product. Evidence has accumulated that acetyl-CoA is formed in a carbonylation reaction with a methyl-corrinoid and an acetyl-corrinoid as intermediates.

INTRODUCTION

Most methanogenic bacteria can reduce CO_2 with H_2 to CH_4 and couple this reaction with the synthesis of ATP from ADP and Pi. Studies on the metabolism of H_2 and CO_2 have mainly been performed with *Methanobacterium thermoautotrophicum* strain ΔH or strain Marburg. This methanogen grows at 65°C on H_2 and CO_2 as sole energy and carbon sources and does not form methane from acetate, methanol, methylamines or formate (Zeikus and Wolfe, 1972; Balch et al., 1979; for differences in strains see Brandis et al., 1981).

41

This chapter summarizes what is known about the catabolism and anabolism of *M. thermoautotrophicum*. There is accumulating evidence that most of the results obtained with this organism can, in principal, be generalized for all methanogens. For most recent reviews see Escalante-Semerena et al. (1984a), Fuchs and Stupperich (1984), Zeikus (1983), and Daniels et al. (1984).

CATABOLISM

The reduction of CO_2 with H_2 to CH_4 is an exergonic process:

$$4\ H_2 + CO_2 \longrightarrow CH_4 + 2\ H_2O \qquad \Delta G^{o\prime} = -131\ kJ/mol$$

The reaction is coupled with the synthesis of ATP as evidenced by growth of methanogens on H_2 and CO_2 as sole energy source. From the free energy change associated with the reaction it can be estimated that about 1 mol of ATP can be formed per mol methane (Thauer et al., 1977; Thauer and Morris, 1984). First the mechanism of CO_2 reduction to CH_4 and then the mechanism of coupling will be discussed.

Mechanism of CO_2 Reduction to CH_4

Cell suspensions of exponentially grown *M. thermoautotrophicum* (td = 2 hours) mediate the reduction of CO_2 with H_2 at a rate of approximately 6-7 $\mu mol\ min^{-1}\ mg\ protein^{-1}$ (Perski et al., 1982). Formate, formaldehyde and methanol, alone or in the presence of H_2, are not converted to CH_4 by the cells at significant rates. Since the cytoplasmic membrane is probably permeable to these C_1 units, this finding indicates that CO_2 reduction to methane proceeds via carrier bound intermediates. This conclusion was first drawn by Barker in 1956 (Barker, 1956).

Cell extracts of the organism catalyze the reduction of CO_2 with H_2 at approximately 1% of the rate of cell suspensions (Gunsalus and Wolfe, 1977; Romesser and Wolfe, 1982). Even upon addition of enzymes, coenzymes and intermediates presently thought to be involved in methanogenesis from CO_2, the rate of the reaction does not increase significantly. The reason for this is not understood. The finding indicates that our understanding of the mechanism of CO_2 reduction to methane is still incomplete.

From studies in cell free systems it became evident that at least five novel coenzymes are involved in CO_2 reduction to methane; methanofuran (formerly named carbon dioxide reducing factor) (Leigh, 1983; Leigh and Wolfe, 1983; Leigh et al., 1984; Escalente-Semerena et al., 1984a); tetrahydromethanopterin (THMP) (identical with formaldehyde activating factor) (van Beelen et al., 1984; Vogels et al., 1984; Escalante-Semerena and Wolfe, 1984; Escalente-Semerena et al., 1984 b and c); coenzyme M (Taylor and Wolfe, 1974); factor F430 (Pfaltz et al., 198s; Livingston et al., 1984; Thauer et al., 1984); and component B (Gunsalus and Wolfe, 1980; Nagle and Wolfe, 1984). With the exception of component B, the structure of these coenzymes have been elucidated (Fig. 1). (There is still a debate whether factor F430 contains covalently bound 1 mol of coenzyme M and 1 mol of 6,7-dimethyl-8-ribityl-5,6,7,8-tetrahydrolumazine (Keltjens et al., 1982; 1983 a and b)). Formylmethanofuran, methenyl-THMP, methylene-THMP, methyl-THMP and methyl coenzyme M have been identified as intermediates in CO_2 reduction to methane (Vogels et al, 1984; Escalente-Semerena et al., 1984a) (Fig. 2). Factor F430 is tightly bound to methyl CoM reductase (Ellefson et al., 1982; Ankel-Fuchs et al., 1984; Hausinger et al., 1984) which catalyzes the reduction of methyl CoM to methane. Component B is a cofactor in this reaction (Nagle and Wolfe, 1983).

Almost nothing is known about the electron carriers involved in CO_2 reduction to methane (Lancaster, 1982). Factor F420 (a 5-deazaflavine derivative) (Eirich et al., 1978) was once thought to be the coenzyme of the methyl CoM reductase (Ellefson and Wolfe, 1980). Recent results have shed doubt on this interpretation. Instead, FAD has now been shown to be required for methyl CoM reduction to methane (Nagle and Wolfe, 1983). From the enzymes involved in the electron transport chain two hydrogenases have been purified. However, it is still not known whether both or only one of the two enzymes participate in CO_2 reduction in methane. Both hydrogenases contain nickel (Graf and Thauer, 1981; Albracht et al., 1982; Jacobson et al., 1982; Kojima et al., 1983; Lindahl et al., 1984; Tan et al., 1984).

Coupling of CO2 Reduction to Methane with the Synthesis of ATP

From the reactions involved in CO_2 reduction to methane (Fig. 2) only the reduction of methyl CoM ($\hat{=}$ CH_3OH) with H_2 to methane ($\Delta G^{0\prime}$ = - 112 kJ/mol) is exergonic enough to drive the synthesis of ATP. The methyl CoM reductase system is

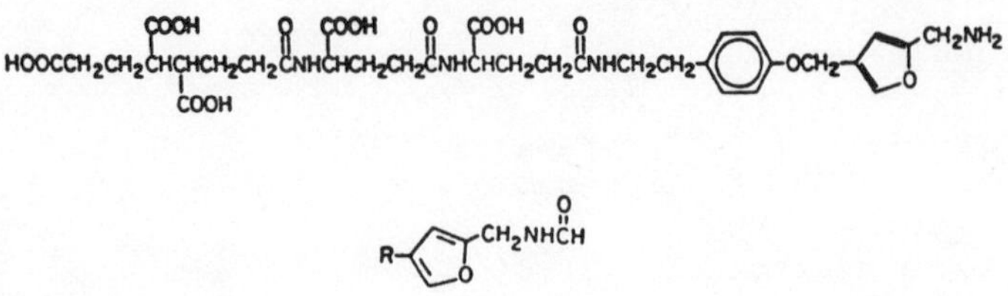

Methanofuran

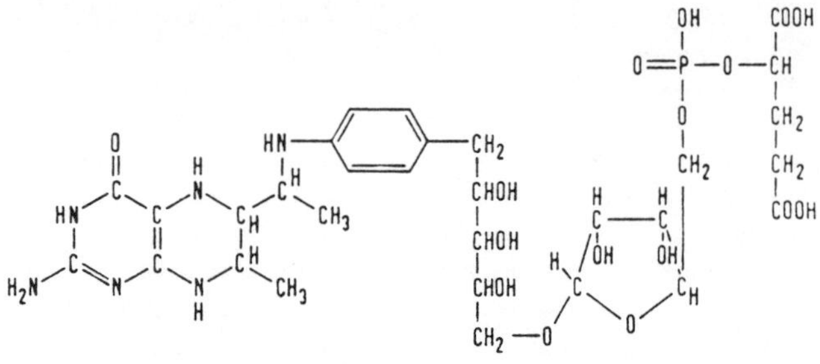

Tetrahydromethanopterin

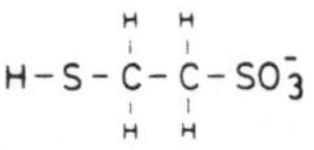

Coenzyme M

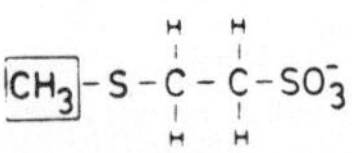

Methylcoenzyme M

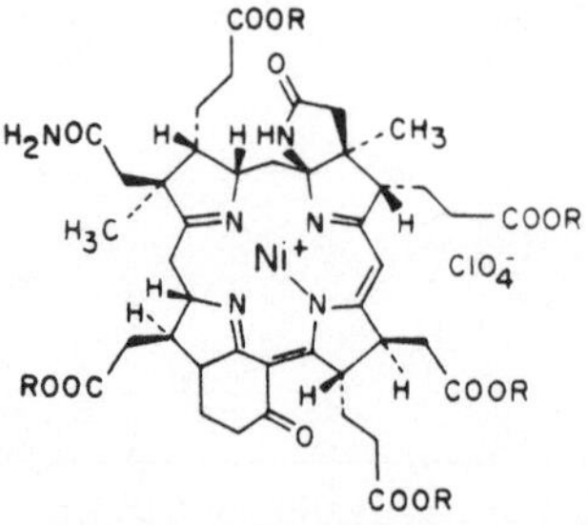

Factor F$_{430}$

Figure 1.   Structures of methanofuran (Leigh et al., 1984; Escalante-Semerena et al., 1984a), tetrahydromethanopterin (van Beelen et al., 1984; Escalante-Semerena et al., 1984b), coenzyme M (Taylor and Wolfe, 1974), and of factor F430 (Pfaltz et al., 1982; Livingston et al., 1984).

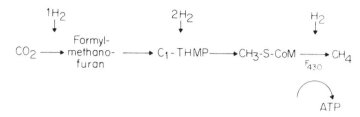

Figure 2. Proposed pathway of methane formation from H_2 and
CO_2 in *Methanobacterium thermoautotrophicum*.
THMP = Tetrahydromethanopterin; C_1-THMP = methenyl-
THMP, methylene-THMP and methyl-THMP. This is a
simplified scheme not considering all the experimental data available. For a more sophisticated
version see Escalante-Semerena et al. (1984a).

therefore generally considered to be the coupling site in
methanogenesis. The mechanism of coupling has not yet been
elucidated. Substrate level phosphorylation can be excluded
since H_2 is the electron donor in the reaction (Thauer et al.,
1977). Therefore a chemiosmotic mechanism has been envisaged.
In favor of this hypothesis are the following findings: (i)
Cells of *M. thermoautotrophicum* that form methane from CO_2
show across the cytoplasmic membrane an electrochemical proton potential in the order ot 200 mV ($\Delta\psi$ = 200 mV, inside
negative; ΔpH = 0) (Jarrell and Sprott, 1981; Butsch and
Bachofen, 1984; Schönheit et al., 1984). $\Delta\mu H^+$ collapses and
ATP synthesis stops when methanogenesis is inhibited. (ii)
Cells of *M. thermoautotrophicum*, in the absence of methane
formation, synthesize ATP when a ΔpH (inside alkaline)
(Doddema et al., 1978) or a $\Delta\psi$ (inside negative) (Schönheit
and Perski, 1983) is artificially imposed to the cells.
There is no evidence, however, that methanogenesis from CO_2
primarily generates an electrochemical proton potential which
then drives the synthesis of ATP. If the membrane associated
ATPase operates reversibly, the data are also consistent
with the build-up of an electrochemical proton potential at
the expense of the hydrolysis of ATP which is formed during
methanogenesis by a yet unknown mechanism. Attempts to dis-
criminate between the two possibilities have failed until now
since the ATPase/ATP synthase of *M. thermoautotrophicum* is
not susceptible to inhibition with dicyclohexylcarbodiimide
(DCCD) or other inhibitors (Sprott and Jarrell, 1982). The
respective enzyme of *Methanosarcina barkeri* (strain Fusaro)

appears, however, to be inactivated by this reagent. With this organism it has recently been demonstrated that in the presence of DCCD methane formation from methanol and H_2 leads to the generation of an electrochemical proton potential in the apparent absence of a measurable phosphorylation potential (Blaut and Gottschalk, 1984). This finding suggests that also in *M. thermoautotrophicum* the formation of the electrochemical proton potential rather than the synthesis of ATP is the primary event in coupling of methanogenesis with the synthesis of ATP (for a review see Daniels et al., 1984).

It has recently been observed that, at low concentrations, protonophores enhance rather than inhibit methanogenesis from H_2 and CO_2 in cell suspensions of *M. thermoautotrophicum*. Under these conditions an electrochemical proton potential was no longer measurable. Surprisingly, however, ATP was still formed. This finding appears to exclude a chemiosmotic mechanism of coupling (P. Schönheit, unpublished).

The finding that sodium ions are required for both methanogenesis and ATP synthesis in *M. thermoautotrophicum* is also of importance (Perski et al., 1981; Perski et al., 1982; Schönheit and Perski, 1983). Whether sodium plays a direct or indirect role in the mechanism of coupling remains to be elucidated.

ANABOLISM

Methanobacterium thermoautotrophicum requires for growth only CO_2 as carbon source, i.e., the bacterium is an autotroph. Evidence has been provided that autotrophic CO_2 fixation in this methanogen does not proceed via the Calvin cycle (for reviews see Fuchs and Stupperich, 1982, 1984). In a series of labelling experiments backed up by enzymatic studies it was shown that acetyl-CoA (formed from 2 CO_2 in a non-noncyclic process) is an early intermediate in CO_2 fixation. From acetyl-CoA pyruvate is formed by reductive carboxylation. Until recently the mechanism of acetyl-CoA formation from 2 CO_2 remained obscure.

A breakthrough in the understanding was the finding that [14]CO is specifically incorporated into C_2 of pyruvate (which is derived from the caroxyl-group of acetyl-CoA) when *M. thermoautotrophicum* is growing on H_2, CO_2, and [14]CO (Stupperich et al., 1983). It had already been shown earlier that methanol is specifically incorporated into the methyl-group of pyruvate (which is derived from the methyl-group of acetyl-CoA) in methanol-growing *Methanosarcina barkeri* (Kenealy and Zeikus, 1982). These observations indicate that, as shown in Fig. 3, methanol (in a carrier-bound form) and CO are intermediates in acetyl-CoA synthesis from 2 CO$_2$.

The scheme in Fig. 3 is supported by the following findings: (i) CO is formed from CO$_2$ and H$_2$ by growing cells and by cell suspensions of *M. thermoautotrophicum*. CO formation is inhibited by cyanide (Conrad and Thauer, 1983; Eikmanns et al., 1984). The cells contain an active carbon monoxide:methyl viologen oxidoreductase which contains nickel (Hammel et al., 1984) and is inactivated by cyanide (Daniels et al., 1977). (ii) Cyanide affects acetyl-CoA formation from 2 CO$_2$ but not the formation of the thioester from CO$_2$ and CO (Stupperich and Fuchs, 1983). In the presence of cyanide the carboxyl-group of acetyl-CoA is exclusively derived from added CO (Stupperich and Fuchs, 1984 a and b).

In Fig. 3 propyl iodide is shown to inhibit CO$_2$ fixation by inhibiting the formation of methylated corrinoid enzyme which is probably involved in the carbonylation reaction. The evidence for this is based on the finding that propyl iodide inhibits growth rather than CH$_4$ and CO formation and that growth inhibition can be reverted by light (Fuchs, unpublished results). The presence of corrinoids in *M. thermoautotrophicum* has been demonstrated (Krzycki and Zeikus, 1980; Scherer and Sahm, 1980). The involvement of an acetyl corrinoid as intermediate is deduced from the observation that methyl B$_{12}$ reacts with CO to acetyl B$_{12}$ in a photoinduced carbonylation reaction (Kräutler, 1984).

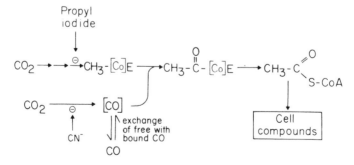

Figure 3.   Proposed pathway of autotrophic CO$_2$ fixation in *Methanobacterium thermoautotrophicum*. [Co] E = Corrinoid containing enzyme.

Acetogenic bacteria which mediate a total synthesis of acetate from 2 CO_2 have recently been shown to form CO (Diekert et al., 1984a). Both CO formation and acetate synthesis were inhibited by cyanide. CO is incorporated by these bacteria into the carboxyl group of acetate (Hu et al., 1982; Diekert and Ritter, 1983; Kerby et al., 1983). A corrinoid enzyme was found to be essential component of the enzyme system catalyzing the carbonylation reaction and a methylated corrinoid enzyme to be an intermediate (Pezacka and Wood, 1984). Thus acetyl CoA synthesis in autotrophic methanogens and in acetogenic bacteria appears to be mechanistically similar in many respects (Diekert et al., 1984b).

ACKNOWLEDGMENTS

This work was supported by a grant from the Deutsche Forschungsgemeinschaft (Bonn-Bad Godesberg) and by the Fonds der Chemischen Industrie.

REFERENCES

Albracht, S.P.J., E.-G. Graf, and R.K. Thauer, *FEBS Lett.*, *140*, 311-313 (1982).

Ankel-Fuchs, D., R. Jaenchen, N.A. Gebhardt, and R.K. Thauer, *Arch. Microbiol.*, In press (1984).

Balch, W.E., G.E. Fox, L.J. Magrum, C.R. Woese, and R.S. Wolfe, *Microbiol. Rev.*, *43*, 260-296 (1979).

Barker, H.A., Bacterial Fermentations, John Wiley and Sons, New York (1956).

van Beelen, P., A.P.M. Stasse, J.W.G. Bosch, G.D. Vogels, W. Guijt, and C.A.G. Haasnoot, *Eur. J. Biochem.*, *138*, 563-571; *139*, 359-365 (1984).

Blaut, M. and G. Gottschalk, *Eur. J. Biochem.*, *141*, 217-222 (1984).

Brandis, A., R.K. Thauer, and K.O. Stetter, *Zbl. Bakt. Hyg.*, *I. Abt. Orig. C 2*, 311-317 (1981).

Butsch, B.M. and R. Bachofen, *Arch. Microbiol.*, *138*, 293-298 (1984).

Conrad, R. and R.K. Thauer, *FEMS Microbiol. Lett.*, *20*, 229-232 (1983).

Daniels, L., G. Fuchs, R.K. Thauer, and J.G. Zeikus, *J. Bacteriol.*, *132*, 118-126 (1977).

Daniels, L., R. Sparling, and G.D. Sprott, *Biochim. Biophys. Acta*, *786*, 113-163 (1984).

Diekert, G. and M. Ritter, *FEMS Microbiol. Lett.*, *17*, 299-302 (1983).

Diekert, G., M. Hansch, and R. Conrad, *Arch. Microbiol.*, *138*, 224-228 (1984a).

Diekert, G., G. Fuchs, and R.K. Thauer, "Microbial Gas Metabolism - Mechanistic, Metabolic and Biotechnical Aspects, (R.K. Poole and C.S. Dow, eds.), Society for General Microbiology-Special Publication, Academic Press, In press (1984b).

Doddema, H.J., T.J. Hutten, C. Van der Drift, and G.D. Vogels, *J. Bacteriol., 136,* 19-23 (1978).

Eikmanns, B., G. Fuchs, and R.K. Thauer, *Eur. J. Biochem.,* Submitted (1984).

Eirich, J.D., G.D. Vogels, and R.S. Wolfe, *Biochemistry, 17,* 4583-4393 (1978).

Ellefson, W.L. and R.S. Wolfe, *J. Biol. Chem., 255,* 8388-8389 (1980).

Ellefson, W.L., W.B. Whitman, and R.S. Wolfe, *Proc. Natl. Acad. Sci. USA, 79,* 3707-3710 (1982).

Escalante-Semerena, J.C. and R.S. Wolfe, *J. Bacteriol., 158,* 721-726 (1984).

Escalante-Semerena, J.C., J.A. Leigh, and R.S. Wolfe, "Microbial Growth in C₁ Compounds; Proceedings of the 4th International Symposium," (R.L. Crawford and R.S. Hanson, eds.), American Society for Microbiology, Washington, DC, pp. 191-198 (1984a).

Escalante-Semerena, J.C., K.L. Rinehart, Jr., and R.S. Wolfe, *J. Biol. Chem.,* In press (1984b).

Escalante-Semerena, J.C., J.A. Leigh, K.L. Rinehardt, Jr., and R.S. Wolfe, *Proc. Natl. Acad. Sci. USA, 81,* 1976-1980 (1984c).

Fuchs, G. and E. Stupperich, *Zbl. Bakt. Hyg., I. Abt. Orig., C 3,* 277-288 (1982).

Fuchs, G. and E. Stupperich, "Microbial Growth in C₁-Compounds; Proceedings of the 4th International Symposium," (R.L. Crawford and R.S. Hanson, eds.), American Society for Microbiology, Washington, DC, pp. 199-202 (1984).

Graf, E.-G. and R.K. Thauer, *FEBS Lett., 136,* 165-169 (1981).

Gunsalus, R.P. and R.S. Wolfe, *Biochem. Biophys. Res. Commun., 76,* 790-795 (1977).

Gunsalus, R.P. and R.S. Wolfe, *J. Biol. Chem., 255,* 1891-1895 (1980).

Hammel, E.H., K.L. Cornwell, G.B. Diekert, and R.K. Thauer, *J. Bacteriol., 157,* 975-978 (1984).

Hausinger, R.P., W.H. Orme-Johnson, and C. Walsh, *Biochemistry, 23,* 801-804 (1984).

Hu, S.-I, H.L. Drake, and H.G. Wood, *J. Bacteriol., 149,* 440-448 (1982).

Jacobson, F.S., L. Daniels, J.A. Fox, C.T. Walsh, and W.H. Orem-Johnson, *J. Biol. Chem., 257,* 3385-3388 (1982).

Jarrel, K.F. and G.D. Sprott, *Can. J. Microbiol., 27,* 720-728 (1981).

Keltjens, J.T., C.G. Caerteling, A. van Kooten, H.F. van Dijk, and G.D. Vogels, *Biochim. Biophys. Acta, 743,* 351-358 (1983a).

Keltjens, J.T., C.G. Caerteling, A.M. van Kooten, H.F. van Dijk, and G.D. Vogels, *Arch. Biochem. Biophys., 223,* 235-253 (1983b).

Keltjens, J.T., W.B. Whitman, C.G. Caerteling, A.M. van Kooten, R.S. Wolfe, and G.D. Vogels, *Biochem. Biophys. Res. Commun., 108,* 495-503 (1982).

Kenealy, W.R. and J.G. Zeikus, *J. Bacteriol., 151,* 932-941 (1982).

Kerby, R., W. Niemczura, and J.G. Zeikus, *J. Bacteriol., 155,* 1208-1218 (1983).

Kojima, N., J.A. Fox, R.P. Hausinger, L. Daniels, W.H. Orme-Johnson, and C. Walsh, *Proc. Natl. Acad. Sci. USA, 80,* 378-382 (1983).

Kräutler, B., *Helv. Chim. Acta, 67,* 1053-1059 (1984).

Krzycki, J. and J.G. Zeikus, *Curr. Microbiol., 3,* 243-245 (1980).

Lancaster, Jr., J.R., "Biomembranes," (L. Packer, ed.), Methods in Enzymology 88, Academic Press, New York, pp. 412-427 (1982).

Leigh, J.A., Doctoral Thesis, University of Illinois, Urbana (1983).

Leigh, J.A. and R.S. Wolfe, *J. Biol. Chem,, 258,* 7536-7540 (1983).

Leigh, J.A., K.L. Rinehart, Jr., and R.S. Wolfe, *J. Am. Chem. Soc., 106,* 3636-3639 (1984).

Lindahl, P.A., N. Kojima, R.P. Hausinger, J.A. Fox, B.K. Teo, C.T. Walsh, and W.H. Orme-Johnson, *J. Am. Chem. Soc., 106,* 3062-3064 (1984).

Livingston, D.A., A. Pfaltz, J. Schreiber, A. Eschenmoser, D. Ankel-Fuchs, J. Moll, R. Jaenchen, and R.K. Thauer, *Helv. Chim. Acta, 67,* 334-351 (1984).

Nagle, D.P. and R.S. Wolfe, *Proc. Natl. Acad. Sci. USA, 80,* 2151-2155 (1983).

Perski, H.-J., J. Moll, and R.K. Thauer, *Arch. Microbiol., 130,* 319-321 (1981).

Perski, H.-J., P. Schönheit, and R.K. Thauer, *FEBS Lett., 143,* 323-326 (1982).

Pezacka, E. and H.G. Wood, *Arch. Microbiol., 137,* 63-69 (1984).

Pfaltz, A., B. Jaun, A. Fässler, A. Eschenmoser, R. Jaenchen, H.H. Gilles, G. Diekert, and R.K. Thauer, *Helv. Chim. Acta, 65,* 828-865 (1982).

Romesser, J.A. and R.S. Wolfe, *J. Bacteriol., 152,* 840-847 (1982).

Scherer, P. and H. Sahm, "Anaerobic Digestion," (D.A. Stafford and B.I. Wheatley, eds.), Proceedings of 1st Internation-al Symposium on Anaerobic Digestion, University College Cardiff, 1979, Scientific Press, Cardiff, p. 45 (1980).

Schönheit, P. and H.-J. Perski, *FEMS Microbiol. Lett.*, *20*, 263-267 (1983).

Schönheit, P., D. Beimborn, and H.-J. Perski, *Arch. Microbiol.*, In press (1984).

Sprott, G.D. and K.F. Jarrell, *Can. J. Microbiol.*, *28*, 982-986 (1982).

Stupperich, E. and G. Fuchs, *FEBS Lett.*, *156*, 345-348 (1983).

Stupperich, E., K.E. Hammel, G. Fuchs, and R.K. Thauer, *FEBS Lett.*, *152*, 21-23 (1983).

Stupperich, E. and G. Fuchs, *Arch. Microbiol*, In press (1984a).

Stupperich, E. and G. Fuchs, *Arch. Microbiol*, In press (1984b).

Tan, S.L., J.A. Fox, N. Kojima, C.T. Walsh, and W.H. Orem-Johnson, *J. Am. Chem. Soc.*, *106*, 3064-3066 (1984).

Taylor, C.D. and R.S. Wolfe, *J. Biol. Chem.*, *249*, 4879-4885 (1974).

Thauer, R.K., K. Jungermann, and K. Decker, *Bacteriol. Rev.*, *41*, 100-180 (1977).

Thauer, R.K. and J.G. Morris, "The Microbe 1984," (D.P. Kelly and N.G. Carr, eds.), 36 Symposium, Society for General Microbiology, Cambridge Univeristy Press, pp. 123-128 (1984).

Thauer, R.K., D. Ankel-Fuchs, G. Diekert, H.-H. Gilles, R. Jaenchen, J. Moll, and P. Schönheit, "Microbial Growth on C₁-Compounds, Proceedings of the 4th Symposium," (R.L. Crawford and R.S. Hanson, eds.), American Society for Microbiology, Washington, DC, pp. 188-190 (1984).

Vogels, G.D., P. van Beelen, J.T. Keltjens, and C. van der Drift, "Microbial Growth on C₁ Compounds, Proceedings of the 4th International Symposium," (R.L. Crawford and R.S. Hanson, eds.), American Society for Microbiology, Washington, DC, pp. 191-198 (1984).

Zeikus, J.G. and R.S. Wolfe, *J. Bacteriol.*, *109*, 707-713 (1972).

Zeikus, J.G., *Adv. Microb. Physiol.*, *24*, 215-299 (1983).

5

EFFECT OF HYDROGEN CONCENTRATION ON POPULATION DISTRIBUTION AND KINETICS IN METHANOGENESIS OF PROPIONATE

Perry L. McCarty and Daniel Smith

Department of Civil Engineering
Stanford University
Stanford, California

ABSTRACT

The recent recognition of the significance of acetogenic dehydrogenation as a major stage in methanogenesis of complex substrates has led to questions about current understanding of bacterial growth and the kinetics of substrate utilization. As indicated by the presented evaluation of propionate utilization in methanogenesis, the complexity of the ecology that results from separate hydrogen production and utilization within the system requires a new approach to modeling these complex systems. Important to recognize is the effect of hydrogen partial pressure on the yields of the different bacterial species involved. Also, better information is needed on the substrate affinity (K_m) of the hydrogen utilizing methanogens because of the significance this has on the effective partial pressure of hydrogen in the system. These are important avenues for further exploration in this interesting and important ecological system.

INTRODUCTION

A most significant advance in the understanding of the ecology of methanogenesis occurred with the discovery by Bryant and co-workers (2) that the conversion of ethanol to methane requires the combined activity of three separate species of bacteria. This led to the concept of a new stage in the overall conversion of complex substrates to methane, now termed acetogenic dehydrogenation. A close association or symbiotic relationship between the molecular hydrogen

producing and consuming species is dictated by the small
energy available from the overall reaction. The importance
of this close relationship in the operation and control
of anaerobic systems is yet far from being well understood.
Many questions remain unanswered. Findings in this area
are likely to overturn many of the traditional concepts
concerning bacterial growth and process kinetics, as well
as the operation and control of anaerobic systems.

In this paper, some of the implications of the third
stage in methanogenesis are explored. This is done through
consideration of the growth and kinetics associated with
the conversion of propionate, a most important intermediate,
to methane. This conversion requires the association of
three separate species of bacteria in a manner similar to
that for ethanol. This was demonstrated by Boone and
Bryant (1) who identified *Syntrophobacter wolinii,* a species
capable of propionate conversion to acetate and hydrogen.

SIGNIFICANCE OF PROPIONATE

Figure 1 illustrates the three stages of methanogenesis
of complex organics together with the flow of electrons
through various substrates. Especially incorporated into
this figure is propionate, an intermediate that is formed
during the hydrolysis and fermentation of complex substrates
such as carbohydrates, proteins, and fats. About 30% of
the electrons flow through propionate as an intermediate
(12), thus its utilization is important in the overall process. An understanding of the interactions that occur with
propionate utilization is useful for reaching an understanding of the overall process.

ENERGY FROM SUBSTRATE UTILIZATION

The importance of the relationship between the three
species of bacteria required for overall methanogenesis of
propionate has been described (9,13,23). This is illustrated by the following three equations together with thermodynamic values summarized by Thauer, Jungermann, and Decker
(21):

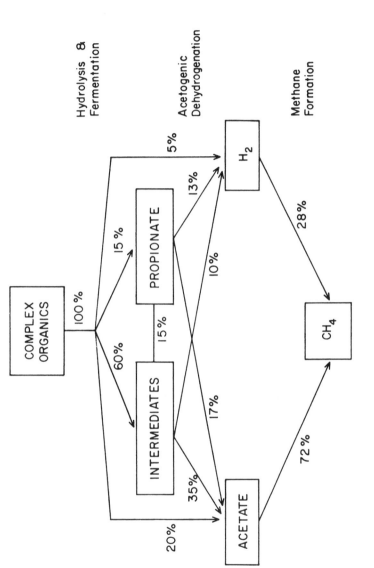

ELECTRON FLOW IN METHANOGENESIS

Figure 1.   The three stages of methanogenesis illustrating electron flow and the significance of propionate in the conversion of complex substrates.

$$\frac{\Delta G_r^{\circ\prime}}{kJ}$$

Propionate

$$CH_3CH_2COO^- + 2H_2O = CH_3COO^- + 3H_2 + CO_2 \qquad 71.67 \qquad (1)$$

Hydrogen

$$3H_2 + \frac{3}{4} CO_2 \qquad = \frac{3}{4} CH_4 + \frac{3}{2} H_2O \qquad -98.06 \qquad (2)$$

Acetate

$$CH_3COO^- + H_2O \qquad = CH_4 + HCO_3^- \qquad -31.01 \qquad (3)$$

Net

$$CH_3CH_2COO^- + \frac{3}{2} H_2O = \frac{7}{4} CH_4 + \frac{1}{4} CO_2 + HCO_3^- \quad -57.40 \qquad (4)$$

The value $\Delta G_r^{\circ\prime}$ represents the standard Gibbs free energy available from substrate catabolism at unit activity and pH 7. Under such standard conditions, propionate conversion is not possible since energy can only be obtained when $\Delta G_r^{\prime}$ is negative. However, the true free energy available is a function of the actual activities of the reactants and products as given by

$$\Delta G_r^{\prime} = \Delta G_r^{\circ\prime} + RT \sum_{i=1}^{m} \nu_{ik} \ln a_i \qquad (5)$$

where ν_{ik} is the stoichiometric coefficient for component i in reaction k and a_i the activity of the component (10).

Figure 2 illustrates the Gibbs free energy available from conversion of each of the three substrates as a function of hydrogen partial pressure or concentration (based upon a gas law constant for hydrogen 1331 atm-liter/mole). Activities for the reactants and products were taken to be $[CH_3CH_2COO^-] = [CH_3COO^-] = 10^{-3}$, $[HCO_3^-] = 0.053$, $p_{CH4} = 0.65$ and $P_{CO2} = 0.35$.

Figure 2 illustrates that propionate conversion in an operating treatment system can occur only when the log P_{H2} (atm) resides between -3.9 and -5.5. Above this value, energy is not available from propionate conversion and below this value it is not available from hydrogen conversion.

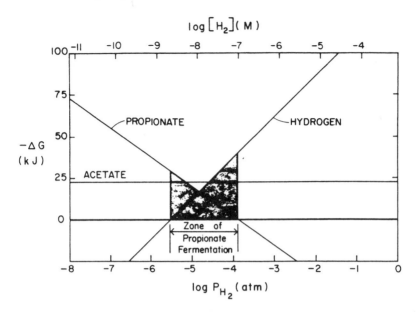

Figure 2.   The effect of hydrogen concentration or partial
            pressure on Gibbs free energy available from
            utilization of three substrates in methanogenesis.

The average log P_{H_2} reported by Heyes and Hall (3) in chemostats operated with 8.3 and 14.5 days residence times with
propionate as the sole substrate for methanogenesis were
-4.2 and -4.4, respectively. These values are within the
range expected from Figure 2, lending validity to the energetic concepts. We have measured similar partial pressures
with propionate utilization in our laboratory.

BACTERIAL YIELDS

     In models commonly used to describe bacterial growth,
coefficients of yield (Y) are generally taken to be constant. However, with hydrogen consuming methanogens, the
yield would be expected to vary considerably with P_{H_2}. The
formulation presented by McCarty (10,11) was used to evaluate this effect.

     Figure 3 represents a summary of the computed yields
for the three bacterial species as a function of hydrogen
partial pressure. There are close parallels between yield
coefficient and reaction energy (Figure 2). Figure 4 compares the computed yield coefficients for hydrogen utilizing

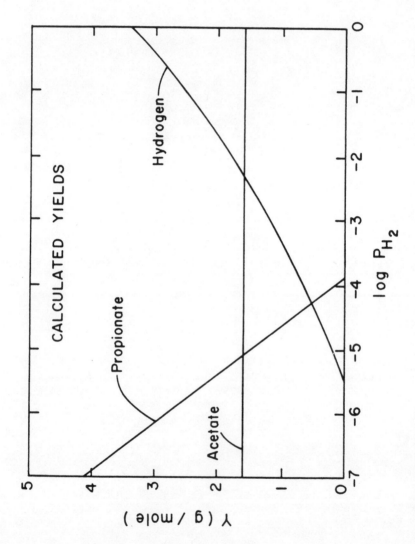

Figure 3.   Computed bacterial yields from acetate, hydrogen, and propionate as a function of hydrogen partial pressure.

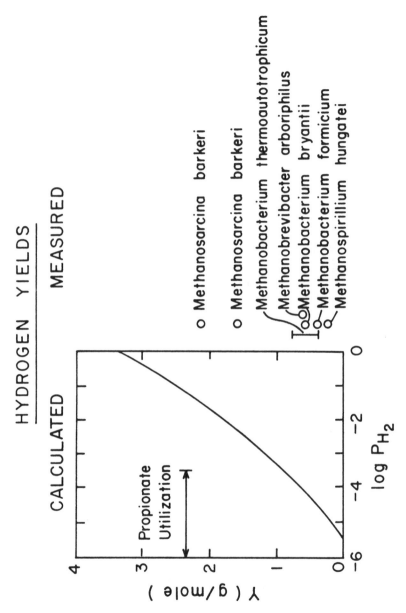

Figure 4. Comparison between predicted yields and measured yields from conversion of hydrogen to methane. References for measured yields are: 14, 15, 17, 19, 22, and 24.

methanogens with measured values reported in the literature. This comparison indicates there is a great variation in the reported yield coefficients for hydrogen utilizers, but they tend to cover the range of the predicted yields. When making yield measurements, little attention is generally paid to the actual hydrogen concentration experienced by the bacteria while growing. Measured partial pressures in the gas phase may bear little resemblance to the partial pressures surrounding the bacteria in the liquid phase because of mass transfer effects. It is obvious from the information displayed in Figure 4 that more attention to this detail is required. It may be that the yield for each hydrogen using methanogen actually varies with hydrogen partial pressure, or that each species has a given yield and will tend to dominate only in environments where the hydrogen partial pressure is appropriate for the organism's yield.

In Figure 5 another comparison is presented. Here the respective yield calculated for each species in the overall conversion of propionate to methane as a function of hydrogen partial pressure is illustrated. Also shown are data from Lawrence (7), who operated a series of chemostats, some with propionate as the sole organic substrate and others with acetate. The measured yields indicated were for retention times in the range of 8 to 12 days. A reasonable comparison between calculated and measured yields is indicated. While the total yield is similar for the range of partial pressures where propionate utilization is possible, the respective yield of hydrogen and propionate using bacteria varies significantly. The results also add validity to the concept that yields are a function of system energetics.

SUBSTRATE AFFINITY

A final area in which a better understanding is needed is in substrate affinity (K_s or K_m), as generally used in equations for bacterial growth or substrate utilization rates:

Bacterial Growth Rate

$$\mu = \mu_m \frac{S}{K_s + S} - b \tag{6}$$

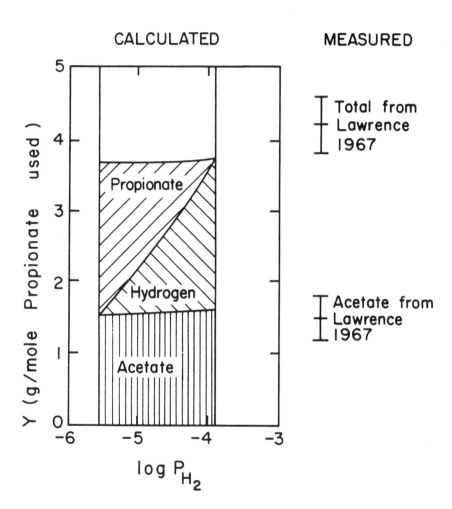

Figure 5.   Comparison between overall predicted yields for propionate conversion to methane and measured yields.

Substrate Utilization

$$- \frac{dS/dt}{X} = k \, \frac{S}{K_m + S} \tag{7}$$

Here, S is substrate concentration; μ and μ_m are net specific and maximum specific growth rates, respectively; b is decay rate; k is specific substrate utilization rate; and X is organism concentration. Generally, K_s is taken to equal K_m, but Robinson and Tiedje (16) have made a distinction between the two. This may be appropriate because of the varying yield rate anticipated as a function of hydrogen partial pressure. Most values of substrate affinity reported for hydrogen consuming methanogens represent K_m, a summary of which is contained in Table 1. In most cases, the values were determined from measured hydrogen partial pressure in the gas phase when hydrogen gas was the substrate. Here, mass transfer problems can be significant as suggested by Robinson and Tiedje (16).

The substrate affinity values reported, while generally quite low, are nevertheless much higher than seem to be required by thermodynamics and the measured rates of substrate utilization with propionate. Hydrogen concentrations in reactors where propionate is utilized must lie between 0.003 and 0.1 μM (Figure 2), generally 1 to 3 orders of magnitude below reported K_m values. With maximum growth rates for hydrogen using methanogens of 1.1 to 1.4 day^{-1} at 35°C (15,18,22,24) the hydrogen concentration should be no more than an order of magnitude lower than K_m when the retention time in a chemostat is about 10 days. Thus, reported values for K_m appear high.

The kinetics of substrate utilization in a chemostat when propionate is the sole organic substrate for three species conversion to methane was modeled using Monod kinetics and thermodynamics of bacterial growth as performed previously (10). The effect of retention time on the concentration of propionate in the reactor effluent as a function of K_m is illustrated in Figure 6. With a K_m of 1 μM, washout occurs at a detention time of about 25 days. A value in the range of 0.01 to 0.1 is associated with washout between 5 to 9 days, which is closer to the range for 35°C that has been experimentally determined by Lawrence with propionate.

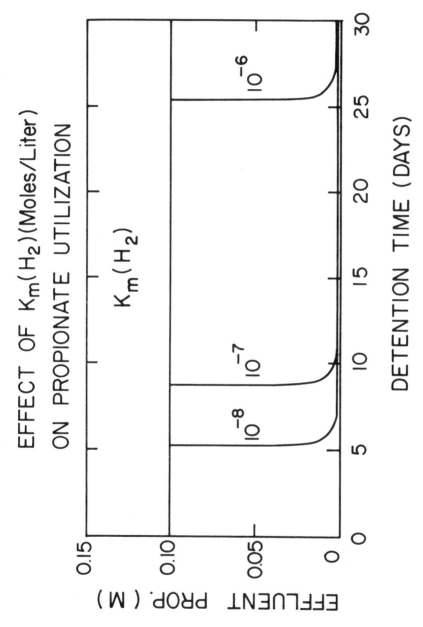

EFFECT OF $K_m(H_2)$ (Moles/Liter) ON PROPIONATE UTILIZATION

$K_m(H_2)$

10^{-6}

10^{-7}

10^{-8}

EFFLUENT PROP. (M)

DETENTION TIME (DAYS)

Figure 6. Computed effect of K_m (H_2) on propionate utilization in a CSTR.

Table 1.   Reported K_m Values for H_2 Utilization

Date	K_m µM	Ref	Date	K_m µM	Ref
1968	560	18	1982	6	16
1970	1	4	1982	6	16
1970	1*	4	1982	7	16
1978	80	5	1982	5	8
1978	3	20	1984	13*	15
1980	150*	17	1984	2*	15
1982	7*	6			

The above calculations indicate that much lower K_m values are required for hydrogen using methanogens than have commonly been reported. This discrepancy may be the result of one or more factors. The various kinetic parameters and interrelationships may not be sufficiently well understood for modeling purposes, including reported thermodynamic values. Hydrogen using methanogens associated with propionate utilization may have much lower substrate affinity than associated with bacteria studied to date. Measured K_m values may also be too high because of the difficulties resulting from mass transfer effects. Better experimental evidence is required in order to sort out these possibilities.

SUMMARY

Propionate is a useful substrate to help evaluate the complex ecological and energetic interrelationships that occur in methane fermentation. Important areas for research are the effect of substrate and product concentrations on yield and substrate affinity.

ACKNOWLEDGMENT

This research has been supported by a research grant from the U.S. National Science Foundation, Division of Civil and Environmental Engineering.

REFERENCES

1. Boone, D.R. and M.P. Bryant, *Appl. and Envir. Micro.*, *40*, 626 (1980).
2. Bryant, M.P., E.A. Wolin, M.J. Wolin, and R.S. Wolfe, *Archiv für Mikrobiologie, 59,* 20 (1967).
3. Heyes, R.H. and R.J. Hall, *Appl. and Envir. Micro.*, *46*, 710 (1983).
4. Hungate, R.E. et al., *J. Bact., 107*, 389 (1970).
5. Kaspar, H.F. and K. Wuhrmann, *Appl. and Envir. Micro.*, *36*, 1 (1978).
6. Kristjansson, J.K., P. Schönheit, and R.K. Thauer, *Arch. Microbiol., 131,* 278 (1982).
7. Lawrence, A.W., "Kinetics of Methane Fermentation in Aerobic Waste Treatment," Ph.D. Dissertation, Stanford University (1967).
8. Lovley, P.R., D.F. Dwyer, and M.J. Klug, Abstract 194, Amer. Soc. for Microbiology (1982).
9. McCarty, P.L., *Anaerobic Digestion 1981,* p.3, (E. Hughes, et al., eds.), Elsevier Biomedical Press B.V. (1981).
10. McCarty, P.L., *Anaerobic Biological Treatment Processes,* 91, Advances in Chemistry Series No. 105, American Chemical Society, Washington, DC (1971).
11. McCarty, P.L., "Energetics and Bacterial Growth," *Organic Compounds in Aquatic Environments,* p. 495, (S.D. Faust and J.V. Hukter, eds.), Marcel Dekker, Inc., New York (1971).
12. McCarty, P.L., *Public Works, 95,* 107 (Sept. 1964).
13. McInerney, M.J. and M.P. Bryant, *Fuel Gas Production from Biomass,* Vol., 1, p. 19, (D.L. Wise, ed.), CRC Press, Inc., Boca Raton, FL (1981).
14. Mylroie, R.L. and R.E. Hungate, *Canad. J. Microbiol., 1,* 55 (1954).
15. Robinson, J.A. and J.M. Tiedje, *Arch. Microbiol., 137,* 26 (1984).
16. Robinson, J.A. and J.M. Tiedje, *Appl. and Envir. Micro.*, *44*, 1374 (1982).
17. Schönheit, P., J. Moll, and R.K. Thauer, *Arch. Microbiol., 127,* 59 (1980).
18. Shea, T.G. et al., *Water Research, 2,* 833 (1968).
19. Smith, M.R. and R.A. Mah, *Appl. and Envir. Micro., 36,* 870 (1978).
20. Strayer, R.F. and J.M. Tiedje, *Appl. and Envir. Micro., 36,* 330 (1978).
21. Thauer, R.K., K. Jungermann, and K. Decker, "Energy Conservation in Chemotrophic Anaerobic Bacteria," *Bacteriol. Rev., 41,* 100 (1977).

22.   Weimer, P.J. and J.G. Zeikus, *Arch. Microbiol., 119,*
      49 (1978).
23.   Zehnder, A.J.B., "Ecology of Methane Formation," *Water
      Pollution Microbiology, Vol. 2,* p. 349, (R. Mitchell,
      ed.), John Wiley & Sons, Inc., New York (1978).
24.   Zehnder, A.J.B. and K. Wuhrmann, *Arch. Microbiol., 111,*
      199 (1977).

6
ENERGETICS OF H$_2$-PRODUCING SYNTROPHIC BACTERIA

P.S. Beaty, M.J. McInerney, and N.Q. Wofford

Department of Botany and Microbiology
University of Oklahoma
Norman, Oklahoma

ABSTRACT

The complete anaerobic degradation of organic matter to CO_2 and CH_4 requires the concerted action of four major metabolic groups of bacteria. Fermentative bacteria hydrolyze the substrate polymers and ferment the products to volatile acids, CO_2 and H_2. The H_2- and acetate producing syntrophic bacteria degrade propionate and longer-chain fatty acids and some aromatic acids to acetate, CO_2 and H_2. The acetogenic bacteria produce acetate and sometimes butyrate from H_2/CO_2, CO, CH_3OH and methyoxy moieties of aromatic compounds. Finally, the methanogens use the H_2 produced by the other groups to reduce CO_2 and CH_4, and some species cleave acetate to CH_4 and CO_2.

Little is known about the physiology of the H_2- and acetate-producing syntrophs. These bacteria grow very slowly and only grow in coculture with H_2- using bacteria. Methods have been developed to mass-culture *Syntrophomonas wolfei* in coculture with *Methanospirillum hungatei* and to obtain cell-free extracts of *S. wolfei* with minimal contamination from cellular components of the methanogen by lysozyme treatment. *S. wolfei* extracts contained high specific activities of the β-oxidation enzymes indicating that fatty acids are degraded by this pathway. Theoretical calculations indicate that end-product excretion may be an important route for energy production in this organism.

INTRODUCTION

H.A. Barker (1) traces the history of the methane fermentation from A. Volta's initial discovery of the presence of methane in eutrophic lake sediments in 1776 to the development of the first systematic classification of methane bacteria. Popoff, Toppeiner, Hoppe-Seyler and Omelianskii used cellulolytic enrichment cultures to show that the methane fermentation was a two stage process. Cellulose was degraded to acetate, butyrate, CO_2, and H_2 which then served as primary substrates for methanogenic bacteria. Methods were developed to cultivate obligate anaerobes such as the methanogenic bacteria and physiological studies of various isolates led to the first systematic classification of methanogens (1). Methanogens were believed to degrade the acids, alcohols, H_2 and CO_2 formed by fermentative bacteria in the first stage to CH_4 and CO_2.

In 1967, Bryant et al. (5), found that the ethanol fermentation carried out by *Methanobacillus omelianskii* was actually carried out by a syntrophic association of two bacterial species. One species (S organism) produced acetate and H_2 from ethanol only when the second species (*Methanobacterium bryantii*) was present to use the H_2 to make CH_4. Under standard conditions, the conversion of ethanol to acetate and H_2 is thermodynamically unfavorable ($\Delta G^{\circ '}$ = +9.6 kJ/reaction) and pure cultures of S organism grow poorly on ethanol (19). Reddy et al. (20), showed that cell-free extracts of S organism contain a pyridine nucleotide-linked, ferredoxin-dependent hydrogenase activity which uses the electrons generated in ethanol oxidation to reduce protons to H_2. Wolin's calculations (26) show that the unfavorable dehydrogenations involved in ethanol oxidation become energetically favorable when the partial pressure of H_2 (pH_2) is kept below 10^{-3} atm. Methanogens have a high affinity for H_2 use which maintains a low H_2 concentration in the system which allows H_2 production from high (-24 mv) potential electrons to be energetically favorable. This kind of interaction between H_2-producing bacteria and H_2-using bacteria such as methanogens is called interspecies H_2-transfer (5,26). Recently, other syntrophic cocultures have been obtained where interspecies H_2-transfer is obligately required for the degradation of the substrate molecule. *Syntrophomonas wolfei* (14) oxidizes C_4 to C_8 fatty acids to acetate and H_2 or acetate, propionate and H_2. *Syntrophobacter wolinii* (4) oxidizes propionate to acetate, CO_2 and H_2, and *Syntrophus buswellii* (16) oxidizes benzoate to acetate and H_2 when grown in coculture with H_2-using bacteria.

Thus, compounds such as ethanol, propionate and longer-chained fatty acids, and certain aromatic acids are not degraded by methanogens per se, as was originally believed, but are degraded by H_2-producing syntrophic bacteria.

Anaerobic digestion is now considered to be a three-stage process where fermentative bacteria hydrolyze polysaccharides, and proteins and ferment the products mainly to volatile fatty acids, CO_2 and H_2. The H_2-producing syntrophic bacteria degrade propionate and longer-chain fatty acids, alcohol and certain aromatic acids to acetate, H_2 and, in some cases, CO_2. The methanogens use acetate and H_2/CO_2 to produce CH_4. Volatile fatty acid degradation is often the rate-limiting step in methanogenesis with the build up of propionate as the first sign of digestor failure.

The coupling of H_2 production by the syntrophic bacteria to CH_4 formation by methanogens theoretically explains how free energy is conserved by the syntrophic bacteria. As an example, the degradation of butyrate with methane formation results in a $\Delta G^{o'}$ of -39.4 kJ/reaction. This corresponds to the amount of free energy needed to form one mole of ATP from ADP and inorganic phosphate via substrate-level phosphorylation with an energy conversion efficiency of 90 to 100%. Such high energy conversion efficiencies are incompatible with the entropy requirements of metabolism (23). Also, the free energy must be partitioned between both organisms. Thus, it seems that the above analysis of the energetics of fatty acid degradation cannot account for ATP formation in the syntroph. The objective of this paper is to separate the syntrophic metabolic process from that of the methanogen and consider mechanisms by which the syntroph can conserve energy.

METHODS OF CALCULATION

S. wolfei is used as a model system to develop a bioenergetic hypothesis for syntrophic growth.

Calculation of $\Delta G'$

Molar activities (mole reactant per mole product) are used to calculate the $\Delta G'$ as shown for the hypothetical reaction, $aA + bB \longrightarrow cC + dD$:

$$\Delta G' = \Delta G^{o'} - RT \ln \frac{[\frac{C}{A+B+C+D}]^c \quad [\frac{D}{A+B+C+D}]^d}{[\frac{A}{A+B+C+D}]^a \quad [\frac{B}{A+B+C+D}]^b}$$

Calculation of the kJ Per mg (Dry Weight)

The kJ per mg (dry weight) of cells was calculated using the following equations:

a. aerobic cultures:

$$kJ/mg = (mol\ O_2/ml)\,(475\ kJ/mol\ O_2)\,(mg\ (dry\ wt))^{-1}$$

b. anaerobic cultures:

$$kJ/mg\ (dry\ wt) = [(\Sigma\ moles\ catabolic\ end\ products/ml)$$
$$(-56.0\ kJ/mol\ pyruvate\ formed) + (mol\ lactate/ml)$$
$$(-43.1\ kJ/mol) + (mol\ acetate/ml)\,(-47.1\ kJ/mol)+$$
$$(mol\ ethanol/ml)\,(-56.9\ kJ/mol)]\times[mg\ dry\ wt/ml]^{-1}$$

The sum of the lactate, ethanol, and acetate was used to calculate the energy release from glucose conversion to pyruvate. The energy released from the conversion of pyruvate to either lactate, acetate, or ethanol was calculated independently.

ENERGETICS OF SYNTROPHIC GROWTH

The efficiency of energy conversion into biogenic power is a function of the availability of potential energy, the flux of monomer pools, and the rate of biosyntehsis. Fundamental generalizations can be made by comparing the energy economy of several bacteria grown with different substrates (Fig. 1). *Escherichia coli* is able to rapidly release large amounts of energy when growing aerobically on glucose. As the growth rate increases, the amount of energy expended per mg (dry weight) of cells increases. Thus, one of the trade offs that occurs at fast growth rates is that the efficiency of energy conversion is low. Similar patterns are observed when acetate or mannitol are the energy sources or with *Streptococcus bovis* growing anaerobically with glucose. Growth with acetate as the sole carbon and energy source is limited by the availability of carbon rather than energy

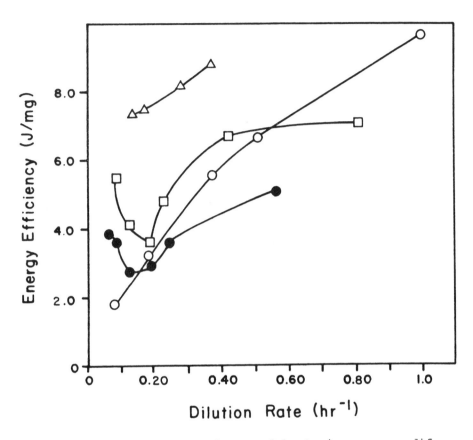

Figure 1.   Energy economy of several bacteria grown on different substrates. The calculations are as described in the Methods of Calculation Section.
O—O *Escherichia coli* grown aerobically on glucose.
●—● *E. coli* grown aerobically on mannitol. △—△ *E.
coli* grown aerobically on acetate (9). ▯—▯ *Streptococcus bovis* in the transition period from homolactic fermentation (D > 0.40) (22).

and this causes a decrease in the efficiency of energy conversion. At growth rates greater than 0.4 per hour, *S. bovis*
exhibits a homolactic fermentation and the full energy potential of the system is not realized due to the effect that a
more efficient system would have on the rate of energy production. As growth rate slows, the system approaches energy
limitation and metabolism shifts to a heterolactic fermentation where more of the potential energy in glucose
can be conserved. As the growth rate decreases, energy
conversion efficiency increases to a maximum. At slow
growth rates (D < 0.1 per hour), energy conversion efficiency
decreases rapidly. In this region large amounts of energy
are expended with little production of cellular material.
Calculations from the data in Verseveld et al. (24), show
that as the growth rate of *E. coli* decreases from 0.15 to
0.07 per hour, approximately 40 kJ of energy is expended
per mg (dry weight) of cells. At growth rates of 0.05
to 0.015 and less than 0.01 per hour, the energy expenditures
are approximately 53 kJ/mg and 72 kJ/mg, respectively. Thus,
at extremely slow growth rates, a large energy expenditure
relative to the available energy is required to displace
the system from equilibrium. Slow growing bacteria can be
considered to be systems near equilibrium where the rate
of biosynthesis is very slow and energy is spent to maintain viability.

    S. wolfei can be considered to be a system at or near
equilibrium. The fastest growth rates observed for *S.
wolfei* were 0.012 per hour for the coculture with the sulfate reducer and 0.007 per hour for the coculture with the
methanogen. Theoretical calculations based on the stoichiometry of butyrate degradation and the concentrations
of reactants and products often observed during growth show
that the ΔG' for butyrate degradation is about -22.1 kJ/mole
of butyrate. D.F. Dwyer, D.R. Shelton and J.M. Teidje
(*Abst. Ann. Meet. Amer. Soc. Microbiol.*, p. 126, 1984)
experimentally found that the ΔG' for butyrate degradation
during exponential phase of growth was about -17 kJ/mole
of butyrate. Thus, this system is near equilibrium. Since
S. wolfei grows very slowly, it must be assumed that there
is a high energy requirement for organizational maintenance (7,24).

    S. wolfei contains high levels of the β-oxidation
enzymes (Wofford, Beaty and McInerney, unpublished data).
Butyrate oxidation via the β-oxidation pathway requires +48.1
kJ/mol under standard conditions (Eq. (1), Table 1).
Inspection of this process reveals that the energy intensive

Table 1. *Table of Equations*[a]

Reaction	$\Delta G^{o'}$ (kJ/mol)[a]
1. *Butyrate* + $2H_2O$ ----> 2 *Acetate* + H^+ + $2H_2$	+48.1
2. *Butyrate* ----> *Butyryl-CoA*	+35.6
3. *Butyryl-CoA* ----> *Crotonyl-CoA*	+75.0
4. *Crotonyl-CoA* ----> β-*hydroxybutyryl-CoA*	+ 8.2
5. β-*hydroxybutyryl CoA* ----> *Acetoacetyl-CoA*	+25.7
6. *Acetoacetyl-CoA* ----> 2 *Acetate* + *2CoA*	-96.7
7. $FADH_2$ ----> *FAD* + H_2	+37.4
8. *NADH* ----> *NAD* + H_2	+18.1

[a] $\Delta G^{o'}$ *taken from Thauer et al. (23).*

reactions are the activation of butyrate to butyryl-Coenzyme A (CoA), the dehydrogenation of butyryl-CoA, and the dehydrogenation of β-hydroxybutyryl-CoA (Eqs. 2, 3, and 5, respectively, Table 1). The production of H$_2$ from electrons generated in these dehydrogenations are endergonic suggesting that H$_2$ produced from reduced NAD or FAD (assuming that these are the in vivo electron carriers) by reverse electron transport would require energy. The formation of acetyl-CoA in β-oxidation suggests that energy can be conserved by the generation of acetyl phosphate from acetyl-CoA. However, the energy released in the cleavage of acetoacetyl-CoA to acetate and CoA (Eq. 6, Table 1) may be needed to displace the unfavorable equilibria involving butyrate activation and H$_2$ production. Thus, it seems that energy conservation by substrate-level phosphorylation may not be energetically feasible in *S. wolfei*.

If the H$_2$ concentration is maintained at a low level (about 0.5 μM in these calculations), the degradation of syntrophic substrates such as ethanol, lactate, propionate, butyrate or benzoate is thermodynamically favorable over a wide range of initial substrate concentrations (Fig. 2). Theoretically, H$_2$ levels must be maintained below 10 μM for butyrate degradation to be energetically favorable. Methanogens rapidly use H$_2$ which maintains low H$_2$ levels, around 1 μM (11,21), in methanogenic ecosystems. Although interspecies H$_2$-transfer reactions explain how butyrate can be degraded in syntrophic systems, it does not explain how energy is conserved by the syntroph. As seen in Figure 2, the change-in-free-energy for butyrate degradation is much less than that required to synthesize ATP from substrate-level phosphorylation. Butyrate oxidation generates electrons in the conversion of butyryl-CoA to crotonyl-CoA ($E_O^{'}$ = -25mV) and in the conversion of β-hydroxybutyrl-CoA to acetoacetyl-CoA ($E_O^{'}$ = -280 mV). This suggests that energy is required to drive reverse electron flow to product H_2. Using the data of Mackie and Bryant (12), an electron flow of 120 μmol of electrons per cell per min must be maintained to account for the observed rate of butyrate degradation in sludge. This would require an energy expenditure of 0.0323 mV per cell per min when the H_2 concentration is 0.1 μM. Colje et al (6), showed that the generation of reducing power needed for nitrogen fixation requires a membrane potential to change the redox potential of the flavodoxin oxidore-ductase. It seems that a membrane potential may be re-quired to drive reverse electron transport in *S. wolfei*.

A comparison of the rates of substrate use by different growth conditions shows that the rates of propionate and butyrate degradation in digestor sludge are much faster than the rates of aerobic metabolism of *E. coli*, *E. coli* using H_2 and fumarate, formate use by *Wolinella succinogenes* or anaerobic glucose metabolism by *S. bovis* (Table 2). This suggests that the flux of mass through the syntroph cell is very large. This suggests that the flux of mass may be important for energetics of syntrophic bacteria. Michels et al. (15), postulated that the build up and excretion of metabolic end products occurs in symport with protons and this generates a proton motive force. Otto et al. (18), showed that *Streptococcus cremoris* contains a system for the electrogenic efflux of lactate. The efflux of lactate from cells generated a proton motive force. Dissipating the lactate gradient by adding increasing amounts of lactate to the medium decreased the molar growth yield of *S. cremoris*. With acetate, a pH near neutrality would be needed to reduce the passive diffusion of the undissociated form of acetate across the membrane. At pH's less than 6.5, the permeability of acetate across the membrane would in-crease making it difficult to maintain an acetate gradient across the membrane. It is interesting to note that the optimal pH range for volatile fatty acid degradation in sludge is between 6.8 to 7.4 (13) which corresponds to the pH range required to maintain an acetate gradient.

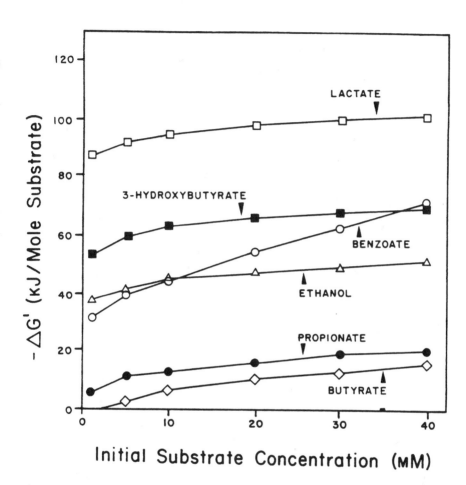

Figure 2. The Gibbs free energy available from the catabolism of various syntrophic substrates at 0.50
μM H$_2$. Calculations were performed as described
in Methods of Calculation based on the ΔG$^{o'}$
values from Thauer et al. (23), assuming the
following product concentrations: acetate,
20 mM; H$_2$, 0.5 μM; and HCO$_3$, 50 mM at 30°C.

Table 2.   *Comparison of Substrate Degradation Rates per ml by Different Bacteria in Varying Environments*[a]

Organism	Substrate	e^- Acceptor	$D(hr^{-1})$	μmol substrate/ ml/hr
Escherichia coli	Glucose	O_2	0.090	0.459
			0.190	0.969
			0.510	2.600
	Mannitol	O_2	0.063	0.371
			0.573	1.670
	Acetate	O_2	0.141	1.410
	H_2	Fumarate	---	4.880
Wolinella succinogenes	Formate	NO_3^-	---	16.000
Streptococcus bovis	Glucose	Organic	0.088	4.200
			0.195	4.600
			0.423	4.630
			0.807	5.290
Animal Waste Digestor	Butyrate	H^+/H_2	0.004	60.000
	Propionate	H^+/H_2	0.004	180.000

[a]The data on Escherichia coli was calculated from Hempfling and Mainzer (9) by the equation (mMol O_2 consumed/ml/hr)(0.167 mmol glucose/μmol O_2) = mmol glucose/ml/hr. Data for E. coli grown on H_2 and formate comes from Bernhard and Gottschalk (2). Data for Wolinella succinogenes from Bokranz et al. (3). Data for Streptococcus bovis from Tussel and Baldwin (22). Calculation method as described in Methods of Calculation using 1/2 total moles end product as moles glucose. The animal waste digestor data is from Mackie and Bryant (12).

Calculations using a butyrate degradation rate of 1 μmol/min/ml obtained from Mackie and Bryant and a partition coefficient of 3.4×10^{-3} cm/sec for acetate show that high internal concentrations of acetate can be generated (Fig. 3). Thus, the rate of butyrate degradation is much more rapid than the rate of acetate diffusion and a large chemical gradient of acetate can be generated inside of the cell. Calculations show that even when the internal acetate concentration is set at 200 mM and the concentrations of H_2, bicarbonate and butyrate are 0.5 μM, 50 mM, and 5 mM, respectively, butyrate degradation is still favorable ($\Delta G'$ of -5 kJ per mole). Evidence that *S. Wolfei* obtains energy from acetate excretion is shown in Figure 4. Increasing the acetate concentration in the medium decreased the amount of biomass formed by the *S. wolfei* - *Methanospirillum hungatei* coculture. The ionic strength was kept constant by added appropriate amounts of NaCl. The amount of butyrate degraded in each case was similar suggesting that high acetate concentrations did not inhibit butyrate metabolism but decreased the amount of energy available for biosynthesis. Other experiments where protein concentrations were monitored instead of absorbance gave similar results. The electrogenic efflux of acetate by *S. wolfei* would provide a mechanism for generating a membrane potential which is necessary to drive reverse electron transport to H_2.

The fermentation of branched-chain amino acids by a marine spirochete is possibly an analogous system where electrogenic efflux of end products is important. Leucine, isoleucine and valine are oxidatively deaminated with the formation of the respective branched-chain volatile fatty acid (isovaleric, 2-methylbutyric, and isobutyric), CO_2, NH_3, and 2 H_2. The reaction is energetically unfavorable at standard conditions ($\Delta G^{o'} = +4.2$ kJ/mole of leucine) and the $\Delta G'$ is dependent on pH_2. The spirochete contains phosphotransacetylase and branched-chain fatty acid kinase activities so ATP could be generated from the respective acyl phosphate. The spirochete does not use branched-chain amino acids as the sole energy source nor does the addition of these compounds affect the molar growth yield of the spirochete on glucose. The use of branched-chain amino acids by starving cells maintained the viability of the culture. Cell viability decreased rapidly after the excretion of branched-chain fatty acids stopped. Thus, the spirochete uses a near equilibrium reaction to create an energy flux which is quite similar to that discussed for syntrophic bacteria.

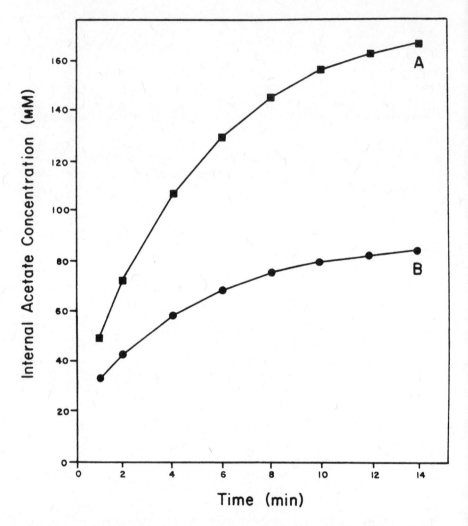

Figure 3.   The intracellular concentration of acetate.
Assuming a cell volume of 10 pl and using the
specific activity for butyrate catabolism (12) of
1 μmol/ml/min an intracellular rate of acetate
production of 0.020 μ/cell/min was calculated.
Based on the permeation of acetate across lipid
bilayers ($P = 3.40 \times 10^{-3}$ cm/sec) as measured by
Walter et al. (25), the intracellular concentration of acetate was calculated using the equation,
$C_t = C_o e^{-Pt}$ where C_o is the initial concentration
and C_t is the calculated concentration at time t.

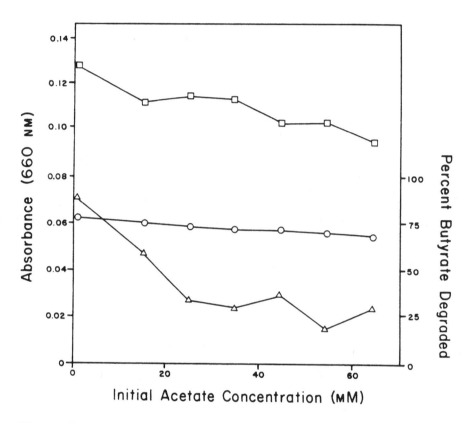

Figure 4.   The effect of extracellular acetate concentration
on the growth of a *Syntrophomonas wolfei-Methano-
spirillum hungatei* coculture. The culture was
grown in a butyrate basal media as described by
McInerney et al. (14), and growth was followed as
absorbance at 660 nm. Butyrate was analyzed by
gas chromatography. Symbols: Δ, change in absorbance; ⊓ maximum absorbance; and O, percent
butyrate degraded.

Washed cell suspensions of *Staphylococcus epidermidis* when starved under anaerobic conditions rapidly degrade intracellular protein and excrete large amounts of amino acids into the medium (10). Viability of the culture declined rapidly after the efflux of amino acids stopped. This is another example where efflux of mass from the cell may be used to generate energy.

CONCLUSIONS

The bioenergetics of syntrophic organisms have been treated as open systems. The calculations of Gibbs free energy changes have been used only to describe processes in terms of their environments. Basically, syntrophic growth is a near equilibrium process ($\Delta G \approx 0$). The analysis of the Gibbs free energy equation ($\Delta G = \Delta H = T \Delta S$, where ΔH is the change in enthalpy, ΔS is the change in entropy and T is the temperature) reveals that either large changes in enthalpy or large changes in entropy result in large changes of free energy. In syntrophic growth, we propose that free energy is conserved by taking advantage of the entropy generated by the system. Specifically, this is done by acetate production. Gibbs free energy calculations are based on a closed system assumption which considers the systems mass to be static. By allowing changes in mass as a result of metabolism and imposing a mass flow boundary in the form of a cell membrane, this system becomes capable of producing a potential difference in the form of a mass gradient. The use of an electrogenic membrane-bound porter system allows the conversion of this chemical potential into a useful form of energy for the cell. As H.J. Morowitz said (17), "It is not energy per se that makes life go, but the flow of energy through the system."

ACKNOWLEDGMENTS

This research was supported by U.S. Department of Energy contract DE-AS05-83ER13053, the Research Council and the College of Arts and Sciences, University of Oklahoma.

REFERENCES

1. Barker, H.A., *Bacterial Fermentations,* John Wiley and Sons, New York (1956).
2. Bernhard, T. and G. Gottschalk, "Cell Yields of *Escherichia coli* During Anaerobic Growth on Fumarate and Molecular Hydrogen," *Arch. Microbiol., 116,* 235-238 (1978).
3. Bokranz, M., J. Katz, J. Schröder, A. Roberton, and A. Kröger, "Energy Metabolism and Biosynthesis of *Vibrio succinogenes* Growing with Nitrate or Nitrite as Terminal Electron Acceptor," *Arch. Microbiol., 135,* 36-41 (1983).
4. Boone, D.R. and M.P. Bryant, "Propionate-degrading Bacterium *Syntrophobacter wolinii* sp. nov., gen. nov. from Methanogenic Ecosystems," *Appl. Environ. Microbiol., 33,* 1162-1169 (1980).
5. Bryant, M.P., E.A. Wolin, M.J. Wolin, and R.S. Wolfe, "*Methanobacillus omelianskii,* A Symbiotic Association of Two Species of Bacteria," *Arch. Microbiol., 59,* 20-31 (1967).
6. Colje, L., W. Klone, W.N. Konings, H. Haaker, and C. Veeger, "The Involvement of the Membrane Potential in N₂ Fixation by Bacteroids of *Rhizobium leguminosum,*" *FEBS Lett., 103*(2), 328-332 (1979).
7. Forrest, W.W. and D.J. Walker, "Change in Entropy During Bacterial Growth," *Nature, 201,* 49 (1964).
8. Harwood, C.S. and E. Canale-Parola, "Branched-chain Amino Acid Fermentation by a Marine Spirochete: Strategy for Survival," *J. Bacteriol., 148,* 109-116 (1981).
9. Hempfling, W.P. and S.E. Mainzer, "Effects of Varying the Carbon Source Limiting Growth on Yield and Maintenance Characteristics of *Escherichia coli* in Continuous Culture," *J. Bacteriol., 123,* 1076-1087 (1975).
10. Horan, N.J., M. Midgley, and E.A. Dawes, "Effect of Starvation on Transport, Membrane Potential and Survival of *Staphylococcus epidermidis* under Anaerobic Conditions," *J. Gen. Microbiol., 127,* 223-230 (1981).
11. Hungate, R.E., "Hydrogen as an Intermediate in the Rumen Fermentation," *Arch. Microbiol., 59,* 158-164 (1967).
12. Mackie, R.I. and M.P. Bryant, "Metabolic Activity of Fatty Acid-oxidizing Bacteria and the Contribution of Acetate, Propionate, Butyrate, and CO₂ to Methanogenesis in Cattle Waste at 40 and 60°C," *Appl. Environ. Microbiol., 41,* 1363-1373 (1981).

13. McCarty, P.L., "Anaerobic Waste Treatment Fundamental, II. Environmental Requirements and Control," *Public Works, 95*(10), 123-126 (1964).
14. McInerney, M.J., M.P. Bryant, R.B. Hespell, and J.W. Costerton, "*Syntrophomonas wolfeii* gen. nov., sp. nov., An Anaerobic,Syntrophic, Fatty Acid-oxidizing Bacterium," *Appl. Environ. Microbiol., 41,* 1029-1039 (1981).
15. Michels, P.A.M., J.P.S. Michels, J. Boonstra, and W.N. Konings, "Generation of an Electrochemical Proton Gradient in Bacteria by the Excretion of Metabolic End Products," *FEMS Microbiol. Lett., 5,* 357-364 (1979).
16. Mountfort, D.O., W.J. Brulla, L.R. Krumholz, and M.P. Bryant, "*Syntrophus buswelli* gen. nov., sp. nov.: A Benzoate Catabolizer from Methanogenic Ecosystems," *Inter. J. Syst. Bacteriol., 34,* 216-217 (1984).
17. Morowitz, H.S., *Energy Flow in Biology,* Academic Press, New York (1968).
18. Otto, R., A.S. Sonnenberg, H. Veldkamp, and W.N. Konings, "Generation of an Electrochemical Proton Gradient in *Streptococcus cremoris* by Lactate Efflux," *Proc. Natl. Acad. Sci. (USA), 77,* 5502-5506 (1980).
19. Reddy, C.A., M.P. Bryant, and M.J. Wolin, "Characteristics of S Organism Isolated from *Methanobacillus omelianskii*," *J. Bacteriol., 109,* 539-545 (1972).
20. Reddy, C.A., M.P. Bryant, and M.J. Wolin, "Ferredoxin and Nicotinamide Adenine Dinucleotide-dependent H_2 Production from Ethanol and Formate in Extracts of S Organism Isolated from *Methanobacillus omelianskii*," *J. Bacteriol., 110,* 126-132 (1972).
21. Robinson, J.A., R.F. Strayer, and J.M. Teidje, "Method for Measuring Dissolved Hydrogen in Anaerobic Ecosystems: Application to the Rumen," *Appl. Environ. Microbiol., 41,* 545-548 (1981).
22. Russell, J.B. and R.L. Baldwin, "Comparisons of Maintenance Energy Expenditures and Growth Yields Among Several Rumen Bacteria Grown in Continuous Culture," *Appl. Environ. Microbiol., 37,* 537-543 (1979).
23. Thauer, R.K., K. Jungermann, and K. Decker, "Energy Conservation in Chemotrophic Anaerobic Bacteria," *Bacteriol. Rev., 47,* 100-180 (1977).
24. Van Verseveld, H.W., W.R. Chesbro, M. Braster, and A.H. Stouthammer, "Eubacteria Have Three Growth Modes Keyed to Nutrient Flow. Consequence for the Concept of Maintenance and Maximal Growth Yield," *Arch. Microbiol., 137,* 176-184 (1984).

25. Walter, A. and J. Gutknelt, "Monocarboxylic Acid Penetration Through Lipid Bilayer Membranes," *J. Memb. Biol., 77*, 255-264 (1984).

26. Wolin, M.J., "Metabolic Interactions Among Intestinal Microorganisms," *Am. J. Clin. Nutr., 27*, 1320-1328 (1974).

7
THERMODYNAMICS OF CATABOLIC REACTIONS IN THE ANAEROBIC DIGESTOR

David R. Boone

Environmental and Occupational Health Sciences
School of Public Health
University of California
Los Angeles, California

ABSTRACT

Anaerobic digestion of organic matter takes place in a stepwise manner with interacting physiological groups of organisms which catalyze successive reactions. The products of one reaction are the substrates for the next, and consequently concentration of any intermediate depends on a balance between its rate of production and rate of degradation. Since many of the reactions of anaerobic digestion yield very small quantities of free-energy, the concentrations of both products and reactants are important. These concentrations can determine whether and to what extent a reaction is exergonic, and thus whether organisms catalyzing that reaction can thrive in the ecosystem. Inefficient degradation of a given fermentation product by subsequent metabolic groups causes that product to accumulate. Elevated concentrations of the product, for instance a fermentation acid, may make its formation relatively less favorable, and other fermentation reactions gain preference. Molecular hydrogen is an important intermediate in this respect; because of its low concentration in digestors it has a very rapid turnover, and its concentration changes rapidly and drastically with small changes in the rate of its production or removal. Hydrogen is an important intermediate in other respects as well: it is a product of many different fermentations, and is the precursor to about 30% of the methane formed in anaerobic digestors. Where hydrogen production is rapid and

localized, for instance in organic particles, it may accumulate and inhibit hydrogen-producing reactions. Measurement of hydrogen in digestor systems may not reflect
this type of concentration gradient.

Complete anaerobic digestion of complex organic matter
is accomplished by many interacting groups of organisms.
The organic matter is degraded in a stepwise manner, the
products of one group of organisms are substrates for subsequent reactions catalyzed by other groups. One physiological group of organisms may depend on other groups for
the production of its substrate, and still other groups
for removal of its products. The obligate proton-reducing
acetogenic bacteria described in the previous talk are a
good example of this phenomenon. The interdependence of
several different physiological groups makes anaerobic
digestors very complex ecosystems.

In contrast to the stepwise nature of anaerobic digestion, compounds degraded in aerobic digestors are usually
completely mineralized by a single species of bacteria
using O_2 as electron acceptor. The reason why these degrada-
tions do not occur with the same stepwise division of labor
exemplified by organisms of anaerobic digestion is unknown.
It is possible that stepwise aerobic digestion did not
evolve because there are not suitable extracellular intermediates which could be produced during aerobic digestion.
Figure 1 shows various groups of compounds and the free
energy in each which could be released by complete aerobic
mineralization. The numbers on the ordinant show the free
energy which it would take to produce the compounds from
CO_2 and H_2O, and are analogous to the free energy of formation
(stoichiometrically required amounts of O_2 are included as
necessary, and standard concentrations are 1 mM rather than
1 M). Thus these numbers indicate the amount of free energy
which would be released by the complete oxidation of those
compounds, and may be called the "aerobic free-energy
content." The figure shows that very little of the energy
available in a mole of glucose is released by fermentation
alone. Even further conversion of those acids via methanogenesis or sulfate reduction releases only a small fraction
of the available energy. Almost all of this energy is
released in the last steps of aerobic digestion, during
which the organic matter is completely oxidized to CO_2 and

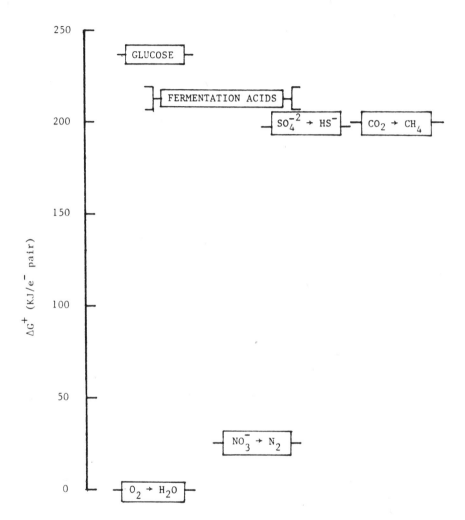

Figure 1.   Various groups of compounds and their free energy
which could be released by complete aerobic mineralization.   Calculations were made for the following concentrations: soluble compounds, 1 mM;
N_2, 1 atm.; O_2, 0.2 atm.; pH = 7.0

O_2 acts as electron acceptor. Thus it is not surprising that organisms do not extravagantly release as products compounds which still contain the vast majority of energy of the substrate, but prefer to complete the oxidation.

Under anaerobic conditions a different situation exists. Here complete oxidation to CO_2 is not an option; for complete anaerobic degradation to take place a part of the CO_2 must act as an electron acceptor in methanogenesis. Under anaerobic conditions the thermodynamically most stable configuration of carbon, hydrogen and oxygen atoms is as the compounds CH_4, CO_2 and H_2O. In examining reactions by which organic compounds are degraded anaerobically, it is more constructive to consider thermodynamics in terms of free energy required for formation of those compounds from the normal products of anaerobic digestion (CH_4, CO_2 and H_2O). This number is equal to the free energy released during complete methanogenic digestion, and may be called the "anaerobic free-energy content." The left side of Figure 2 shows the anaerobic free-energy content of several compounds. At the right are shown the anaerobic free-energy contents of glucose (at the top) and of several other groups of compounds which can be produced from a single mole of glucose. For instance, in the heterolactic fermentation, one mole of lactate, one mole of ethanol, and one mole of CO_2 may be stoichiometrically produced from a mole of glucose. The box labelled "Heterolactic" shows the sum of the free-energy contents of this collection of products, and the placement of the box shows the amount of energy released by the fermentation of glucose to these products as well as the amount of free energy remaining to be released by complete conversion of these compounds to CH_4 and CO_2.

This figure shows that there are a number of potential intermediates in anaerobic fermentation whose production releases significant quantities of free energy relative to the total available free energy. Bacteria growing in the absence of O_2 thus have the option of excreting these intermediates after obtaining a large portion of that available free energy, and these fermentation acids are the major extracellular intermediates in anaerobic digestors.

I should digress here for a moment and qualify that last statement, that fermentative bacteria have the option of excreting fermentation acids rather than completing their conversion to CH_4. I do not mean to imply that there are

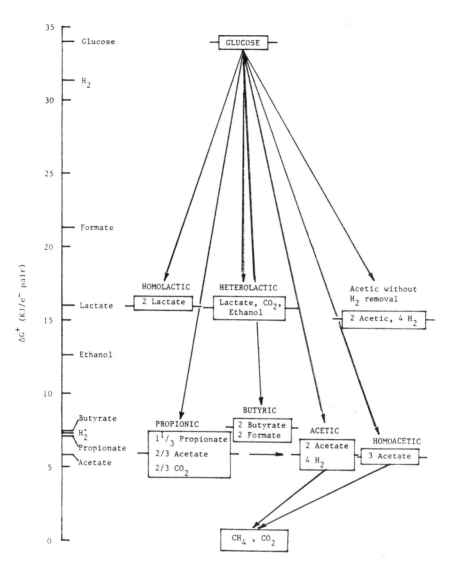

Figure 2. The anaerobic free-energy of glucose and of several groups of compounds which can be produced
from a single mole of glucose. Concentrations are
the same as Fig. 1. H_2 is H_2 at 5×10^{-5} atm.

fermentative bacteria which can produce CH_4, for rather than if it were more efficient for a single bacterium to perform this entire reaction sequence, such an organism would have evolved. I would predict that a successful attempt to manufacture such an organism may increase understanding of anaerobic digestion, but that such an organism added to anaerobic digestors will not increase their efficiency.

Classically, anaerobic digestion has been divided into two stages. The fermentation acids which are produced as intermediates are mainly acetic and propionic, with smaller amounts of butyric. The remainder of this paper will deal with the second phase: reactions which convert these extra-cellular intermediates to the terminal products, CH_4 and CO_2. I would like to describe a continuous, mixed-culture fermentor which was operated to mimic these reactions of anaerobic digestors. First I will quickly review the reactions (Fig. 3).

Methane is produced directly from H_2-CO_2 or acetate by pure cultures of methanogens. The degradation of longer fatty acids is accomplished by a group of bacteria known as the obligate proton-reducing acetogens, which were discussed in the previous paper. Propionate represents this class of compounds in this figure, and is the most difficult of these intermediates for the microflora to degrade. During periods of stress it may be the first intermediate to accumulate, and it may be toxic when it accumulates. Figure 4 shows the free-energy change which accompanies propionate oxidation at various concentrations of H_2. One can see that only at very low H_2 concentrations is propionate oxidation exergonic. CH_4 production from H_2 and CO_2 must accompany propionate oxidation because methanogenesis is the mechanism for H_2 removal. Thus methanogenesis must also be exergonic. There is only a very small range of H_2 concentrations which allows both of these reactions to occur, and this is the range in which H_2 concentration must fall in anaerobic digestors, or any methanogenic environment where propionate is degraded. Propionate-degrading bacteria have no other reaction from which to derive energy, and so in order to survive they must couple energy from this reaction to ATP synthesis, as described in the previous talk. Thus the propionate-degrading reaction diagrammed here is not the complete reaction: the synthesis of some amount of ATP from ADP and inorganic P must be added. This would make the complete reaction even less favorable than shown (the line would be higher). Likewise, CH_4 production is the energy-yielding

Organic matter

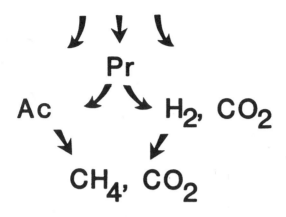

Figure 3. Main reactions in an anaerobic digestor. Ac = acetic acid; Pr = propionic acid.

reaction for methanogens, and ATP synthesis must be coupled to that reaction as well. This leaves an even more narrow range of H_2 concentrations which would allow both propionate oxidation and methanogenesis.

The main difference between the two-stage model and the more complex one of Figure 3 is the segregation of compounds which are degraded via interspecies H_2-transfer. The most important of these are the volatile organic acids longer than acetate, especially propionate. There may also be "reverse reactions" such as acetate production from CO_2 and H_2 or butyrate production from acetate and H_2. The "forward" reactions of anaerobic digestion must be exergonic because there are bacteria making a living by carrying them out. Thus the "reverse reactions" must be endergonic, requiring an input of energy by bacteria, or there must be micro-environments in anaerobic digestors, perhaps with higher H_2 concentrations, in which these reactions occur exergonically. I believe the latter is more likely, and I will present some data later which support this idea.

In order to understand the reactions leading to CH_4 and CO_2 from the products of the fermentative bacteria it is helpful to simplify, using a model compound to represent those degraded by interspecies H_2-transfer. Propionate was

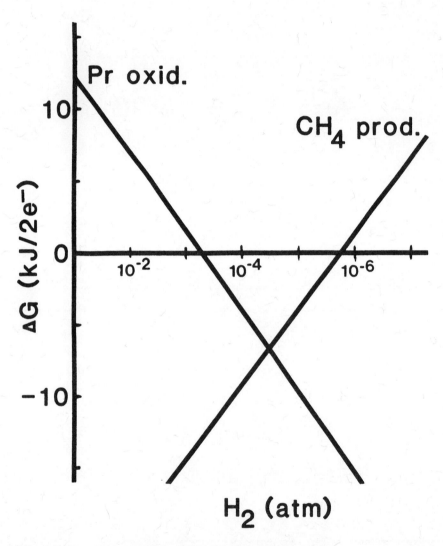

Figure 4.   Free-energy changes during propionate oxidation
at various H_2 concentrations.

chosen because it is quantitatively the most important and because the thermodynamics for its degradation are the most demanding. The three reactions I will use in modeling the methanogenic phase of anaerobic digestion are the aceto-clastic reaction, methanogenesis from H_2-CO_2, and propionate oxidation:

$$CH_3COO^- + H^+ \rightarrow CH_4 + CO_2$$

$$4H_2 + CO_2 \rightarrow CH_4 + 2H_2O$$

$$CH_3CH_2COO^- + 1/2H_2O \rightarrow CH_3COO^- + 1/4CO_2 + 3/4CH_4$$

To measure the turnover rates of the important inter-mediates (H_2, acetate and propionate), both the concen-trations and degradation rates must be quantified. This is relatively easy for acetate and propionate. H_2 concen-tration cannot, however, be measured by routine gas chro-matographic analysis. One can determine the range of allowable H_2 concentrations by using the thermodynamic calculations described above.

When this anaerobic digestor was examined, acetate and propionate concentrations were 1.2 and 0.7 mM (respectively), but H_2 concentration fell in the range of 1.2 to 450 nM. The turnover-rate constants for acetate and propionate were 1.33 and 0.46 h^{-1}, but H_2 turnover rate was 1.6 to 600 s^{-1}. The turnover rate of H_2 is extremely rapid because of the very low concentration. The range of H_2 concentrations shown here is based on the maximum allowable range of concentrations, based on thermodynamics. Because of the necessity of coupling the propionate-degrading and methanogenic reactions to microbial growth, the range is much more narrow, and the actual H_2 concentration is pro-bably fairly close to the middle of the range. That would mean that dissolved H_2 in digestors turns over about 300 times per second.

I would like to describe a continuous, mixed-culture fermentor which I operated to mimic the degradation of these three compounds in anaerobic digestors. This work has been published (*Appl. Environ. Microbiol., 48,* 112-126). The continuous, mixed-culture fermentor was operated at the same pH, temperature, and solids and liquid retention time as a lab scale anaerobic cattle-waste digestor on which the fermentor was modeled.

The simplified anaerobic degradation reactions shown in Figure 3 were used to develop the continuous-culture fermentor. These reactions were first measured in the cattle-waste digestor. Then a medium for the continuous-culture fermentor was derived which added these substrates at a rate matching the cattle-waste digestor. The simplest way to operate the continuous-culture fermentor was to add these intermediates (acetate, propionate, and H_2-CO_2). However, it is difficult to get gases such as H_2 to go into solution at the rapid rate at which they are utilized. Therefore gases were not added as substrates, but rather compounds were added from which these gases were rapidly produced. Figure 5 shows the actual reactions which were occurring in the anaerobic fermentor. Formic acid and ethanol were added as sources of H_2 and CO_2. Neither formic acid nor ethanol was ever detected in the fermentor and they were assumed to be instantaneously converted to products: formic acid to one mole each of H_2 and CO_2 and ethanol to one mole of acetic acid and two moles of H_2.

When this medium was used to operate the continuous mixed-culture fermentor propionate was not efficiently removed, and accumulated. When propionate was removed from the medium and only the methanogenic reactions were used in the model, CH_4 was produced in the fermentor at the same

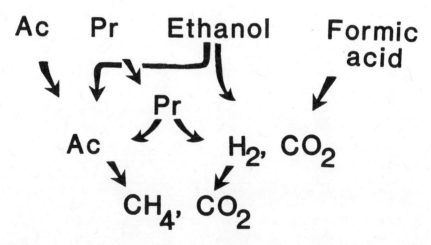

Figure 5. Actual reactions occurring in the anaerobic fermentor.

rate as the digestor. When intermediate levels of propionate were added that propionate accumulated to an intermediate level, still much higher than that of the anaerobic
digestor, even at very low rates of propionate addition.
When H_2 addition rates were decreased there were substantial
decreases in propionate concentration, even with full
addition rates of propionate. When H_2 addition rates were
decreased by only about 25% propionate levels were similar
to the digestor; it seemed that the fermentor acted like
the animal waste digestor only when the H_2-addition rate
to the continuous-culture fermentor was reduced. This
finding led to a hypothesis that a compartmentalization or
segregation of reactions occurred in the digestor.

Figure 6 shows a model which illustrates the hypothesis. I examined the substrate of the animal waste digestor
and found that 80% of the biodegradable organic matter was
contained in particulate matter. Fermentative reactions may
be localized in or on the surface of particles. H_2 may be a
product of these reactions, and because it turns over very
rapidly in digestors, may be catabolized before it can
diffuse from the particle. Rapid production of H_2 there
would cause its concentration to elevate, and stimulate
rapid methanogenesis. Volatile organic acids such as acetic
and propionic turn over much more slowly, and would diffuse
from the particle. Thus the microenvironment associated
with particles may have a fermentation pattern analogous to
the rumen, where also volatile acids accumulate and the
methanogenic substrate is predominantly H_2 produced by fermentative bacteria.

The volatile acids diffusing from the particles may be
catabolized by bacteria in the interparticle fluid. Because
most of the fermentatively produced H_2 is degraded before it
diffuses from particles, H_2 production may be low in interparticle fluid, and the concentration need not be high to
allow its degradation. Low H_2 concentration is a requisite
for the degradation of volatile organic acids. Thus it is
the reactions of the interparticle fluid which the continuous mixed-culture fermentor mimicked.

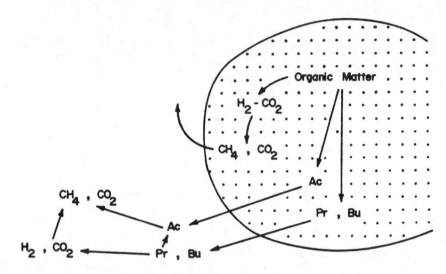

Figure 6.   A model illustrating organic matter biomethanation and localization of reactions in particles. Ac = acetic acid; Pr = propionic acid; Bu = butyric acid.

8

A LINK BETWEEN HYDROGEN AND SULFUR METABOLISMS IN *METHANOSARCINA BARKERI* (DSM 800)

G. Fauque,* M. Teixeira,** I. Moura,**, A.R. Lino,** P.A. Lespinat,*
A.V. Xavier,** D.V. Dervartanian,*** H.D. Peck, Jr.,*** J. Legall,*
and J.J.G. Moura**

A. R. B. S.　　　　　　　　　　**Centro de Quimica Estrutural*
E. C. E.　　　　　　　　　　　　*Lisbon, Portugal*
CEN-Cadarache
Saint-Paul Lez Durance, France

***Department of Biochemistry*
University of Georgia
Athens, Georgia

ABSTRACT

The hydrogenase of the acetoclastic methanogenic bacterium *Methanosarcina (M.) barkeri* strain DSM 800 grown on methanol has been purified to homogeneity. This soluble hydrogenase has a high molecular weight (800 kDa) and a final specific activity of 270 μmoles of H_2 evolved per min. per mg of protein. It shows an absorption spectrum typical of a non-heme iron protein with a ratio A400/A275 = 0.29.

It contains 8-10 iron atoms, 0.6-0.8 nickel atoms and 1 FMN per subunit of molecular weight of 50 kDa. Under hydrogen the purified hydrogenase can reduce the F_{420} cofactor (either free or protein-bound) and the ferredoxin isolated from *M. barkeri* as well as cytochromes c_{553} and c_3 from *Desulfovibrio* and c_7 from *Desulfuromonas*. *M. barkeri* presents also sulfite, thiosulfate and trithionate

97

reductases activities. The *M. barkeri* sulfite reductase
has been isolated (P_{590}); it contains 0.9 moles of sirohaem
and 4.9 g atom of iron per subunit of molecular weight =
23 KDa.

The EPR spectrum of P_{590} exhibits characteristics of
low spin ferrihaem with g values at 2.40, 2.30 and 1.88.

INTRODUCTION

The methanogenic bacteria are a diverse group of strictly
anaerobic microorganisms that produce methane from a limited
number of substrates. *Methanosarcina barkeri* belongs to a
group of methanogens that are characterized by their ability
to grow not only chemoautotrophically on H_2 + CO_2 but also
chemoorganotrophically with acetate, methanol or methylamines as energy sources (29).

The methanogenic bacteria have recently been reclassi-
fied as Archaebacteria in recognition of the fact that they
are only distantly related in the evolutionary scale to
eucaryotes (1) and the strictly anaerobic bacteria such as
the sulfate reducing organisms and Clostridia (16). Several
electron carriers and factors unique to methanogenic bacteria have been isolated from these organisms such as: coenzyme M (19), F_{420} (5), F_{342} (10) and F_{430} (29,6) which
may be isolated in a protein-bound form (20). The metabolism
of sulfur in methanogenic bacteria remains obscure although
it is of special importance because of these organisms synthesize large amounts of coenzyme M, which contains both
reduced and oxidized sulfur.

In this paper we report on a link between hydrogen and
sulfur metabolisms in *M. barkeri* (DSM 800).

MATERIALS AND METHODS

*Growth of the Microorganism and Preparation of the Crude
Extract*

M. barkeri (strain DSM 800) was grown at 37°C in a
methanol-containing medium as previously described (3). The
cells were suspended in 10 mM Tris-HCl buffer pH 7.6. DNase
was added and the extract was ruptured once in a French press
at 62 MPA under N_2 atmosphere. The extract was centrifuged
at 12000 rpm for 30 min and the supernatant constituted the
crude cell extract.

Assays

The sulfite reductase activity was measured by a mano-
metric assay as described by Schedel et al. (24). It requires the generation of reduced methylviologen by an excess
of hydrogenase under hydrogen atmosphere. The reduced dye
then serves as electron donor to the sulfite reductase.
The thiosulfate and trithionate reductases activities were
also determined manometrically by the same technique.

The sulfite reductase (P_{590}) has been isolated from
M. barkeri as previously described (21). The siroheme content of P_{590} was analyzed according to the method of Siegel
et al. (25). The hydrogenase of *M. barkeri* has been purified as previously described (7). Three methods were used
for the determination of hydrogenase activity: H_2 evolution with sodium dithionite (15 mM) as electron donor and
methylviologen (1 mM) as mediator in 50 mM Tris-HCl buffer
(23); H_2 uptake was followed by the reduction under H_2 of
benzyl-viologen (10 mM) at pH 8.0; and D $-H^+$ exchange activity was followed directly in the liquid phase in a reaction
vessel connected to a mass spectrometer via a teflon membrane allowing the diffusion of dissolved gases to the ion
source (2).

The hydrogenase specific activities are expressed as
moles of H_2 or HD produced and H_2 consumed per minute per
mg of protein. Total iron was determined by the TPTZ method
(8) and nickel by atomic absorption using a Perkin-Elmer
model 403.

Protein was determined by the Lowry (18) and biuret
(17) methods. The molecular mass of the native proteins
was estimated by gel filtration on a column of Sephadex
G-200 (30) and the subunit structure was determined on SDS
polyacrylamide gel electrophoresis (28). The coupling
effect of hydrogenase on the reduction of different electron
carriers was assayed by a spectrophotometric method following the reduction of the chromophores at 553 nm (cytochromes), 420 nm (F_{420}) and 400 nm (ferredoxins) in the
presence of hydrogenase under a hydrogen atmosphere (1
atm., 25°C).

Spectroscopic Instrumentation

The ultraviolet and visible spectra were recorded on a Beckman model 35 spectrophotometer. Electron paramagnetic resonance spectroscopy (EPR) was carried out on a Bruker 200-tt spectrometer, equipped with an ESR-9 flow cryostat and a Nicolet 1180 computer with which mathematical manipulations were performed.

RESULTS

Properties of the Hydrogenase from M. barkeri

The hydrogenase of the acetoclastic methanogenic bacterium *M. barkeri* (strain DSM 800) grown on methanol has been purified to homogeneity (7). This soluble hydrogenase has a high molecular weight (800 kDa) and a final specific activity of 270 µmoles of H_2 evolved per min. per mg of protein in the dithionite reduced methylviologen assay. The protein is rather stable to high temperature and to exposure to air at 4°C. It shows an absorption spectrum typical of a non-heme iron protein with maxima at 275, 380 and 403 nm and a ratio A400/A275 = 0.29 (Figure 1). It contains 8-10 iron atoms, 0.6-0.8 nickel atoms and 1 FMN per subunit of molecular weight of 60 kDa.

The electron paramagnetic resonance (EPR) spectrum of the native enzyme shows a rhombic signal with g values at 2.24, 2.20 and $\sim$ 2.0 probably due to nickel which is optimally measured at 40 K. In the reduced state, using molecular hydrogen or dithionite as reductants, at least two types of g = 1.94 EPR signals, due to iron-sulfur centers, could be detected and differentiated on the basis of power and temperature dependence.

Under hydrogen the purified hydrogenase can reduce the F_{420} cofactor (either free or protein-bound) and the ferredoxin isolated from *M. barkeri* as well as cytochromes c_{553} and c_3 from *Desulfovibrio* and c_7 from *Desulfuromonas* (Table 1). In the same conditions, this protein can also reduce methyl and benzyl viologens.

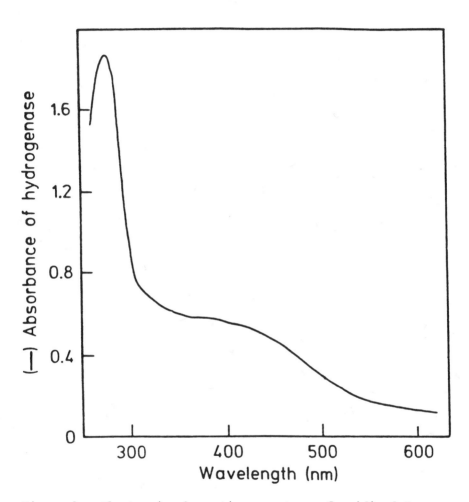

Figure 1. Electronic absorption spectrum of oxidized *M. barkeri* hydrogenase (0.62 mg/ml) at 25°C.

Table 1.   Direct Coupling Between Electron Transfer Proteins and Hydrogenase from Methanosarcina barkeri (DSM 800)

Protein	Active Center	Reduction by Hydrogenase Under H_2
Cytochrome c_{553} D. Desulfuricans (Berre Eau)	1 Heme C (Met, His)	Yes
Cytochrome c_7 Desulfuromonas Acetoxidans	3 Heme C (His, His)	Yes
Cytochromes c_3 * (M.W. 13000)	4 Heme C (His, His)	Yes
Ferredoxin M. Barkeri	[3 Fe – X S]	Yes
Ferredoxin II D. Desulfuricans (Berre Eau)	[4Fe-4S] + [3Fe-XS]	Partial
Factor F_{420} M. barkeri	5 – Deazaflavin Mononucleotide	Yes

* From D. gigas, D. desulfuricans Berre Eau, D. baculatus strain 9974.

A comparative study of two (Fe-Ni) hydrogenases (from *D. gigas* and *M. barkeri*) and one (Fe-) hydrogenase (from *D. vulgaris* Hildenborough) has been made (Table 2). The periplasmic hydrogenase from *D. vulgaris* Hildenborough is the most active in H_2 evolution, H_2 uptake and D_2-H^+ exchange

Sulfur Metabolism in M. barkeri

The *M. barkeri* crude extract exhibits sulfite, thiosulfate and trithionate reductases activities (Table 3). The sulfite reductase (P_{590}) has been purified from this strain as previously described (21). The optical spectrum of *M. barkeri* P_{590} shown in Figure 2 exhibits absorption bands at 590, 543, 395 and 275 nm. There is no band around 715 nm as is usually seen in other sulfite reductases; this band is characteristic of high spin Fe^{3+} complexes of isobacteriochlorins (26). The lack of this band in P_{590} is probably indicative that the siroheme is in a different spin state. The EPR spectrum of *M. barkeri* P_{590} shown in Figure 3 exhibits characteristics of low spin ferriheme with g values at 2.40, 2.30 and 1.88. Spin quantitation of this EPR signal yields a value close to 1 spin/siroheme. When P_{590} reacts with cyanide under reducing conditions (in the presence of methyl viologen) an EPR spectrum characteristic of reduced $[4Fe-4S]^{1+}$ center is observed (our unpublished results).

The chemical analysis of iron and siroheme together with EPR analysis show that P_{590} must contain one siroheme and probably one [4Fe-4S] center per mole of protein.

DISCUSSION

The specificity of the soluble hydrogenase from *M. barkeri* (DSM 800) for the reduction under H_2 of several electron carriers has been investigated. This protein is able to reduce some monoheme and multiheme cytochromes c isolated from sulfate and sulfur reducing bacteria. This hydrogenase reduces also completely the ferredoxin of *M. barkeri* and partially the *D. sulfuricans* (Berre Eau) ferredoxin. This protein can finally reduce the cofactor F_{420} from *M. barkeri* (free or protein-bound). We should note that the factor F_{420} is not reduced by the other (Fe-Ni) hydrogenase from *D. gigas* (G. Fauque and J. LeGall , unpublished).

Table 2. Comparison of Specific Activities of 2 Fe-Ni Hydrogenases (*Desulfovibrio Gigas* and *Methanosarcina Barkeri*) and 1 Fe-Hydrogenase (*D. vulgaris* Hildenborough)

Organism	H_2 Evolution (1 mM M.V. + 1/5 mM $Na_2S_2O_4$) pH 7.6	H_2 Uptake (10 mM BV) pH 8.0	HD Production (20% D_2 in Ar) pH 7.6
D. gigas	440	1 500	147
M. barkeri	270	960	35
D. vulgaris	3 800	37 000	607

The specific activities are expressed as μmoles of gases evolved or consumed per minute per milligram of protein.

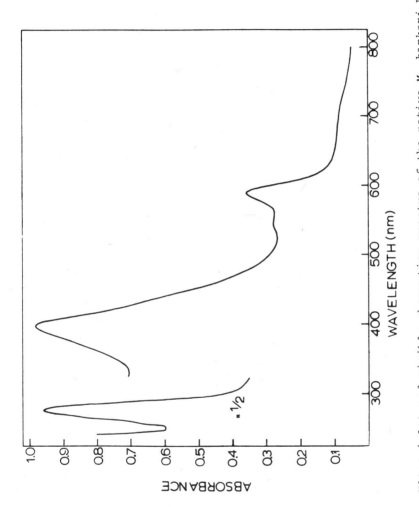

Figure 2. Ultraviolet and visible absorption spectra of the native *M. barkeri* P$_{590}$ (sulfite reductase).

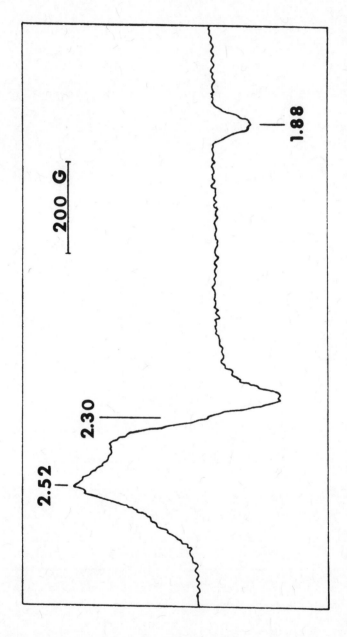

Figure 3.   EPR spectrum of native *M. backeri* P$_{590}$.   Temperature 9.6 K, microwave power 2mW, gain 5 x 10^5.

*Table 3.   Reduction Under Hydrogen of Sulfite, Thiosulfate
and Trithionate by a Crude Extract of M. barkeri*

Additions	*Activity*
	$\mu moles\ H_2\ min^{-1}\ mg^{-1}$
None	*0*
Na_2SO_3	*18*
$Na_2S_2O_3$	*5*
$Na_2S_3O_6$	*8*
Na_2SO_4	*0*
plus ATP	

*All reaction mixtures contained crude extract of M. barkeri
(DSM 804) plus methyl viologen (5 mM), temp. 30°C; gas phase
H_2. Na_2SO_3, 20 mM; phosphate buffer, pH 6.5, 0.2 M; $Na_2S_2O_3$,
20 mM; phosphate buffer, pH 7.6, 0.2 M; $Na_2S_3O_6$, 20 mM; Tris
buffer, pH 7.5, 0.1 M; Na_2SO_4, 20 mM; $MgCl_2$, 20 mM, ATP,
20 mM, phosphate buffer, pH 7.0, 0.2 M.*

Recently, EPR and Mössbauer studies of assimilative
sulfite reductase isolated from *D. vulgaris* Heldenborough
firmly establish that the siroheme is also in a low spin
heme complex (12). Mössbauer spectroscopy demonstrated that
the siroheme is exchange-coupled to the [4Fe-4S] center.
The siroheme-[4Fe-4S] unit is a common prosthetic group
found in *Escherichia coli* sulfite reductase and spinach
nitrite reductase. However, *D. vulgaris* sulfite reductase
and *M. barkeri* P_{590} are the only known examples where the
native state of the enzyme contains a low-spin ferric siroheme.

CONCLUSIONS

The presence of sulfite, thiosulfate and trithionate
reductase activities in *M. barkeri* shows that its sulfur
metabolism is far more complex than was first suspected.
Since sulfide ions are most probably predominant over oxidized sulfur compounds at the low redox potentials which are
necessary for the growth of methanogenic organisms, it is
not clear why these bacteria would need to reduce these

compounds. It has already been suggested (21) that *M.
barkeri* sulfite reductase could actually function in a
reverse direction, thus allowing the production of oxidized
sulfur which is necessary for the biosynthesis of coenzyme
M. Another attractive hypothesis is that a sulfur cycle
might function when *M. barkeri* grows in association with
sulfate reducing bacteria: a reoxidation of sulfide to
sulfite would allow the latter organisms to obtain extra
energy through oxidative phosphorylation coupled to methane
formation and the constant cycling of sulfide and sulfite
might prevent the accumulation of toxic levels of sulfide.
A study of *Desulfovibrio vulgaris* growing through interspecies hydrogen and acetate transfer with *M. barkeri* has
shown that the growth yield of the sulfate reducing bacterium is much higher than expected based on the loss of
ATP originating from oxidative phosphorylation (27). The
presence of a sulfite/sulfide cycle, eliminating the loss
of ATP for the activation of sulfate and allowing the production of ATP through oxidative phosphorylation during
sulfite reduction would explain the unexpected high growth
yield of the sulfate reducing bacteria.

In contrast to the F_{420} hydrogenase isolated from
Methanobacterium formicicum (22,13,14), *M. barkeri* hydrogenase appears to be extremely stable as far as its reactivity toward F_{420} is concerned. As seen in Table 1, its
reactivity is extremely broad and is not limited to two
electron acceptors since it can reduce the monoheme cytochrome c_{553} from sulfate reducing bacteria.

The sulfite reductase of *M. barkeri* appears similar
to the so-called assimilatory sulfite reductase from *D.
vulgaris* strain Hildenborough (15,12) in its molecular weight,
heme and nonheme iron content and in the spin state of its
siroheme. This is the second protein from a methanogenic
"archaebacterium" that is shown to bear similarity with a
protein from a sulfate reducing bacterium. With 26 direct
homologies, the ferredoxin from *M. barkeri* (11) is more
closely related to ferredoxin II from *D. desulfuricans*
Norway 4 (9) than the later protein is to *D. gigas* ferredoxin (19 direct homologies) (4).

ACKNOWLEDGMENTS

This work was made possible thanks to a collaborative agreement between UGA, CNRS and CEA. Financial support from INIC, INICT and NATO are gratefully acknowledged, also Solar Energy Research Institute Contract XD-1-1155-1, and AID Grant No. 936-5542-G-SS-4003-00.

LITERATURE CITED

1. Balch, W.E., G.E. Fox, L.J. Magrum, C.R. Woese, and R.S. Wolfe, "Methanogens: Reevaluation of a Unique Biological Group," *Microbial. Rev., 43,* 260-296 (1979).
2. Berlier, Y.M., G. Fauque, P.A. Lespinat, and J. LeGall, "Activation, Reduction and Proton-deuterium Exchange Reaction of the Periplasmic Hydrogenase from *Desulfovibrio gigas* in Relation with the Role of Cytochrome c_3," *FEBS Lett., 140,* 185-188 (1982).
3. Blaylock, B.A. and T.C. Stadtman, "Methane Biosynthesis by *Methanosarcina barkeri* - Properties of the Soluble Enzyme System," *Arch. Biochem. Biophys., 116,* 138-152 (1966).
4. Bruschi, M., "Amino Acid Sequence of *Desulfovibrio* Ferredoxin: Revisions," *Biochem. Biophys. Res. Comm., 91,* 623-628 (1979).
5. Cheeseman, P., A. Tomswood, and R.S. Wolfe, "Isolation and Properties of a Fluorescent Compound, Factor$_{420}$ from *Methanobacterium* Strain M.O.H.," *J. Bacteriol, 112,* 527-531 (1972).
6. Diekert, G., H.H. Gilles, R. Jaenchen, and R.K. Thauer, "Incorporation of 8 Succinate Per Mol Nickel into F_{430} by *Methanobacterium thermoautotrophicum,*" *Arch. Microbiol., 128,* 256-262 (1980).
7. Fauque, G., M. Teixeira, I. Moura, P.A. Lespinat, A.V. Xavier, D.V. DerVartanian, H.D. Peck, Jr., J. LeGall, and J.J.G. Moura, "Purification, Characterization and Redox Properties of Hydrogenase from *Methanosarcina barkeri* (DSM 800)," *Eur. J. Biochem., 142,* 21-28 (1984).
8. Fischer, D.S. and D.C. Price, "A Simple Serum Iron Method Using the New Sensitive Chromogen Tripyridyl-s-triazine," *Clin. Chem., 10,* 21-25 (1964).
9. Guerlesquin, F., M. Bruschi, G. Bovier-Lapierre, J. Bonicel, and P. Couchoud, "Primary Structure of the Two (4Fe-4S) Clusters Ferredoxin from *Desulfovibrio desulfuricans* (strain Norway 4)," *Biochimie, 65,* 43-47 (1983).

10. Gunsalus, R.P. and R.S. Wolfe, "Chromophoric Factors F_{342} and F_{430} of *Methanobacterium thermoautotrophicum*," *FEBS Lett., 3,* 191-193 (1978).

11. Hausinger, R.P., I. Moura, J.J.G. Moura, A.V. Xavier, H. Santos, J. LeGall, and J.B. Howard, "Amino Acid Sequence of a 3Fe-3S Ferredoxin from the Archebacterium *Methanosarcina barkeri*," *J. Biol. Chem., 257,* 14192-14194 (1982).

12. Huynh, B.H., L. Kang, D.V. DerVartanian, H.D. Peck, Jr., and J. LeGall, "Characterization of a Sulfite Reductase from *Desulfovibrio vulgaris:* Evidence for the Presence of a Low-Spin Siroheme and an Exchange-coupled Siroheme - 4Fe-4S Unit," *J. Biol. Chem.,* In press (1985).

13. Jin, S.L., D.K. Blanchard, and J.S. Chen, "Two Hydrogenases with Distinct Electron Carrier Specificity and Subunit Composition in *Methanobacterium formicium*," *Biochim. Biophys. Acta, 748,* 8-20 (1983).

14. Kojima, N., J.A. Fox, R.P. Hausinger, L. Daniels, W.H. Orme-Johnson, and C. Walsch, "Paramagnetic Centers in the Nickel-containing, Deazaflavin Reducing Hydrogenase from *Methanobacterium thermoautotrophicum*," *Proc. Natl. Acad. Sci., USA, 80,* 378-382 (1983).

15. Lee, J.P. and H.D. Peck, Jr., "Purification of the Enzyme Reducing Bisulfite to Trithionate from *Desulfovibrio gigas* and its Identification as Desulfoviridin," *Biochem. Biophys. Res. Comm., 45,* 583-589 (1971).

16. LeGall, J., J.J.G. Moura, H.D. Peck, Jr., and A.V. Xavier, "Hydrogenase and Other Iron Sulfur Proteins from the Sulfate-Reducing and Methane Forming Bacteria," *Iron-Sulfur Protein,* (T.G. Spiro, ed.), pp. 177-248, John Wiley and Sons, New York (1982).

17. Levin, R. and R.W. Brauer, "The Biuret Reaction for the Determination of Proteins. An Improved Reagent and its Application," *J. Lab. Clin. Med., 38,* 474-479 (1951).

18. Lowry, O.H., N.J. Rosebrough, A.L. Farr, and R.J. Randall, "Protein Measurement with the Folin Phenol Reagent," *J. Biol. Chem., 193,* 265-275 (1951).

19. MacBride B.C. and R.S. Wolfe, "A New Coenzyme of Methyl Transfer, Coenzyme M.," *Biochemistry, 10,* 2317-2324 (1971).

20. Moura, I., J.J.G. Moura, H. Santos, A.V. Xavier, G. Burch, H.D. Peck, Jr., and J. LeGall, "Proteins Containing the Factor F_{430} from *Methanosarcina barkeri* and *Methanobacterium thermoautotrophicum*," *Biochim. Biophys. Acta, 742,* 84-90 (1983).

21. Moura, J.J.G., I. Moura, H. Santos, A.V. Xavier, M. Scandellari, and J. LeGall, "Isolation of P_{590} from *Methanosarcina barkeri*: Evidence for the Presence of Sulfite Reductase Activity," *Biochem. Biophys. Res. Comm.*, *108*, 1002-1009 (1982).

22. Nelson, M.J.K., D.P. Brown, and J.G. Ferry, "FAD Requirement for the Reduction of Coenzyme F_{420} by Hydrogenase from *Methanobacterium formicium*," *Biochem. Biophys. Res. Comm.*, *120*, 775-781 (1984).

23. Peck, H.D., Jr., and H. Gest, "A New Procedure for the Assay of Bacterial Hydrogenases," *J. Bacteriol*, *71*, 70-80 (1956).

24. Schedel, M., J. LeGall, and J. Baldensberger, "Sulfur Metabolism in *Thiobacillus denitrificans*. Evidence for the Presence of a Dissimilatory Sulfite Reductase," *Arch. Microbiol.*, *105*, 339-341 (1975).

25. Siegel, L.M., M.Y. Murphy, and H. Kamin, "Siroheme: Methods of Isolation and Characterization," *Methods Enzymol.*, *52*, 436-447 (1978).

26. Stolzenbach, A.M., S.H. Strauss, and R.H. Holm, "Iron (II,III)-Chlorin and-Isobacteriochlorin Complexes. Models of the Heme Prosthetic Groups in Nitrite and Sulfite Reductases: Means of Formation and Spectroscopic and Redox Properties," *J. Am. Chem. Soc.*, *203*, 4763-4778 (1981).

27. Traore, S.A., C. Gaudin, C.E. Hatchikian, J. LeGall, and J.P. Belaich, "Energetics of Growth of a Defined Mixed Culture of *Desulfovibrio vulgaris* and *Methanosarcina barkeri*: Maintenance Energy Coefficient of the Sulfate Reducing Organism in the Absence of and Presence of its Partner," *J. Bacteriol.*, *155*, 1260-1264 (1983).

28. Weber, K. and M. Osborn, "The Reliability of Molecular Weight Determination of Dodecyl Sulfate Polyacrylamide Gel Electrophoresis," *J. Biol. Chem.*, *244*, 4406-4412 (1969).

29. Weimer, P.J. and J.G. Zeikus, "One Carbon Metabolism in Methanogenic Bacteria," *Arch. Microbiol.*, *119*, 49-57 (1978).

30. Whitaker, J.R., "Determination of Molecular Weights of Proteins by Gel Filtration of Sephadex," *Anal. Chem.*, *35*, 1950-1953 (1963).

31. Whitman, W.B. and R.S. Wolfe, "Presence of Nickel in Factor$_{430}$ from *Methanobacterium bryantii*," *Biochem. Biophys. Res. Comm.*, *92*, 1196-1201 (1980).

9

THE ASSIMILATORY REDUCTION OF N_2 AND NO_3^- BY
METHANOCOCCUS THERMOLITHOTROPHICUS

Negash Belay, Richard Sparling and Lacy Daniels

Department of Microbiology
University of Iowa
Iowa City, Iowa

ABSTRACT

   Methanococcus thermolithotrophicus growing on formate as
an energy source used dinitrogen (N_2) as the nitrogen source,
as evidenced by increase in culture turbidity, protein concentration and methanogenesis. No growth occurred in control
cultures lacking N_2 or added NH_4^+. Acetylene reduction assays
for nitrogenase using 0.3% acetylene with cells that were producing methane from formate under N_2-fixing conditions showed
the production of ethylene; cells grown on NH_4^+ as the nitrogen source did not reduce acetylene. *Mc. thermolithotrophicus*
also grew using NO_3^- as the nitrogen source. This organism
is the first methanogen, archaebacterium or thermophile growing above 60°C that has been demonstrated to carry out dinitrogen fixation.

INTRODUCTION

   The methane-producing bacteria (methanogens) are
presently the largest group of the archaebacteria (1,2). They
are characterized by their ability to make methane as an
obligatory and energy-generating end product of their metabolism. Most can produce methane from H_2/CO_2, and some can
produce methane and carbon dioxide from formate, methanol,
methylamine or acetate (1,2). Most media for the growth of
these organisms contain ammonium as the nitrogen source. For
example, ammonium was demonstrated as a good nitrogen source
for *Methanobrevibacter (Mbr.) ruminantium* (strains M1 and

PS) and *Methanobacterium (Mb.) bryantii,* but amino acids and
peptides could not be used (3). *Methanococcus voltae* does
not use amino acids, nitrate, nitrite, urea, taurine or
methylamine as sole nitrogen sources, but can transport several amino acids (4,5). *Methanobacterium (Mb.) thermoautotrophicum* strain ΔH cannot use nitrate, nitrite, dinitrogen
or amino acids as nitrogen sources (L. Daniels, Master's Thesis, Univ. of Wisconsin, Madison, 1974). *Methanobacillus
omelianskii,* which is now known to have been a mixture of at
least two organisms (the methanogen *Mb. bryantii* and the "S"
organism;6), was reported to fix dinitrogen (N_2) as its sole
nitrogen source (7); due to it being an impure culture, it is
unknown whether the methanogen was the true nitrogen fixer.
Recently, Belay et al. (8), and Murray and Zinder (8a) have
reinvestigated the phenomenon of dinitrogen fixation by methanogens, and have demonstrated that *Methanococcus (Mc.)
thermolithotrophicus* grown on H_2/CO_2) and *Methanosarcina
barkeri* strain 227, respectively, can fix dinitrogen as the
sole source of nitrogen. We report here our further investigations of nitrogen fixation and nitrate reduction by *Mc.
thermolithotrophicus.*

MATERIALS AND METHODS

   Mc. thermolithotrophicus was a gift of Dr. K.O. Stetter
(9), and was grown at 65°C using the techniques of Balch and
Wolfe (10) and the previously described medium (11). All
gases for the gas mixtures were of research grade, 99.9995%
pure (Air Products; N_2O <0.1 molar ppm), except for acetylene, which was generated from calcium carbide. All cultures
were grown under a H_2/CO_2 (80:20, v/v) gas phase unless otherwise stated.

   For preparation of inocula for experiments demonstrating
N_2 reduction, 50 ml of medium supplemented with 0.2 µM tungstic
acid and 0.2 µM sodium selenate but lacking NH_4^+ was prepared
in 250 ml pressure bottles (Wheaton Scientific). These bottles
were inoculated with 4 ml of culture grown in the presence of
16.8 mM NH_4Cl, resulting in a carryover of 1.2 mM NH_4^+. This
concentration of ammonium is sufficient to promote full growth
of cells in this first transfer. From these bottles, 5 ml of
inocula were transferred to 250 ml bottles containing 50 ml of
ammonium-free media prepared as indicated below.

   For adapting cells to nitrogen-fixation, the gas phase
in the bottles was replaced before inoculation with a 95%
$N_2/5\%$ CO_2 gas mixtures, by gassing thoroughly with entry and
exit needles and pressurized to 35 kPa above atmospheric

(1 atm = 105 kPa). Sterile sodium sulfide (0.5 ml of a 100 mM solution) was readded to replace lost H$_2$S. The bottles were then pressurized to a final overpressure of 140 kPa with H$_2$/CO$_2$ (80:20, v/v), thus there were about 140 kPa of N$_2$ and 105 kPa of H$_2$/CO$_2$ in the bottles. The bottles were then inoculated. After a lag of several weeks the cultures grew and were transferred several more times to medium with N$_2$ as the nitrogen source using the above mentioned procedure.

For adapting cells to NO$_3^-$ reduction, ammonium-free bottles containing 11 mM NaNO$_3$ were inoculated from the N$_2$ fixing culture. Growth occurred after several weeks. Both N$_2$ fixing and nitrate-reducing cultures showed faster growth and very short lags after several transfers and were ready to be used for comparative growth studies. Cultural purity was assured by plating on Gelrite as described previously (8).

A nitrogen-fixing culture was adapted to growth on formate in the above described ammonium free medium supplemented with 294 mM sodium formate, 1 μM sodium selenate and 1 μM tungstic acid. The pH of the medium was adjusted to 6.2 with Na$_2$CO$_3$ while gassing with a 50% CO$_2$/50% N gas mixture. The medium was dispensed in 50 ml amounts into 500 ml pressure bottles and made anaerobic, as previously described (11).

For formate growth experiments comparing N$_2$ and NH$_4^+$ as nitrogen sources, bottles of media with either N$_2$ or NH$_4^+$ as the nitrogen source were initially pressurized to 35 kPa of N$_2$/CO$_2$ (50:50), then to 140 kPa above the atmospheric when growth had started (in order to offset the pH increase caused by the utilization of formate). Control cultures were inoculated into formate-containing ammonium-free medium adjusted to pH 6.2 with 100% CO$_2$ and pressurized to 35 kPa above atmospheric under the same gas.

For growth experiments on H$_2$/CO$_2$ comparing NO$_3^-$ and NH$_4^+$ as nitrogen sources, a nitrate-adapted culture was transferred to ammonium free medium (to give a carryover of 1.9 mM NO$_3^-$). After full growth, 4 ml of this stock was used to inoculate 250 ml bottles containing 50 ml of nitrate or ammonium medium. NaNO$_3$ was added to nitrate bottles using a sterile syringe to a final concentration of 17.14 mM from a stock solution that was made anaerobic under an argon gas phase and autoclaved. No ammonium or nitrate was added to control bottles.

Information on the history of the inocula and the source of nitrogen used in the growth experiments is given in the Results section.

Incubation was carried out with the bottles horizontal in a gyratory shaker. Growth was measured by absorbance at 600 mM in a Perkin Elmer 552A spectrophotometer. Methane production and reduction of acetylene to ethylene was measured by gas chromatography (Varian Aerograph 2700), and protein was determined by the Bradford assay (12).

RESULTS AND DISCUSSION

Cultures of *Mc. thermolithotrophicus* growing on H_2/CO_2 and using N_2 as the nitrogen source were originally obtained, as described above, by transferring a 1.2 mM NH_4^+ inoculum into ammonium-free medium. There was no growth in control bottles lacking N_2 and containing 0.1 mM NH_4^+. Repeated plating of cells on Gelrite^R petri plates resulted in isolated colonies able to reduce N_2, which proved that a contaminant was not responsible for the phenomenon; further, microscopic examination revealed no other organism. Growth under nitrogen fixation conditions still occurred in the presence of antibiotics (78 µg/ml each of kanamycin and streptomycin). Furthermore, when normal *Methanococcus* medium (8) with a gas phase of N_2/CO_2 (80:20 v/v) was supplemented with a variety of heterotrophic nutrients (0.2% peptone and 0.2% yeast extract; 0.1% yeast extract and a sugar mix 0.05% each of glucose, sucrose, xylose and glycerol; 10 mM each of lactate and sulfate, 0.02% yeast extract and a vitamin mix), no growth was observed. When grown in H_2/CO_2 with the supplements present, the methanogen grows with no contaminants observed by microscope. We proceeded with experiments to demonstrate N_2 fixation by cells growing on formate, which we describe here.

A N_2-fixing culture grown on H_2/CO_2 was transferred to formate-containing medium under a N_2/CO_2 atmosphere. After a lag of several days, growth occurred. This culture was used, after several transfers in formate medium, in the experiment described in Fig. 1. Both NH_4^+-grown and N_2-grown cultures grew well: there was a consistent rise in A_{600}, protein and methane present. The control culture without N_2 or NH_4^+ did not grow. Note that in Fig. 1, the N_2-fixing culture produces almost as much CH_4 as the NH_4^+ cultures, whereas it produces significantly less protein and A_{600}. This suggests a difference in cell yield (g dry weight of cells/mole CH_4).

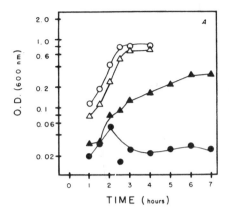

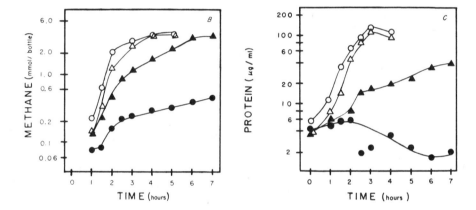

Figure 1.   Growth of *Methanococcus thermolithotrophicus* on
            formate as carbon and energy source in medium con-
            taining either ammonium or dinitrogen as nitrogen
            source. O represent data collected from an ammon-
            ium growth culture inoculated from an ammonium
            culture; Δ, data from an ammonium grown culture
            inoculated from a N$_2$-fixing culture; ▲, data from
            a culture grown on N$_2$ as its sole nitrogen source
            inoculated from a N$_2$-fixing culture; ●, data from
            a fixed nitrogen source free culture under N$_2$/CO$_2$,
            inoculated from a N$_2$-fixing culture. Fig. 1a, ab-
            sorbance at 600 nm of these various cultures was
            monitored. Fig. 1b, methanogenesis was monitored
            as total methane per bottle; each bottle (Wheaton
            Scientific #223952) contained 30 ml of medium pre-
            pared as described in the text, in a total volume
            of 540 ml per bottle. Fig. 1c, protein levels of
            the same cultures were monitored. All data points
            represent the average of duplicates.

Table 1 describes the quantitative differences in yields. Cells grown with NH_4^+ as the nitrogen source gave yields of 1.8 g cell dry weight/mole CH_4, whether or not the inoculum came from N_2 or NH_4^+ grown cultures; these values were 2.5-fold higher than cells grown under N_2-fixing conditions. Since N_2 fixation is an ATP-consuming process, this difference is reasonable: more methane must be made to provide the extra ATP needed for growth. These data are consistent with our previous data with H_2/CO_2 grown cultures (8).

Two 100 ml samples of cells, grown in a 16 ℓ fermentor on formate using N_2 as a nitrogen source, were removed into two 500 ml bottles and acetylene added to a final concentration of 0.3%. At this level both methanogenesis and acetylene reduction to ethylene occurred. In 24 hrs these bottles produced 27.8 mmole $CH_4 \cdot mg$ protein^{-1} and 26.4 μmole ethylene·mg protein^{-1}. Cells exposed to 10% acetylene lysed and did not reduce acetylene; this might be expected since acetylene is known to be an inhibitor of methanogens (13). Cells of *Mc. thermoautotrophicus* (100 ml mid log cells in 250 ml bottles with 0.3% acetylene), when grown under $H_2/CO_2 + N_2$ (8), do not reduce acetylene in the presence 85 μM Br-CoM, an inhibitor of methanogenesis (13a) and therefore an inhibitor of ATP production. In 4 hours in the presence of acetylene 7000 nmol/ml gas phase CH_4 and 2 nmol/ml ethylene were produced compared to 500 nmol/ml CH_4 and no acetylene reduction in the presence of Br-CoM, whether supplemented or not with acetylene. Br-CoM (300 μM) did not however inhibit growth of cultures of sewage sludge mesophilic anaerobic heterotrophs grown in 0.2% dextrose and 0.2% yeast extract suggesting it is specific for methanogens. The evidence presented here (growth curves, growth yields and acetylene reduction assays) conclusively proves that *Mc. thermolithotrophicus* can assimilate dinitrogen reductively as a nitrogen source for growth.

Another source of nitrogen for growth that can be reductively assimilated by some bacteria is nitrate (NO_3^-). We examined this compound with *Mc. thermolithotrophicus*. When NH_4^+ was first replaced by 11 mM NO_3^-, growth occurred after 10-30 days of incubation. Control cultures lacking NO_3^- and containing only 0.1 mM NH_4^+ showed no growth. As in the N_2 experiments described above, the nitrate cultures grew better and with shorter lags after more transfers. Fig. 2 illustrates the ability of *Mc. thermolithotrophicus* to grow on NO_3^- as a nitrogen source. Ammonium is a slightly better nitrogen

TABLE 1.

Nitrogen Source	Inoculum History	Protein Concentrations* (μg/ml)		Yield§ (g cell dry weight/ mmole CH$_4$)	Specific Activity I (μmole CH$_4$/ min. mg protein)
		Initial	Final		
None	N$_2$ grown	3.8	5.4	Δ	——
NH$_4^+$	NH$_4^+$ grown	5.1	122	1.84	19.0
NH$_4^+$	N$_2$ grown	3.55	155	1.83	21.1
N$_2$	N$_2$ grown	3.46	45	0.73	10.7

* Protein concentration determined by the Bradford assay method. The final protein concentration being that at the end of the logarithmic phase of growth.

Δ ____ not detectable

§Yields were calculated as the difference of protein concentrations divided by that of methane concentrations over the period of logarithmic growth. To obtain gram dry weights, one, assuming that proteins comprise 50% of the cell's dry weight, multiplies by two.

ISpecific activity was calculated as the methane produced between two time points within the logarithmic phase of growth over the average protein concentration between these points.

Both growth yields and specific activities may also be calculated from the data shown in the graphs presented herein.

All data presented in this table are averages of duplicate experiments.

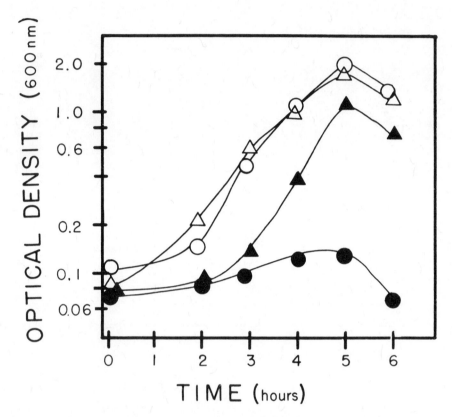

Figure 2.   Growth of *Methanococcus thermolithotrophicus* on
             H_2/CO_2 as energy source in medium containing either
             ammonium or nitrate as nitrogen source. O, repre-
             sent data from an ammonium grown culture inocu-
             lated from an ammonium culture; Δ, data from an
             ammonium grown culture inoculated from a nitrite
             culture; ▲, data from a nitrate grown culture in-
             oculated from a nitrate culture; ●, data from
             cells inoculated into a nitrogen-source free medium
             from a nitrate grown culture; the only nitrogen
             source is that from the carryover nitrate from the
             inoculum. All data points represent the average
             of duplicates.

source, but nitrate works very well; the control culture with less than 0.2 mM NO_3^- and no NH_4^+ did not grow at all. This data proves that *Mc. thermolithotrophicus* can carry out reductive assimilation of nitrate.

We have recently demonstrated that *Mc. thermolithotrophicus* can fix N_2 while growing on H_2/CO_2 (8). Also, P.A. Murray and Steve Zinder (8a) have shown that another methanogen, *Methanosarcina barkeri,* fixes dinitrogen. These are the first reports of dinitrogen fixation by pure cultures of methanogenic bacteria, and the first demonstrations of this phenomenon in any of the archaebacteria. The novelty of these reports and our findings in this paper prompts several questions about the evolution of nitrogenase since archaebacteria are thought to have separated evolutionarily from eubacterial ancesters very early in evolution. *Mc. thermolithotrophicus* is also the first thermophile that can fix N_2 at about 60°C; previously, *Mastigocladus* was reported to fix N_2 at a maximum of about 55°C (14), even though the organism can grow with NH_4^+ at temperatures above 60°C. We expect that the nitrogenase from methanogens could be interestingly different from other known nitrogenases.

The question remains of how widespread nitrogen fixation is among the methanogens. It is worth examining *Mb. bryantii* again, as well as checking the ability of a mesophilic marine *Methanococcus.*

Ecologically, both the reductive assimilation of N_2 and NO_3^- have many implications. In some anaerobic waste digestion systems, e.g., cellulosic conversion to methane, nitrogen could be a factor limiting the growth of organisms. The ability of a variety of obligate and facultative anaerobes to fix N_2 is known (e.g., *Klebsiella* and *Clostridium*): these organisms should provide some input of nitrogen into nitrogen-poor anaerobic ecosystems. Several NO_2^- reducers are also known to occur in similar environments. Our work with *Mc. thermolithotrophicus* suggests that at least one methanogen has similar abilities.

The original isolation site for *Mc. thermolithotrophicus* (9), a shallow submarine hydrothermal vent near the coast of Italy, may be low in ammonium. Thus, nitrogen fixation may be important to its ability to live in that environment, since seawater contains little nitrate or ammonium (15) but has significant levels of N_2 (about 8-15 mg/l at 10°C) (15,16). Also, nitrate and ammonium levels in freshwater ecosystems are often very low (0.005-1.0 mg/l) compared to N_2 (about 16

mg/l) (15). Terrestrial hot springs and lakes containing geo-
logically generated H_2, but with little organic nitrogen
input are other nitrogen-poor environments that might contain dinitrogen-fixing methanogens.

ACKNOWLEDGMENTS

Support during the preparation of this paper was provided by grants from the National Institute of Health (GM-
30868), the National Science Foundation (PCM82-07809) and the
Petroleum Research Fund of the American Chemical Society
(14062-AC5). R.S. was supported by a postgraduate scholarship from the Natural Sciences and Engineering Council of
Canada.

REFERENCES

1. Daniels, L., R. Sparling, and G.D. Sprott, "The Bioener-
    getics of Methanogenesis," *Biochim. Biophys. Acta, 768,*
    113-163 (1984).
2. Nagel, D., Comprehensive Biotechnology, Vol. 1, Ch. 5,
    (N. Moo-Young, ed.), Pergammon Press, Inc., New York
    (1984).
3. Bryant, M.P., S.F. Tzeng, I.M. Robinson, and A.E. Joyner,
    "Nutrient Requirements of Methanogenic Bacteria. Anaero-
    bic Biological Treatment Processes" (Amer. Chem. Soc.),
    Adv. Chem. Ser. No. 105, 23-40 (1971).
4. Whitman, W.B., E. Ankwanda, and R.S. Wolfe, "Nutrition
    and Carbon Metabolism of *Methanococcus voltae," J. Bacter-
    iol, 149,* 852-863 (1982).
5. Jarrell, K.F., S.E. Bird, and G.D. Sprott, "Sodium-
    dependent Isoleucine Transport in the Methanogenic Archae-
    bacterium *Methanococcus voltae," FEBS Lett., 166,* 357-
    361 (1984).
6. Pine, M.J. and H.A. Barker, "Studies on the Methane Bac-
    teria. XI. Fixation of Atmospheric Nitrogen by *Methano-
    bacterium omelianskii," 68,* 589-591 (1954).
7. Bryant, M.P., E.A. Wolin, M.J. Wolin, and R.S. Wolfe,
    "*Methanobacillus omelianskii,* a Symbiotic Association of
    Two Species of Bacteria," *Arch. Mikrobiol., 59,* 20-31
    (1967).
8. Belay, N., R. Sparling, and L. Daniels, "Dinitrogen fixa-
    tion by a Thermophilic Methanogenic Bacterium," *Nature
    (London), 312,* 286-288 (1984).
8a. Murray, P.A. and S.H. Zinder, "Nitrogen Fixation by a
    Methanogenic Archaebacterium," *Nature (London), 312,*
    284-286 (1984).

9. Huber, J.M., M. Thomm, H. Konig, G. Thies, and K.O.
    Stetter, "*Methanococcus thermolithotrophicus,* A Novel
    Thermophilic Lithotrophic Methanogen," *Arch. Microbiol.,*
    132, 47-50 (1982).

10. Balch, W.E. and R.S. Wolfe, "New Approach to the Culti-
    vation of Methanogenic Bacteria: 2-Mercaptoethanesul-
    fonic Acid (HS-CoM)-Dependent Growth of *Methanobacterium
    ruminantium* in a Pressurized Atmosphere," *Appl. Environ.
    Microbiol., 32,* 781-791 (1976).

11. Daniels, L., N. Belay, and B. Mukhopadahyay, "Considera-
    tions for the Use and Large Scale Growth of Methanogenic
    Bacteria," *Biotech. Bioeng. Sixth Sympos. Biotech. Fuels
    and Chemicals,* In press (1984).

12. Bradford, M.M., "A Rapid and Sensitive Method for the
    Quantitation of Microgram Quantities of Protein Utilizing
    the Principle of Protein-dye Binding," *Anal. Biochem.,
    72,* 248-254 (1976).

13. Sprott, G.D., K.F. Jarrell, K.M. Shaw, and R. Knowles,
    "Acetylene as an Inhibitor of Methanogenic Bacteria,"
    J. Gen. Microbiol., 128, 2453-2462 (1982).

13a. Balch, W.E. and R.S. Wolfe, "Specificity and Biological
    Distribution of Coenzyme M (2-Mercaptoethanesulfonic
    Acid)," *J. Bacteriol., 137,* 256-263 (1979).

14. Brock, T.D., Thermophilic Microorganisms and Life at
    High Temperature, Springer-Verlag, New York (1978).

15. Brock, T.D., Principles of Microbial Ecology, Prentice-
    Hall, Inc., Englewood Cliffs, New Jersey (1966).

16. Raymont, J.E.G., Plankton and Productivity in the Oceans,
    Pergamon Press, Inc., New York (1963).

10
A DEFINED STARTER CULTURE FOR BIOMETHANATION OF PROTEINACEOUS WASTES

M.K. Jain

Department of Bacteriology
University of Wisconsin-Madison
Madison, Wisconsin

J.G. Zeikus

Michigan Biotechnology Institute
Michigan State University
East Lansing, Michigan

ABSTRACT

A stable, defined mixed culture was developed which readily degraded gelatin and other proteins into methane and carbon dioxide. The mixture comprised of *Methanosarcina barkeri* and a novel proteolytic, hydrogen consuming acetogen, *Clostridium proteolyticum* sp. nov. *C. proteolyticum* was isolated from a sewage digestor as a prevalent hydrolytic species and its general cellular, growth and metabolic properties are described here. In mono-culture, *C. proteolyticum* produced both protease and collagenase activity and fermented all the amino acids in gelatin, except proline, into acetate and carbon dioxide as the main products with hydrogen, isovalerate and isobutyrate detected in trace products (< 1 mM). In co-culture with *M. barkeri,* gelatin was completely hydrolyzed and transformed into methane and carbon dioxide with varying levels of intermediary acetate formed as a function of incubation time. The data show that complex proteinaceous polymers can be readily transformed into methane and carbon dioxide at 30-40°C by a stable co-culture which does not require exogenous growth factor additions. In addition, the co-culture was readily transferable and methanogenesis initiated rapidly without the need for exogenous pH control.

INTRODUCTION

Biomethanation is becoming increasingly attractive as an alternative to other means of waste disposal (19). The reasons for this include: low biomass production under anaerobic conditions, elimination of energy requiring aerobic treatment processes, the value of methane as a by-product fuel energy source, and new anaerobic digestor designs with improved treatment rates. Proteinaceous materials constitute about 19-27% of sewage sludge wastes. In anaerobic digestion processes, complex substrates (carbohydrates, proteins and lipids) are hydrolysed and degraded into methane and carbon dioxide by a mixed population (26). Acetate and H_2-CO_2 are the major immediate precursors for methanogenic bacteria in industrial digestors. Also, organisms that produce these intermediary metabolites from complex organic matter are an important part of any defined starter culture developed for biomethanation. One requirement for reliable industrial waste treatment is to gain process control over the activity of the mixed microbial population. Define mixed populations of bacteria that convert various saccharides including cellulose into methane and CO_2 have been studied (3,7,8,10,12,16,18,20,22,23). To our knowledge, however, defined co-cultures have not been developed which completely degrade complex proteins into methane and carbon dioxide.

The purpose of the present report is to show that a simple defined co-culture comprised of a novel *Clostridium* species and *Methanosarcina barkeri* readily degrades gelatin and other complex proteins into methane and carbon dioxide.

MATERIALS AND METHODS

Clostridium proteolyticum sp. nov., was isolated from Nine-Spring sewage digestor, Madison, Wisconsin (9,21). *Methanosarcina barkeri* strain MS had been adapted and maintained on 80 mM acetate as carbon and energy source. *M. barkeri* was grown in 158 ml serum bottles containing 50 ml PBB medium (14) with H_2-CO_2 as energy and carbon source under shaking conditions at 37°C. H_2-CO_2 (80:20) was replenished every day until the culture reached late log phase. At this time, the gas phase of the serum bottle was changed with nitrogen. To this, sterilized gelatin (0.5%) was added along with 2.0% inoculum of *C. proteolyticum* and a time course was followed by incubating at 37°C under shaking conditions. When the gelatin was consumed, the gas phase was replaced as described above followed by a new addition of 0.5% substrate. Fatty acids and alcohols were analyzed by flame ionization detection in a Packard 419 gas chromatograph as described by Zeikus et al. (25). All gases were quantified by gas chromatography as described elsewhere (15). Gelatin concentration was analyzed by the modified Bradford's dye binding protein assay (5). Growth was measured by light scattering at 660 nm directly by spectrophotometry. Amino acids were analyzed using a Durram Amino Acid Analyzer model D-500 in the Biophysics Laboratory, University of Wisconsin, Madison, WI. DNA was isolated and purified from cells by the method of Marmur (11). DNA base compositions were calculated by the method of DeLay (4) from DNA thermal denaturation in 0.015 M NaCl and 0.0015 M trisodium citrate as determined in a Gilford model 250 Spectrophotometer equipped with a model 2527 thermoprogrammer. *Escherichia coli* DNA VIII, Lot D-2001 from Sigma Chemical Co., served as a standard. Protease and collagenase activity was assayed by the methods of Blackburn and Hullah (1) and Miyoshi and Rosenbloom (13), respectively, with slight modifications. Filtrated samples were used to determine amino acids released from casein or azocoll by the ninhyridin method (17). One unit of enzyme activity represents the amount of protease or collagenase which under the above conditions, would produce the equivalent of 1 µg of leucine min^{-1} from either casein or azocoll.

RESULTS

Gelatin hydrolyzing bacteria when enumerated by the five tube MPN technique were present at 3.5 x $10^6/ml^{-1}$ in a sewage digestor. The bacterium isolated by enrichment technique was purified and characterized. Based on its cellular growth and metabolic characteristics (Table 1) and the VPI Anaerobe Laboratory Manual (6) the bacterium was identified as *Clostridium* sp. and named as *C. proteolyticum* sp. nov. The isolate grew on gelatin with a doubling time of two hours. It produced acetate and carbon dioxide as the main products. Hydrogen, ethanol, iso-butyrate and iso-valerate were also detected as trace products (< 1 mM). The organism did not grow autotrophically on H_2-CO_2, but it was capable of consuming exogenous H_2 in the stationary growth phase. Hydrogen was not inhibitory to growth at 3 atmospheres. However, cysteine when added as a reducing agent inhibited growth. Figure 1 shows that *C. proteolyticum* produced variable levels

Table 1. Distinguishing characteristics of
Clostridium proteolyticum sp. nov.

Cellular Characteristics

Morphology: single rods with sub-terminal spores
Motility: polar flagellum
Wall: gram-positive
DNA Base Composition: 24.22 mol % G+C

Growth Characteristics

Temperature: 30-37°C optimum
O_2: obligate anaerobe
pH Range: 6-8
Doubling Time: 2 h on gelatin

Metabolic Characteristics

Energy Source: gelatin, serum albumin, collagen, casein,
  meat, various proteins
Fermentation End Products: major--acetate; traces--H_2,
  ethanol, isolvalerate and isobutyrate
Nutrition: does not require growth factors

Habitat

Sewage sludge digestors, 3.5 x 10^6/ml

---

of extracellular protease during growth which were highest
during the lag phase and lowest during the stationary phase.
Activity of extracellular collagenase increased parallel to
growth.

Figure 2 shows a time course for gelatin degradation
into fermentation end products. At the end of growth, about
90% of gelatin was consumed. The medium pH remained near
neutral during the entire fermentation.

Table 2 illustrates that all the amino acids of gelatin are metabolized (61-90% during growth except proline.

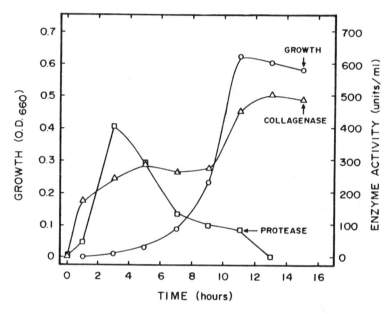

Figure 1.   Protease and collagenase activity during growth
of *C. prteolyticum* on gelatin.   The organism was
grown under a N_2 gas phase in bunged aluminum
crimp-sealed culture tubes that contained 10 mL
PBB medium with 0.5% gelatin.   Substrates for protease and collagenase activity assay were casein
and azocoll, respectively.

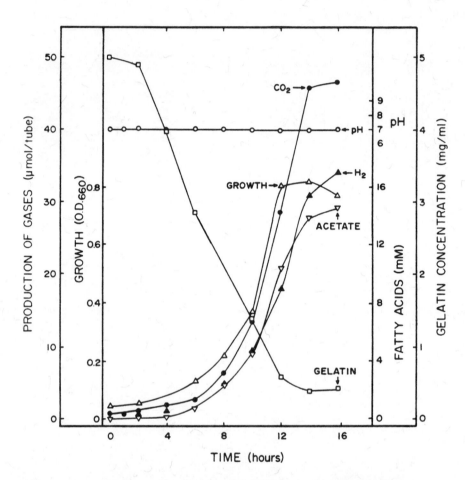

Figure 2.   Fermentation time course of *C. proteolyticum* on
gelatin.   The organism was grown under N_2 gas
phase in sealed serum bottles that contained 50
mL PBB medium with 0.5% gelatin as the sole carbon and energy source.

Table 2. Analysis of the Major Amino Acids in Gelatin
            Fermented by C. proteolyticum[a]

Amino Acid[b]	Concentration (nmoles/40 µl) T_{oh}	At $T_{16}h$	% Metabolized
Aspartate	10.04	3.01	60.06
Glutamic Acid	15.18	2.29	84.91
Proline	27.82	36.11	- -
Glycine	70.82	8.01	88.64
Alanine	22.38	6.83	69.48
Arginine	10.19	1.29	87.34
Hydroxyproline	19.227	1.74	90.95

[a]Experimental Conditions: C. proteolyticum was grown in
culture tubes containing 10 ml PBB medium with 0.5% gelatin.
Samples were withdrawn at t_o and $t_{16}h$ and subjected to acid
hydrolysis (6 N HCl) at 110°C for 20 h, followed by freeze-
drying. Amino acid analysis was done using Durram Amino
Acid Analyzer Model D-500.

[b]Only those A.A. whose initial concentrations are greater
than 10 nmol/40 µl sample are reported. These comprise
greater than 60% of total A.A.

Table 3 shows the effect of different gelatin concentrations on growth and end product formation by C. proteolyticum. Growth was neither affected by high concentration
of gelatin nor by the end products (e.g., 100 mM acetate).
Also pH remained neutral at up to 10.0% gelatin concentration.

Table 4 illustrates the range of protein sources fermented by C. proteolyticum. The organism readily fermented
gelatin, peptone, trypticasetryptone, cooked meat and poly-
pep. However, a prolonged lag phase was required for growth
on casein or serum albumin which were not as readily fermented. Interestingly, the major products from all the protein fermentations were acetate and CO_2 with only traces of

Table 3.   Effect of Different Gelatin Concentrations on Growth and End Product Formation by Clostridium proteolyticum[a]

Gelatin (%)	Growth 16 h (A_{660})	pH	End Products Formed (mM)				
			Acetate (mM)	Propionate (mM)	Isobutyrate (mM)	Isovalerate (mM)
0.5	1.244	6.96	19.64	0	0.31	0.49
1.0	2.774	6.78	36.67	0.036	0.70	0.99
2.5	2.850	6.64	73.33	0.038	1.70	2.41
5.0	2.930	6.63	80.67	1.25	2.32	3.26
10.0	3.255	6.67	100.00	1.60	3.01	4.35

[a]*Experimental Conditions: Products were measured after growth in aluminum sealed culture tubes that contained 10 ml PBB medium with different concentrations of gelatin. The inoculum was grown on gelatin. All tubes were incubated at 37°C under stationary conditions.*

Table 4. Comparison of Different Protein Fermentations by Clostridium proteolyticum[a]

Protein Source	Growth (A_{660})	Final pH	H_2 (µmoles)	Fermentation Products (mM)					
				Acetate	Propionate	Isobutyrate	Butyrate	Isovalerate	Ethanol
Gelatin	0.800	6.96	0.06	19.57	0	0.34	0	0.49	0
Peptone	1.664	7.10	0.87	17.56	0.68	1.08	0.05	0.22	0
Casein	0.210	7.21	23.06	3.72	0.53	0.54	0	1.07	0.41
Trypticase- Tryptone	0.400	7.23	23.95	3.60	0.63	0.81	1.36	1.67	0
Bovine-Serum Albumin	0.210	---	26.10	5.13	0.75	0.83	0	1.98	0.80
Poly-pep	0.510	6.89	0.08	12.06	0	0.18	0	0.30	0
Cooked Meat	---	7.12	15.08	53.33	6.90	7.18	21.99	16.10	0

[a]Experimental Conditions: Products were measured after growth in aluminum sealed tubes that contained 10 ml PBB medium with 0.5% substrate except cooked meat for which cooked meat medium (dehydrated Difco) was used at 1.25 g 10 ml^{-1}. The inoculum of C. proteolyticum was grown on gelatin. All tubes were incubated at 37°C under shaking conditions.

of iso-butyrate, isovalerate, ethanol, hydrogen and/or propionate. Although acetate was a major end product when the
organism was grown in cooked meat medium (Difco Lab), significant amounts of butyrate were also produced. Poly-pep
(a proprietary protein digest that provides increased stabilization of tissue sections in histochemical studies) was
utilized by this bacterium. However, dipeptide mixtures
(i.e., L-Ala-L-Ala, Glycyl-L-Alanine, Glycyl-L-Aspartic acid,
Glycyl-Glycyl, Glycyl-DL-Phenylalanine, L-Leucyl-Glycine) were
not fermented.

Figure 3 shows a typical time course for gelatin fermentation by the co-culture. At day 5, the vials were
flushed with N_2 and a second feeding with 0.5% gelatin was
performed. After 16 h into the first feeding, about 85.0%
of the gelatin was hydrolyzed, intermediary levels of acetate were at maximal values and H_2 was detected as a trace
metabolite. However, methane production continued lin-
early for the next 3 days in response to acetate consumption. Except for not detecting H_2, the kinetics of gelatin
conversion to methane repeated on the second feeding. On
day 5 of the time course, an inoculum (10.0%) from this
co-culture was transferred into fresh medium containing
8 mM methanol. Figure 4 shows that gelatin was readily
transformed to CH_4 by the co-culture indicating the transferability and high activity of the starter culture.

DISCUSSION

Our results show that proteins such as gelatin are completely hydrolyzed and further degraded into CH_4 and CO_2
by a defined starter culture comprised of *C. proteolyticum*
sp. nov. and *M. barkeri* strain MS (acetate adapted). These
results suggest that an anaerobic microbial food chain required to readily degrade polymeric proteins may consist
of only two species. This is of interest because biomethanation of other simple substrates (i.e., sucrose) required
five different species for complete transformation (7).
It was previously reported that polymeric pectin could be
completely degraded to CH_4 and CO_2 by a co-culture of *C.
butyrium* and *M. barkeri* (18); and, cellulose by a co-culture
of *Acetovibrio cellulolyticus* and *M. barkeri* (8). Worth
noting here, is that proteins unlike the above polysaccharides are comprised of many different amino acids in
lieu of the same repeating sugar monomer.

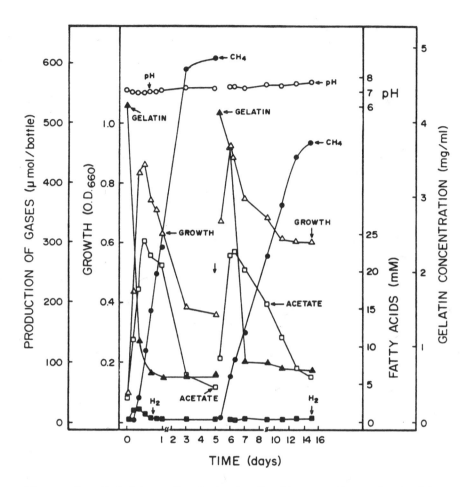

Figure 3.   Relation of substrate feeding on the gelatin biomethanation time course for *C. proteolyticum* and
M. barkeri co-cultures. *M. barkeri* was pregrown
under shaking in serum bottles that contained 50
ml PBB medium and H_2/CO_2. The gas phase was replaced with N_2 and *C. proteolyticum* was added at
2% level along with 0.5% gelatin at 0 day. Where
indicated by down arrow, the gas phase of the
bottle was replaced with N_2 and 0.5% gelatin was
added. Bottle was incubated at 37°C with shaking.

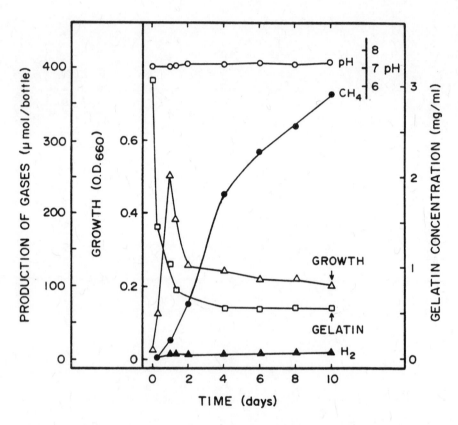

Figure 4.   Relation of co-culture transferability to gelatin
biomethanation.   The co-culture was transferred
(as 10% inoculum from day 5 in Figure 3) to a
fresh serum bottle that contained 50 ml PBB medium
with 0.5% gelatin and 8 mM methanol.   Incubation
was done at 37°C with shaking.

The use of *C. proteolyticum* in defined started culture also limits the number of ancillary species required for effective biomethanation because this species ferments the amino acids in proteins into acetate as the major product and H_2 as a trace product. Both of these metabolites are the substrates for methanogenic species but not for the obligate hydrogen syntrophic species (24). *C. proteolyticum* readily degrades various proteins of animal origin because it contains very active proteases including collagenase.

Based on the properties described here, the proteolytic isolate is unique and the name *C. proteolyticum* has been proposed (Jain and Zeikus, manuscript in preparation). Of the described species, *C. proteolyticum* most closely resembles *C. histolyticum,* a pathogen, in that it produces collagenase; but, it differs in type of flagellation, nutrition, fermentation substrate range, and end products formed (2,6).

CONCLUSIONS

Our data show near complete transformation of protein polymer to CH_4 by a simple mixture of *C. proteolyticum* and *M. barkeri.* This co-culture may have industrial utility as a starter culture for treatment of proteinaceous waste waters because it is stable, transferable and readily active. Furthermore, pH control is not needed even though high acetate levels are produced. These data clearly show that acetate transformation and not protein degradation was rate limiting to biomethanation of proteinaceous wastes.

ACKNOWLEDGMENTS

This research was supported by the College of Agricultural and Life Sciences, University of Wisconsin-Madison, and by a research grant from Institut Pasteur Foundation to J.G.Z. M.K. Jain was on research leave from Haryana Agricultural University, Hissar, India.

LITERATURE CITED

1.   Blackburn, T.H. and W.A. Hullah, "Cell bound protease of *Bacterioides amylophilus,*" *Can. J. Microbiol., 20,* 435-441 (1974).
2.   Buchnan, R.E. and N.E. Gibbons, "Bergey's Manual of Determinative Bacteriology, 8th ed., Williams and Wilkins, Baltimore, MD (1974).

3. Chen, M. and M.J. Wolin, "Influence of CH₄ production by *Methanobacterium ruminantium* on the fermentation of glucose and lactate by *Selenomonas ruminantium*," *Appl. Environ. Microbiol., 34,* 756-759.

4. Deley, J., "Reexamination of the association between melting point, buoyant density and chemical base composition of the deoxyribonucleic acid," *J. Bacteriol., 101,* 738-754 (1970).

5. Duhamel, R.C., E. Meezan, and K. Brendel, "The addition of SDS to the Bradford dye binding protein assay, a modification with increased sensitivity to collagen," *J. Biochem. Biophys. Methods, 5,* 67-74 (1981).

6. Holdeman, L.V., E.P. Cato, and W.E.C. Moore, "Anaerobe Laboratory Manual", 4th ed., V.P.I. Anaerobe Laboratory, The Virginia Polytechnic Institute, Blacksburg, VA (1977).

7. Jack-Jones, W., J.P. Guyot, and R.S. Wolfe, "Methanogenesis from sucrose by defined immobilized consortia," *Appl. Environ. Microbiol., 47,* 1-6 (1984).

8. Khan, A.W., "Degradation of cellulose to methane by a co-culture of *Acetovibrio cellulolyticus* and *Methanosarcina barkeri*," *FEMS Microbiol. Lett., 9,* 233-235 (1980).

9. Krzycki, J.A., R. Wolkin, and J.G. Zeikus, "Comparison of unitrophic and mixotrophic substrate metabolism by an acetate adapted strain of *Methanosarcina barkeri*," *J. Bacteriol., 149,* 247-254 (1982).

10. Laube, V.M. and S.M. Martin, "Conversion of cellulose to methane and carbon dioxide by tri-culture of *Acetovibrio cellulolyticus, Desulfovibrio* sp. and *Methanosarcina barkeri*," *Appl. Environ. Microbiol., 42,* 413-420 (1981).

11. Marmur, J., "A procedure for the isolation of deoxyribonucleic acid from the microorganisms," *J. Molecular Biol., 3,* 208-218 (1961).

12. McInerney, M.J. and M.P. Bryant, "Anaerobic degradation of lactate by syntrophic associations of *Methanosarcina barkeri* and *Desulfovibrio* species and effect on H₂ on acetate degradation," *Appl. Environ. Microbiol., 41,* 346-354 (1981).

13. Miyoshi, M. and J. Rosenbloom, "General proteolytic activity of highly purified preparations of clostridial collagenase," *Connec. Tissue Res., 2,* 77-84 (1974).

14. Moench, T.T. and J.G. Zeikus, "Nutritional growth requirements for *Butyribacterium methylotrophicum* on single carbon substrates and glucose," *Curr. Microbiol., 9,* 151-154 (1983a).

15.   Nelson, D.R. and J.G. Zeikus, "Rapid method for the isotopic analysis of gaseous end products of anaerobic metabolism," *Appl. Microbiol.*, *28*, 258-261 (1974).

16.   Patel, G.B. and L.A. Roth, "Acetic acid and hydrogen metabolism during co-culture of an acetic acid producing bacterium with methanogenic bacteria," *Can. J. Microbiol.*, *24*, 1007-1010 (1978).

17.   Rosen, H., "A modified Ninhydrin colorimetric analysis for amino acids," *Arch. Biochem. Biophys.*, *67*, 10-15 (1957).

18.   Schink, B. and J.G. Zeikus, "Microbial ecology of pectin decomposition in anoxic lake sediments," *J. Gen. Microbiol.*, *128*, 393-404 (1982).

19.   Stafford, D.A., B.I. Wheatley, and D.E. Hughes, "Anaerobic Digestion," Applied Science Publishers, Ltd., London England, pp. 528 (1980).

20.   Weimer, P.J. and J.G. Zeikus, "Fermentation of cellulose and cellobiose by *Clostridium thermocellum* in the absence and presence of *Methanobacterium thermoautotrophicum*," *Appl. Environ. Microbiol.*, *33*, 289-297 (1977).

21.   Weimer, P.J. and J.G. Zeikus, "Acetate metabolism in *Methanosarcina barkeri*," *Arch. Microbiol.*, *119*, 175-182 (1978).

22.   Winter, J.U. and R.S. Wolfe, "Complete degradation of carbohydrate to carbon dioxide and methane by syntrophic cultures of *Acetobacterium woodii* and *Methanosarcina barkeri*," *Arch. Microbiol.*, *121*, 97-102 (1979).

23.   Winter, J.U. and R.S. Wolfe, "Methane formation from fructose by syntrophic association of *Acetobacterium woodii* and different strains of methanogens," *Arch. Microbiol.*, *124*, 73-79 (1980).

24.   Zeikus, J.G., "The biology of methanogenic bacteria," *Bacteriol. Rev.*, *41*, 514-541 (1977).

25.   Zeikus, J.G., P.W. Hegge, and M.A. Anderson, "*Thermobacterium brockii* gen. nov. and sp. nov., a new chemoorganotrophic, caldoactive, anaerobic bacterium," *Arch. Microbiol.*, *122*, 41-48 (1979).

26.   Zeikus, J.G., "Microbial populations in digestors." *Anaerobic Digestion*, (D.A. Stafford, et al., eds.), pp. 61-90, Applied Science Publishers, LTD, London, England (1980).

11
STRATEGIES FOR THE ISOLATION OF MICROORGANISMS
RESPONSIBLE FOR POLYPHOSPHATE ACCUMULATION

N. Suresh,[1] R. Warburg,[1] M. Timmerman,[2] J. Wells,[1] M. Coccia,[1]
M.F. Roberts,[3] and H.O. Halvorson[1]

[1] Rosenstiel Basic Medical Sciences
Research Center
Brandeis University
Waltham, Massachusetts

[2] Air Products and Chemicals, Inc.
Allentown, Pennsylvania

[3] Department of Chemistry
Massachusetts Institute of Technology
Cambridge, Massachusetts

ABSTRACT

Several strategies were used to isolate organisms involved in the uptake and subsequent release of inorganic
phosphate from waste water sludge. These included direct
staining for polyphosphates (polyP), growing in ^{32}P inorganic
phosphate followed by autoradiography, resistance to dicyclohexyl carbodiimide (DCCD), an ATPase inhibitor, and isolation on the basis of the buoyant density of the cell.
Among those microorganisms isolated, three were identified
as *Acinetobacter lwoffii, A. calcoaceticus* and *Pseudomonas
vesicularis*. The *P. vesicularis* culture had 31% of phosphate as polyP. ^{31}P NMR analysis of the whole cells revealed the presence of polyP when the cultures were grown
aerobically to the late stationary phase and its subsequent
loss during anaerobic incubation. Loss of polyP was also
associated with a decrease in buoyant density of the cell.
In the presence of DCCD, there was a decrease in the polyP
peak, but a substantial increase in the sugar phosphates
which is consistent with a hypothesis that polyP is used as
a reserve energy source. *P. vesicularis* cells showed a

two-fold increase in the level of polyphosphatase during
early stationary phase, but a thirty-fold increase in polyphosphate kinase activity during late stationary phase.
This increased enzyme activity is consistent with the increased polyP synthesis during late stationary phase.

INTRODUCTION

Phosphate is an essential component for a variety of
biosynthetic reactions of cells from generating energy from
metabolic reactions to a wide variety of biosynthetic reactions (Harold, 1966; Mühlradt, 1971; Kuleav, 1975).   In
excess, phosphate can lead to major environmental concerns,
such as eutrophication of bodies of water.   The recognition
in recent years that selected populations of microorganisms
can remove excess phosphate has led to interest in their
identification and the mechanism(s) of phosphate uptake and
release (Harold, 1966; Smith et al., 1954; Thomas, 1965).

Excess phosphate accumulation in microorganisms has
been identified with a variety of biological polymers.
These include 4S RNA in *Arthrobacter globiformis* (Shoda et
al., 1980), methanophosphagen in *Methanobacterium thermoautotrophicum* (Kanodia and Roberts, 1983), polyphosphates
associated with capsid in *Neisseria gonorrhoeae* (Nogel and
Gotschlich, 1983) or outside the plasma membrane (Tijssen
and von Steveninck, 1984; Umanov et al., 1975), or as long
chain cytoplasmic reserves in a variety of microorganisms
(Harold, 1966; Kuleav, 1975; Miller, 1984; Cramer and Davis,
1984).   The recent demonstration that inorganic pyrophosphate can be directly used as a source of energy for growth
of *Clostridium* (Varma et al., 1983), suggests that even further
phosphate reserves may exist.

We have chosen to examine biomass from an anaerobic/
aerobic waste water treatment process which is characterized
by dramatic phosphate removal (Deakyne et al., 1984).   In
this process, as shown in Figure 1, an initial anaerobic
feeding zone is followed by a well aerated aerobic zone
(Spector, 1977).   Approximately 1 mole of inorganic phosphate is released to the aqueous phase per mole of glucose
absorbed from the aqueous phase by the biomass (Timmerman,
1979).   Under subsequent aerobic conditions, both the phosphate release in the anaerobic zone and that originally
present in the feed to the system are taken up by the cells
as the glucose is oxidized.   Since the initial anaerobic
feeding phase appears essential for efficient phosphate

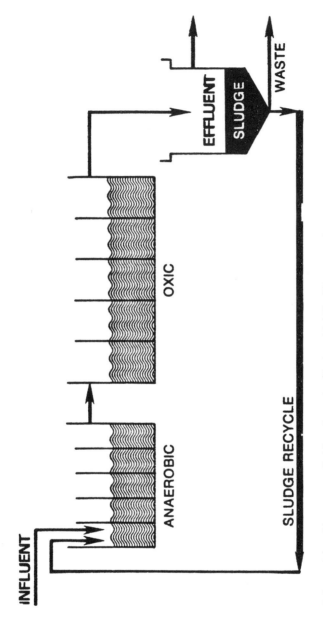

Figure 1. Schematic diagram of the Anoxic/Oxic (A/O) flow system.

removal, a number of interesting questions arise. What
consortium of microorganisms are responsible for this phenomenon? What is the mechanism of phosphate removal? In the
aerobic zone, how does sugar oxidation support phosphate
uptake--an energy dependent process?

The studies on this process are very limited. Fuhs
and Chen (1975) reported that *Acinetobacter* is responsible
for the luxury uptake of phosphate in certain waste water
treatment plants (Millburg et al., 1971; Vacker et al.,
1971; Bargmen, 1970). Fuhs and Chen (1975) found that a
strain, similar to *A. lwoffii,* was able to release inorganic
phosphate upon the addition of acetate to the medium, by
lowering the pH or by both, under anaerobic conditions.
The mechanism of this process is not as yet understood from
the molecular biology point of view.

In this report we examine the bacterial consortium taken
from a functional anaerobic/aerobic waste water treatment
demonstration plant in Baltimore, MD. The nature of the
accumulated phosphate is further defined and new methods
for isolating microorganisms involved in phosphate removal
and release are described. The non-invasive technique of
in vivo ^{31}P NMR spectroscopy was employed to monitor small
phosphorous metabolites and polyphosphates as a function of
cell growth conditions.

MATERIALS AND METHODS

Maintenance of Cultures

The cultures were grown in Luria broth (LB) at 30°C on
a rotary shaker. Log phase cultures were diluted 1:1 with
50% sterile glycerol and were stored at -20°C.

Transformation

trpE27 and *ilv* strains for *Acinetobacter* were provided
by Juni. DNA isolation and transformation was carried out
by the method of Juni (1978). This involved incubation of
a cell paste in 0.05% sodium dodexyl sulphate in standard
saline citrate for 1 hr at 60°C. A loopful of the resulting
"DNA solution" was then used to spread a small amount of auxotrophic bacterial cells on heart infusion medium. After incubation at 30°C for 18 hr, the resulting colony lawns were
replica plated to minimal plates. Any prototrophs resulting
were scored after 18 hr at 30°C.

^{31}P NMR Analysis

Unless otherwise noted, late stationary phase cells grown in LB overnight at 30°C in a rotary shaker to OD_{540} ~2.0 were used for ^{31}P NMR spectroscopy. The cells were harvested by centrifugation in a Sorvall RC 2B at 10,000 g for 10 min. The cell paste was dispersed in 2 ml buffer containing 10 mM potassium phosphate, 100 mM Pipes, 37 mM NH_4Cl, 2 mM $MgCl_2$, 2.6 mM Na_2CO_3, 2 mM Na_2S, pH 7.2 with 10% D_2O added for a lock signal. Resuspended cells (~1.0 x 10^{11} cells/ml) were placed in 10 mm NMR tubes and stored on ice until used. Spectra were obtained at 4°C within 8 hr of sample harvesting. ^{31}P-NMR spectra were obtained at 109.3 MHz on a Bruker 270 spectrometer with ^1H-noise decoupling. Typical parameters for in vivo experiments include 8000 Hz sweep width, 60° pulse angle, 0.5 sec recycle time, 1000 transients, and an exponential multiplication factor at 10 Hz.

Enzyme Assays

Polyphosphatase. Polyphosphatase was assayed by a modification of the method of Harold and Harold (1965). The assay mixture contained 0.05 ml of 2 M KCl, 0.1 ml of 5 mM $MgCl_2$, 0.1 ml of Tris HCl buffer (0.1 M, pH 7.0), 0.05 ml of the endogenously prepared ^{32}P polyP and 0.1 ml of crude cell extract. The mixture was incubated for 30 min at 30°C and the reaction stopped by addition of 0.1 ml of bovine serum albumin solution (10 mg/ml) and 0.5 ml of 0.5 N perchloric acid (PCA). After standing on ice for 15 min the mixture was centrifuged at 4°C for 5 min at 12,000 x g. The ^{32}P Pi released was estimated by counting the soluble radioactivity.

Polyphosphate kinase. PolyP kinase was assayed in the reverse direction by the transfer of P^{32} from polyP to ADP according to Harold and Harold (1965), except that we used the endogenously prepared P^{32} polyP substrate. The assay mixture was the same as used for assay of polyphosphatase with the addition of 20 mM ADP. The difference in the polyP hydrolyzed with and without ADP was defined as polyP kinase activity.

RESULTS AND DISCUSSION

Isolation and Screening for Phosphate Accumulating Bacteria

To provide pure cultures for examining the mechanisms of phosphate uptake and release, we isolated phosphate accumulating bacteria from an anaerobic/aerobic waste water treatment plant. The sludge from the anaerobic/oxic waste water treatment plant was sonicated at low speed for 15 sec to break up clumps and, after dilution in 0.9% saline, plated on Luria broth (Miller, 1972) and allowed to grow for 2 days at 30°C. The colonies were examined microscopically for different morphological varieties, stained with Neisser's stain (Eikelboon and van Buijsen, 1981) for putative poly-phosphate (polyP) granules and those which were positive selected for further testing. Two approaches were used: (a) autoradiography and (b) resistance to an ATPase inhibitor. For autoradiography colonies were transferred by replica plating from LB plates to plates containing 0.5 µCi ^{32}P-inorganic phosphate in LB. After 15 h at 30°C an X-ray film was placed on the resulting colonies for half an hour. After development of the film, production of dark spots was scored. Out of 17 colonies, 10 were positive. Other workers such as Shoda et al. (1980), have used this autoradiography tech-nique to screen for high phosphate accumulating cells. They isolated a strain of *Arthrobacter globiformis* which had a high RNA content but little polyP. To make the selection more specific for bacteria with increased amounts of polyP, we placed sterile 541 Whatman filter paper discs onto fresh colonies on LB plates and allowed them to grow into the filter paper for 4 hr. The discs were subsequently placed onto ^{32}P-inorganic phosphate containing LB plates for 4 h at 30°C. All cellular material, except for polyP, was re-moved from the resulting filter papers by sequential treat-ment with lysozyme/EDTA, Triton-X 100, RNAse, DNAse, pronase and cold TCA. The filter papers were placed onto X-ray film and subjected to autoradiography for half an hour. Only two colonies were positive out of the seventeen tested, but we did not find these to contain substantial amount of polyP compared to the other isolates. It is possible that these colonies may not have been lysed by the lysozyme/EDTA treat-ment resulting in a strong signal on autoradiography.

An alternative technique was developed to screen for colonies which contained more polyP than other bacteria. The rationale was that polyP may serve as an alternative energy source to ATP in these bacteria. One would predict that in the presence of dicyclohexyl carbodiimide (DCCD), an ATPase inhibitor, only those bacteria which accumulate polyP and can utilize it as a source of energy will be able to survive. To test this, the previously isolated colonies were plated on LB plates containing 200 μM DCCD. The bacteria from colonies which grew on these plates not only stained strongly for polyP but also did so when grown on normal LB plates. The colonies also gave strong signals when grown in the presence of ^{32}P inorganic phosphate containing LB plates containing 200 μM DCCD followed by autoradiography. Table 1 shows the correlation between DCCD resistance and autoradiography.

Two of the isolates which stained strongly for polyP and which were resistant to DCCD (up to 800 μM) were identified as *Acinetobacter lwoffii* and *A. calcoaceticus* by the American Type Culture Collection based on their fermentation and other biochemical properties. This identification was supported by the plemorphic morphology of the bacteria and transformation assays with known *Acinetobacter* strains (Table 2) according to Juni (1978). Both strains identified as *Acinetobacter lwoffii* and *A. calcoaceticus* gave poor transformation compared to the strains used by Juni. This indicates that although they are *Acinetobacter* their relationship is distant. Juni (1972) found *trp*E27 to be the most conserved marker amongst all the strains he tested. It is therefore interesting to note that the strain identified as *A. calcoaceticus* gave poor transformation for *trp*E27 but not for *ilv*.

The strain of *A. lwoffii* had 7.5% of its phosphate as polyP compared to 1.8% in the *A. calcoaceticus* strain when measured by the method of Poindexter and Eley (1983). This strain of *A. lwoffii* was studied further and may be the same or similar to that isolated by Fuhs and Chen (10), and will be referred as *A. lwoffii* henceforth.

To test for phosphate uptake and release, a culture of *A. lwoffii* grown in LB for 18 h at 30°C was divided into aliquots. One was preincubated anaerobically and the other aerobically for one hour. They were then subjected to aerobic incubation in the presence of ^{32}P inorganic phosphate for 30 min. The culture subjected to anaerobic preincubation took up twenty times more inorganic phosphate than the aerobic control (Table 3). Thus, anaerobic preincubation induced the uptake of inorganic phosphate during aerobic incubation, reminiscent of sludge behavior in the

Table 1.   Correlation Between DCCD Resistance and Autoradiography

Culture	Autoradiography	Growth on DCCD
A. *lwoffii*	+	+
A. *calcoaceticus*	+	+
Culture #1	+	+
#2	+	+
#3	−	−
#4	+	+
#5	−	−
#6	+	+
#7	+	+
#8	−	−
#9	−	−
#10	+	+
#11	−	−
#12	−	−
#13	+	+
#14	+	+
#15	+	+
Bacillus subtilis	−	−

Cultures were transferred on toothpicks to LB plates containing 0.5 μCi ^{32}P inorganic phosphate and onto plates containing 200 μM DCCD.   The colonies were allowed to grow at 30°C for 18 hr.   The colonies from the plates containing ^{32}P inorganic phosphate were covered with Saran Wrap and overlaid with Kodak X-ray film for 30 min.   The film was developed and colonies scored "+" or "−" for dark spots. Similarly, the colonies were scored as "+" or "−" for growth on plates containing DCCD.

anaerobic/aerobic waste water treatment plant.   Further, 24-25% of the ^{32}P taken up during the aerobic treatment was released after 15 min of anaerobic incubation (Table 3).

Use of Percoll Density Gradients

Because of the difficulties described above with direct selection for phosphate accumulating bacteria, a new strategy was sought.   PolyP contents as high as 20% of the dry mass have been reported (Harold, 1966; Shoda et al., 1980). Since polyP, and particularly the calcium or magnesium salt of polyP, are considerably denser than the vegetative cell ($\rho=1.01$g.cm^{-1}), one would expect that cells higher in polyP content would be denser than cells with little or no polyP.

Table 2. Transformation of Acinetobacter Cells

| DNA Source | Auxotrophic Marker | | Identification |
	trpE27	ilv	
Cells alone	0,0*	0,0	Control
trpE27	0,0	1000,1100	Acinetobacter
ilv	1000,900	0,0	Acinetobacter
A. lwoffii	30,30	20,30	Acinetobacter
A. calcoaceticus	1,0	50,30	Acinetobacter

*
Results of two independent experiments. Numbers of prototro-
phic recombinants resulting from each cross are given in the
table. trpE27 and ilv strains for Acinetobacter were provided
by Juni. DNA isolation and transformation was carried out by
the method of Juni (1978). This involved incubation of a cell
paste in 0.05% sodium dodexyl sulphate in standard saline
citrate for 1 hr at 60°C. A loopful of the resulting "DNA
solution" was then used to spread a small amount of auxotrophic bacterial cells on heart infusion medium. After
incubation at 30°C for 18 hr, the resulting colony lawns were
replica plated to minimal plates. Any prototrophs resulting
were scored after 18 hr at 30°C.

Table 3. ^{32}P Phosphate Uptake and Release

| Treatment before oxic uptake | Oxic Uptake in 30 mins | | Anaerobic Release in 15 mins | |
	CPM in cell pellet	% Uptake	CPM in cell pellet	% Release
1 hr anaerobic	7414	10	4850	25
1 hr oxic	320	0.5	315	--

A. lwoffii cells grown in LB for 18 hr at 30°C was divided into
two aliquots. One was preincubated anaerobically and the other
aerobically for one hour. They were then subjected to aerobic
incubation in the presence of ^{32}p inorganic phosphate for 30
min followed by anaerobic incubation for 15 min. % Uptake
represents the % of ^{32}P inorganic phosphate in the cell pellet
compared to the total available.

To test this possibility we compared the density of *A. lwoffii* which had a polyP content of about 7.5% of the total phosphate in the cell with *A. calcoaceticus* culture which has only 1.5%. On a 50% Percollgradient (Kubirschek et al., 1983) the *A. lwoffii* banded at a density corresponding to 1.08 $g.cm^{-1}$ whereas *A. calcoaceticus* cells banded on top of the gradient corresponding to a density of 1.01 $g.cm^{-1}$ (Fig. 2). On anaerobic incubation of the *A. lwoffii* culture or by allowing the gradient to stand (and the cells to become anaerobic in this way), the cell band moved to the top of the gradient, corresponding to a density of 1.01 g/cm^{-1}. This finding is consistent with the prediction that the heavier density of the *A. lwoffii* culture is due to its higher polyP content.

The Percoll density gradient technique makes it possible to enrich microorganisms from sludge with very high polyP contents. To explore this possibility we aerated a sludge sample, from an anaerobic/aerobic waste water treatment plant overnight, layered it on top of 50% Percoll and centrifuged for one hour at 15K in a SS34 rotor. The population from the bottom-most part (less than 1% of the total population) of the gradient was collected, grown in LB, and the culture recycled through the enrichment technique once again. The resulting heavy band was plated on LB and isolated individual colonies picked and their Percoll banding pattern examined (see Fig. 3). Cells of one of the isolates (F11) had a density of 1.132 g/cm^{-1} and was chosen for further studies.

When F11 cells were allowed to stand in a Percoll gradient, several bands appeared within 30 min from the original band. On further standing, these bands were in turn lost and a single band, corresponding to a density of 1.01 g/cm^{-1} formed. This is similar to the behavior of *A. lwoffii* cells seen before, however, F11 had 31% of its phosphate as polyP compared to the 7.5% of *A. lwoffii* isolate.

F11 forms motile aerobic Gram negative rods of around 1µm in length. The colonies formed are yellow and oxidase positive. These characters indicate that they belong to the genus *Pseudomonas* (Kreg and Hold, 1984). Since the cells were unable to oxidize glucose, to reduce nitrate or to grow on McConckey agar and were indole, urease and catalase negative, they were further identified as *P. vesicularis*.

Bouyant density (g/ml)

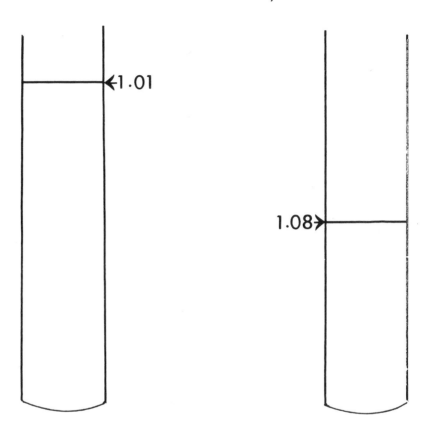

A. calcoaceticus   **A. lwoffii**

Figure 2.   Centrifugation of *A lwoffii* and *A. calcoaceticus*
cells on a 50% Percoll density gradient.   *A. lwoffii* and *A. calcoaceticus* were grown aerobically
at 30°C in LB medium to stationary phase.   After
collecting by centrifugation, the cells were
layered on the top of 50% Percoll and centrifuged
for one hr at 15,000 rpm in a SS34 rotor.   Marker
beads (Pharmacia Chemicals) were included to
standardize the density of the gradient.

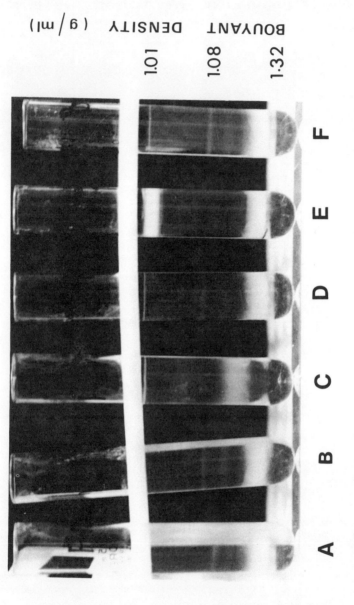

Figure 3. Survey of the densities of bacteria isolated by the Percoll method. As described in the text, a number of bacteria were isolated from the dense fraction present in sludge from an anaerobic/aerobic waste water treatment plant. Individual cultures were centrifuged on 50% Percoll as described in the legent to Fig. 2. (A) Fll, (B) Fl2, (C) FlY, (D) Bll, (E) BlY, (F) Bl4.

^{31}P *NMR Analysis*

^{31}P NMR spectroscopy of whole cells is an excellent non-invasive technique for following polyP and other phosphate containing metabolites in cells (Ugurbil et al., 1979; Kanodia and Roberts, 1983). Aerobically grown cultures of *A. lwoffii* (Fig. 4A) and *P. vesicularis* (Fig. 5A) show distinct polyphosphate resonances. The relative amount of polyP increases as the cells go from early log to late stationary phase. For example, *P. vesicularis* harvested in early log phase shows very little polyphosphate resonances (Fig. 5B). The relative amount of polyP increases as the cells go from early log to late stationary phase.

The *A. calcoaceticus* showed a less prominent polyP peak (Fig. 6). This is consistent with the higher polyP content in the *A. lwoffii* and *P. vesicularis* cultures compared to *A. calcoaceticus* cultures. In addition to the polyP peak other peaks corresponding to inorganic phosphate, sugar phosphates, phosphodiesters and pyrophosphate esters were readily visible (Fig. 4A). On anaerobic incubation there was a distinct and significant decrease in the visible polyP peak in the above cultures (Figs. 4B and 5C) corresponding to the release of inorganic phosphate. The *A. lwoffii* cells grown in the presence of DCCD showed a decreased polyP peak (Fig. 4C) and a significant increase in the sugar phosphate region. This may indicate that polyP is used to phosphorylate the sugars and is consistent with the idea that polyP is used for energy storage.

Polyphosphate Kinase and Polyphosphatase

Numerous microorganisms have been shown to accumulate polyP when exposed to unbalanced growth conditions. The commonly involved mechanism is the transfer of the terminal phosphoryl group of ATP to a primer polyP by the enzyme polyP kinase. This is a reversible enzyme (Kornberg, 1957) but the hydrolysis of polyP to inorganic phosphate (Pi) is also catalyzed by polyphosphatase. In *Aerobacter aerogenes* these two enzymes are derepressed four to eight-fold by Pi starvation (Harold, 1964).

To further clarify the mechanism of phosphate storage and breakdown we have investigated the levels of polyphosphatase and polyP kinase during the various growth phases of *P. vesicularis*. Previous experiments had shown that in LB media, polyP accumulates at the end of the growth phase.

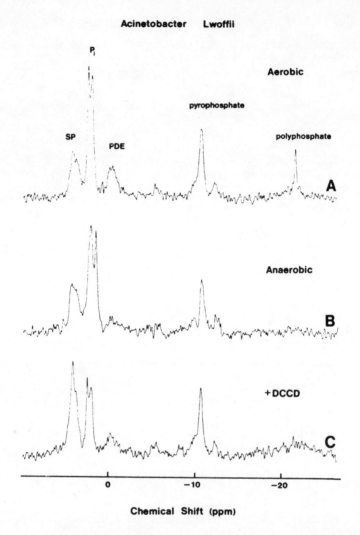

Figure 4. ^{31}P NMR spectra at 109.3 MHz of *A. lwoffii* late
stationary phase cells. Late stationary phase
cells were resuspended to a cell density of ~1
x 10^{11} cells/ml. Chemical shifts are referenced
to external 85% H$_3$PO$_4$. (A) grown aerobically,
(B) aerobically grown culture incubated for 1 hr
anaerobically, (C) grown aerobically in the
presence of DCCD.

Pseudomonas vesicularis

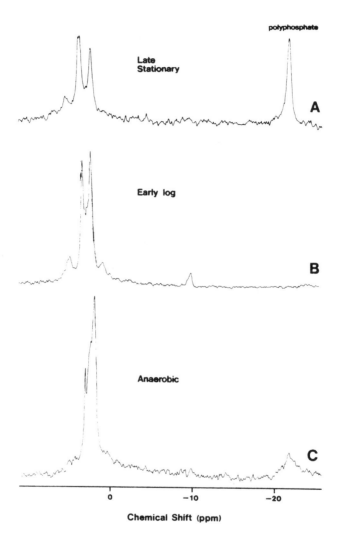

Figure 5. ^{31}P NMR spectra at 109.3 MHz of *P. vesicularis*. (Experimental details same as for Fig. 4). (A) grown aerobically and harvested at late stationary phase. (B) grown aerobically and harvested at early log phase. (C) aerobically grown culture incubated for 1 hr under anaerobic conditions.

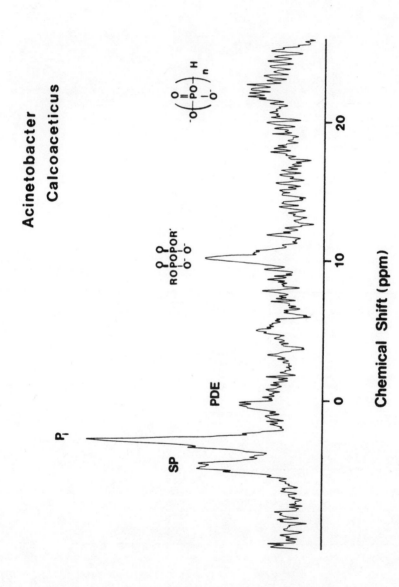

Figure 6. ^{31}P NMR spectrum at 109.3 MHz of *A. calcoaceticus* grown aerobically and harvested at late stationary phase.   (Experimental details same as for Fig. 4.)

A modification of the method of Poindexter and Eley (1983) was used to prepare the endogenous ^{32}P polyP substrate for the enzyme assays. The *P. vesicularis* culture which produced high levels of polyP was grown for 18 hr at 30°C in 100 ml LB media containing 200 µCi inorganic ^{32}P phosphate. The cells were centrifuged and washed at 4°C with LB media. The cell pellet was suspended in 20 ml of "Chlorox" bleach adjusted to pH 9.8, for one hr at 20°C. The polyP was precipitated by centrifugation at 15,000 rpm for 15 mins in a SS-34 rotor and the pellet was washed with 1.5 M NaCl containing 5 mM Na$_3$ EDTA and recentrifuged as before. The resulting pellet was dissolved in 5 ml 0.9% NaCl and stored at 4°C.

As seen in Fig. 7, during the aerobic phase of growth there was a peak of the polyphosphatase levels in early stationary phase, where its activity is almost double that of other growth stages. In contrast, the levels of polyP kinase showed a tenfold increase in early stationary phase compared to log phase cells and in stationary phase there was a thirtyfold increase compared to the levels during the log phase. These findings show that polyP kinase derepression parallels polyP accumulation.

CONCLUSIONS

We have isolated a variety of organisms capable of accumulating inorganic phosphate. A few of these have been identified and shown unambiguously by NMR analysis to contain polyP which is broken down under anaerobic conditions. The use of Percoll gradients in conjunction with NMR analysis will enable us to identify the active members in the population and to compare their properties to that of the population. One isolate, identified as *P. vesicularis,* although a minor component (less than 1%) of the system from which it was derived, is highly active in phosphate uptake and release. It provides a good model for other organisms in the population to study the mechanisms of such phosphate metabolism. The presence of highly active phosphate metabolizing enzymes means that their regulatory controls may be more easily studied and the derivation of mutants defective in such enzymes more simply derived. Further the potential of genetic studies in such *Pseudomonas* species will facilitate the isolation of genes involved.

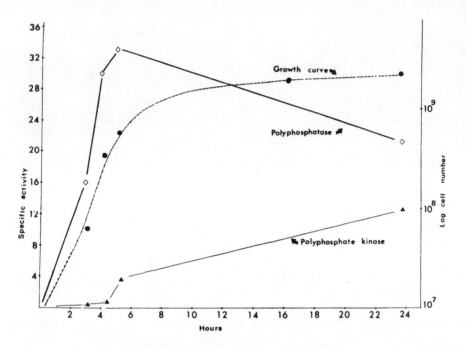

Figure 7.   Polyphosphatase and polyphosphate kinase activity
            as a function of the growth phase of *P. vesicularis.*
            P. vesicularis was grown aerobically at 30°C in LB
            medium.   Growth was followed by measuring the opti-
            cal density at 540 nm in a Backman spectrophotometer.
            At intervals samples were removed and measured for
            polyphosphatase and polyphosphate kinase activities
            as described in the text and plotted as specific ac-
            tivity (units per mg protein) as a function of time.
            One unit of polyphosphate kinase was defined as
            that amount of enzyme which transfers 1 p mole of
            phosphorus from polyphosphate to ADP in 1 min.   One
            unit of polyphosphatase was defined as that amount
            of enzyme which catalyzes the hydrolysis of 1 p
            mole of phosphate per min.

ACKNOWLEDGMENTS

This research was supported by a grant from Air Products and Chemicals, Inc., Allentown, PA.

REFERENCES

Bargmen, R.D., "Nitrogen-phosphorous Relationships and Removals Obtained by Treatment Processes at the Hyperion Treatment Plant," *Advances in Water Pollution Research,* Pergamon Press, Oxford, England (1970).

Cramer, C.L. and Davis, R.H., "Polyphosphate-cation Interaction in the Amino Acid-containing Vacuole of *Neurospora crassa," J. Biol. Chem., 259,* 5152-5257 (1984).

Deakyne, C., M. Patel, and D. Krichten, "Pilot Plant Demonstration of Biological Phosphate Removal," *J. Water Pollut. Control Fed.,* In press (1984).

Eikelboom, D.H. and H.J.J. van Buijsen, *Microscopic Sludge Investigation Manual,* p. 45, TNO Research Institute for Environmental Hygiene, Delft, Netherlands (1981).

Fuhs, G.W. and Min. Chen, "Microbiological Basis of Phosphate Removal in the Activated Sludge Process for the Treatment of Waste Water," *Microbial Ecology, 2,* 119-138 (1975).

Harold, F.M., "Enzymic and Genetic Control of Polyphosphate Accumulation in *Aerobacter aerogenes," J. Gen. Microbiol., 20,* 482-495 (1964).

Harold, F.M. and R.L. Harold, "Degradation of Inorganic Polyphosphates in Mutants of *Aerobacter aerogenes," J. Bacteriol., 89,* 1262-1270.

Harold, F.M., "Inorganic Polyphosphates in Biology: Structure, Metabolism and Functions," *Bacteriol. Rev., 30,* 772-794 (1966).

Juni, E., "Interspecies Transformation of *Acinetobacter:* Genetic Evidence for a Ubiquitous Genus," *J. Bacteriol., 12,* 917-931 (1972).

Juni, E., "Genetics and Physiology of *Acinetobacter," Ann. Rev. Microbiol., 32,* 349-371 (1978).

Kanodia, S. and M.F. Roberts, "Methanophosphagen: Unique Cyclic Pyrophosphate Isolated from *Methanobacterium thermoautotrophicum," Proc. Natl. Acad. Sci. USA, 80,* 5217-5221 (1983).

Kornberg, S.R., "Adenosine Triphosphate Synthesis from Polyphosphate by an Enzyme from *Escherichia coli," Biochim. Biophys. Acta, 26,* 294-300 (1957).

Kreg, N.R. and J.G. Hold, *Bergey's Manual of Systematic Bacteriology, Vol. I* (1984).

Kubitschek, H.E., W.W. Baldwin, and R. Craetger, "Buoyant
    Density Constancy During the Cell Cycle of *Escherichia
    coli*," *J. Bacteriol.*, *155*, 1027-1032 (1983).
Kuleav, I.S., "Biochemistry of Inorganic Polyphosphates,"
    Rev. Physiol. Biochem. Pharmacol., *73*, 131-158 (1975).
Millbury, W.F., D. McCauley, and C.H. Hawthorne, "Operation
    of Conventional Activated Sludge for Maximum Phosphorous
    Removal," *J. Water Pollut. Control Fed.*, *43*, 1890-1901
    (1971).
Miller, J.H., *Experiments in Molecular Genetics*, p. 433,
    Cold Spring Harbor, NY (1972).
Miller, J.J., "In vitro Experiments Concerning the State of
    Polyphosphate in the Yeast Vacuole," *Can. J. Microbiol.*,
    30, 236-246 (1984).
Mühlradt, P.F., "Synthesis of High Molecular Weight Poly-
    phosphate with a Partially Purified Enzyme from *Sal-
    monella*," *J. Gen. Microbiol.*, *65*, 115-122 (1971).
Noegel, A. and E.C. Gotschlich, "Isolation of a High Mole-
    cular Weight Polyphosphate from *Neisseria gonorrhoeae*,"
    J. Exptl. Med., *157*, 2049-2060 (1983).
Poindexter, J.S. and L.F. Eley, "Combined Procedure for
    Assays of Poly-β-hydroxybutyric Acid and Inorganic
    Polyphosphate," *J. Microbiol. Methods*, *1*, 1-17 (1983).
Shoda, M., T. Ohsumi, and S. Udaka, "Screening for High
    Phosphate Accumulating Bacteria," *Agric. Biol. Chem.*,
    44, 319-324 (1980).
Smith, I.W., J.F. Wilkinson, and J.P. Duguid, "Volutin Pro-
    duction in *Aerobacter aerogenes* Due to Nutrient Im-
    balance," *J. Bacteriol.*, *68*, 450-463 (1954).
Spector, M., "Production of Non-Bulking Activated Sludge,
    U.S. Patent 4056 465, November (1977).
Thomas, E.A., "Phosphat-Elimination inde Belebtschlammanlage
    von Männedorf und Phosphate-Fixation in See und
    Klärschlamm," *Vierteljahresschr. Naturf. Ges. Zürich*,
    110, 419-434 (1965).
Tijssen, J.P.F. and J. von Steveninck, "Detection of a yeast
    Polyphosphate Fraction Located Outside the Plasma Mem-
    brane by the Method of Phosphorous-31 Nuclear Magnetic
    Resonance," *Biochem. Biophys. Res. Comm.*, *119*, 47-51
    (1984).
Timmerman, W.M., "Biological Phosphate Removal from Domestic
    Waste Water Using Anaerobic/Aerobic Treatment," *Develop-
    ments in Industrial Microbiology*, *20*, 285-298 (1979).
Ugurbil, K., R.G. Shulman, and T.P. Brown, *Biological Appli-
    cations of Magnetic Resonance*, (R.G. Shulman, ed.),
    pp. 537-589, Academic Press, New York (1979).

Umnov, A.M., A.G. Steblyak, N.S. Umnova, S.E. Mansurova, and I.S. Kulaev, "Possible Physiological Role of the High Molecular Weight Polyphosphate and Polyphosphate Phos-phohydrolase System in *Neurospora crassa*," *Mikrobiologia, 44,* 414-421 (1975).

Vacker, P., C.H. Connell, and W.N. Wells, "Phosphate Removal Through Municipal Waste Water Treatment at San Antonio, Texas," *J. Water Pollut. Control Fed., 39,* 759-771 (1967).

Varma, A.K., W. Rigsby, and D.C. Jordan, "A New Inorganic Pyrophosphate Utilizing Bacterium from a Stagnant Lake," *Can. J. Microbiol., 29,* 1470-1474.

12

HYBRIDIZATION OF DNA FROM METHANOGENS WITH EUBACTERIAL PROBES

Lionel Sibold, Dominique Pariot, Lakshmi Bhatnager, Marc Henriquet, and Jean-Paul Aubert

Unité de Physiologie Cellulaire
Département de Biochimie et Génétique Moléculaire
Institut Pasteur
Paris, France

ABSTRACT

By using the Southern hybridization technique, homologies were examined between restricted DNA of *Methanobacterium ivanovi*, *Methanococcus voltae*, *Methanosarcina barkeri*, *Methanobacterium thermoautotrophicum* strain ΔH and several eubacterial probes. No hybridization was observed with the following genes: *Escherichia coli cya* (adenylate cyclase), *E. coli* and *Anabaena glnA* (glutamine synthetase), *Bacillus subtilis gyrAB* (gyrase), and *Rhizobium japonicum hup* (hydrogenase uptake). Positive results were obtained with *B. subtilis* rRNA genes and *Klebsiella pneumoniae* and *Anabaena nifH* genes that code for the subunit of nitrogenase Fe protein.

INTRODUCTION

Since methanogens have been shown to belong to the "third kingdom" of Archaebacteria (Balch et al., 1979), they have generated considerable interest among molecular biologists. Recently, cloned fragments of *Methanosarcina barkeri* and *Methanobrevibacter arboriphilus* DNA were shown to be expressed in *E. coli* mini-cells (Reeve et al., 1982) or maxi-cells (Bollschweiler and Klein, 1982). As a consequence, complementation was obtained between *E. coli* auxotrophic mutations and cloned fragments of *M. voltae* (Wood et al., 1983), *M. barkeri* MS and *Methanobrevibacter smithii* (Hamilton and Reeve, 1984).

Another approach to characterize genes of interest consists in looking for DNA/DNA hybridization with specific probes. In this paper, we report on Southern hybridization studies between DNA of four methanogens and eubacterial genes. Positive results were obtained with *B. subtilis* rRNA genes and structural genes for the nitrogenase Fe protein of *K. pneumoniae* or *Anabaena*.

MATERIALS AND METHODS

Bacterial Strains and Plasmids

Methanobacterium strain *ivanov* now defined as *M. ivanovi* (J.G. Zeikus, personal communication), *Methanobacterium thermoautotrophicum* strain ΔH and *Methanosarcina barkeri* strain MS were obtained from Dr. J.G. Zeikus. *Methanococcus voltae* strain PS was obtained from the German Collection of Microorganisms, Göttingen, FRG. Plasmids used are listed in Table 1.

Culture Conditions

M. ivanovi and *M. barkeri* were grown at 37°C and *M. thermoautotrophicum* at 63°C in the phosphate buffered basal minimal (PBB) medium as described by Bhatnagar et al. (1983). *M. voltae* was grown at 37°C as described by Whitman et al. (1982).

Isolation of Chromosomal DNA

M. voltae and *M. ivanovi* chromosomal DNA was prepared as follows. Bacteria (1 g wet weight) were suspended in 10 ml of 50 mM Tris 20 mM EDTA pH 8.0. Pronase E (Merck, FRG) and sodium dodecyl sulfate (SDS) were added at final concentrations of 250 µg/ml and 2 % respectively. After 3 h of incubation at 37°C, the lysate was extracted twice with phenol and once with phenol-chloroform. The aqueous phase was dialyzed against 10 mM Tris 1 mM EDTA 50 mM NaCl pH 8.0,and RNAse A (Sigma Chemical Co.) was added at a final concentration of 100 µg/ml, followed by a 3-h incubation at 37°C. After one extraction with phenol-chloroform and two with ether, the aqueous phase was dialyzed extensively against 10 mM Tris 1 mM EDTA pH 8.0. DNA was clean enough for endonuclease treatment. In the case of *M. barkeri* and *M. thermoautotrophicum,* the above lysis treatment was ineffective and the cells were disrupted by hand grinding with alumina. *M. thermoautotrophicum* DNA obtained by this method was partially degraded.

Table 1. Plasmids Used as Probes

Name	Cloned Genes	Relevant Features	Source or Reference
pBR322		Cloning vector	Bolivar et al. (1977)
PACYC184		Cloning vector	Chang & Cohen (1978)
p14B8	rRNA genes	5 kbp BamHI fragment containing B. subtilis 5S, 16S and 23S rRNA genes cloned into pBR313	Stewart et al. (1982)
pna1[s]	gyrAB	6.2 kbp EcoRI fragment containing B. subtilis gyrA and gyrB (gyrase) genes cloned into pMK4	K. Bott, unpublished
pHU1	hup	25 kbp EcoRI fragment containing Rhizobium japonicum hup (uptake hydrogenase) gene cloned into pLAFR1	Cantrell et al. (1983)
p812	glnA ntrBC	11.5 kbp HindIII fragment containing E. coli glnA (glutamine synthetase) and ntrBC (nitrogen regulation) genes cloned into pBR322	Tuli et al. (1982)
pAn503	glnA	7.5 kbp HindIII fragment containing Anabaena glnA gene cloned into pBR322	Tuli et al. (1982)
pDIA100[a]	cya	4.1 kbp EcoRI-SalI fragment containing E. coli cya (adenylate cyclase) gene cloned into pBR322	Roy & Danchin (1982)
pPC880	nifJ	2.7 kbp EcoRI-BglII fragment containing Klebsiella pneumoniae nifJ (oxydoreductase) gene cloned into pBR322	This laboratory
pGR113	nifNE	2.2 kbp EcoRI fragment containing K. pneumoniae nifNE (synthesis of the nitrogenase Fe Mo cofactor) genes cloned into pACYC184	Riedel et al. (1983)
pMC71A	nifA	2.8 kbp SalI fragment containing K. pneumoniae nifA (positive regulatory) gene cloned into pACYC184	Buchanan-Wollaston et al. (1981)
pSA30	nifHDK	6.2 kbp EcoRI fragment containing K. pneumoniae nifHDK genes cloned into pACYC184	Cannon et al. (1979)
pPC1201	nifH	0.9 kbp EcoRI-KpnI fragment containing K. pneumoniae nifH (Fe protein) gene cloned into pGV822	This work
pAn154.3	nifH	1.8 kbp HindIII fragment containing Anabaena 7120 nifH (Fe protein) gene cloned into pBR322	Mevarech et al. (1980).

[a] A 2 kbp BamHI-HindIII fragment, internal to the cya coding sequence, was used as a probe. This fragment, prepared by electro-elution, was kindly provided by Mrs. I. Crenon.

Rhizobium ORS 571, *Azospirillum brasilense* sp 7, *K. pneumoniae* M5al and *Anabaena* 7120 DNAs were kindly provided by Dr. C. Elmerich. *B. subtilis* 168 DNA was a gift of Dr. J. Millet and *E. coli* FB8 DNA was a gift of Mrs. I. Crenon.

Plasmid Isolation

Plasmid DNA was purified according to Humphreys et al. (1975). Small-scale plasmid isolation was performed using the alkaline lysis "mini-prep" procedure described by Maniatis et al. (1982).

Restriction Enzyme Cleavage

Restriction enzyme (New England Biolabs, USA or Boeh-ringer, RG) cleavage was performed in the TA buffer of O'Farrell et al. (1980). Agarose gel electrophoresis was performed as described by Maniatis et al. (1982).

Hybridization Techniques

DNA labelled probes were prepared by nick-translation (Rigby et al., 1977). Specific activity of 50-100 µCi/µg DNA were routinely obtained. Hybridization experiments were performed mainly as described by Southern (1977) with the following modifications. Prehybridization and hybridization were performed at 42°C in the presence of formamide to a final concentration of 10%, 30% or 50% (v/v). Depending on the concentration of formamide used, the blots were washed at 42°C (10%), 52°C (30%) or 68°C (50%) with 0.5% SDS 0.1 SSC (1 SSC is 150 mM NaCl 15 mM Na-citrate).

Construction of Plasmid pPC1201

Plasmid pPC1201 was obtained after ligation of the electro-eluted 0.9 kbp *EcoRI-KpnI* fragment of pSA30 with the *EcoRI-KpnI* digested pBR322 derivative pGV822 (Sibold et al., submitted for publication).

RESULTS

As negative controls, we checked twice that no hybri-dization occurred with the cloning vectors pBR322 and pACYC184. Positive controls were included as chromosomal DNA homo-logous to the probe used. With plasmids or DNA fragments carrying the *cya*, *hup* or *gyr* genes, no hybridization was detected. In the case of the *glnA* probes, no hybridization was found with the *E. coli* gene, but a weak hybridization

was found between *M. ivanovi* DNA and the *Anabaena* gene
(data not shown). However this result was not reproducible
and was only observed twice out of four experiments. Positive results were obtained with a rRNA gene probe and *nif*
probes.

Hybridization with B. subtilis rRNA Genes

Results obtained with *M. ivanovi* and *M. voltae* DNA
using pl4B8 as a probe are shown in Figure 1. Patterns of
hybridizing bands in *M. ivanovi* and *M. voltae* DNA suggest
that the rRNA genes are likely to be present in several
copies as in *B. subtilis* (Stewart et al., 1982). In *M.
ivanovi* these copies might have a common 4.5 kbp *Hind*III
fragment. No hybridization was detected with *M. barkeri*
or *M. thermoautotrophicum* DNA.

Hybridization With nif Genes

No band was detected with the *K. pneumoniae nifJ, nifNE*
or *nifA* probes were used. On the contrary, when pSA30 or
the 6.2 kbp *Eco*RI fragment carrying the genes coding for
the nitrogenase complex were used as probes, bands were
detected. In order to determine which part of the probe
was homologous to total DNA, a 0.9 kbp *Eco*RI-*Kpn*I fragment
of the 6.2 kbp *Eco*RI fragment was subcloned into pGV822.
Plasmid pPC1201 contained two-thirds of the *nifH* (Fe protein)
coding sequence plus 300 bp located upstream. From this
plasmid a 0.4 kbp *Ava*II-*Kpn*I internal *nifH* fragment can be
isolated and can be used as a specific *nifH* (intra-*nifH*
probe). In addition to the *K. pneumoniae nifH* probes, an
Anabaena probe was used. Plasmid pAn154.3 contains the
entire *Anabaena nifH* coding sequence surrounded by 750 bp
of non-*nif* DNA upstream to the translational initiation
codon and by the intra-cistronic region between *nifH* and
nifD followed by the first 150 bp of the *nifD* coding sequence. This plasmid was considered in first approximation
as a *nifH* probe. Hybridization results obtained with these
probes and *Hind*III restricted DNA of *M. ivanovi* and *M. voltae*
are shown in Figure 2. Sizes of the bands found are indicated in Table 2 with those found in the case of *M. barkeri*
and *M. thermoautotrophicum*.

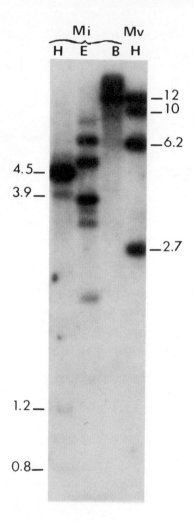

Figure 1.   Southern hybridization between *M. ivanovi* and *M.*
            voltae DNA and *B. subtilis* rRNA genes.   The hy-
            bridization was performed under 10% formamide
            conditions, 0.1% SSC washing.   The probe used was
            p14B8.   Mi: *M. ivanovi,* Mv: *M. voltae.*   E: *Eco*RI,
            B: *Bgl*II, H: *Hin*dIII.   Band sizes (kbp) are indi-
            cated in the case of *M. ivanovi* and *M. voltae* DNA
            restricted by *Hin*dIII.   Band sizes (kbp) found in
            M. ivanovi DNA restricted by *Eco*RI are: 8.2, 6.2,
            4.7, 3.7, 3.5, 3.2, 1.9; in *M. ivanovi* DNA restrict-
            ed by *Bgl*II: 15,12, (these sizes are likely to be
            underestimated).

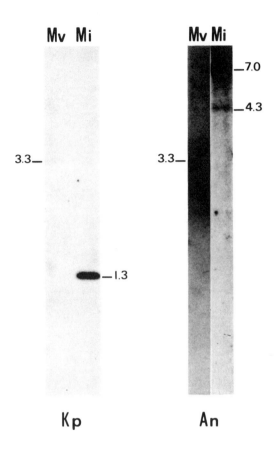

Figure 2. Southern hybridization between *M. ivanovi* and *M. voltae* DNA restricted by *Hind*III and *K. pneumoniae* and *Anabaena nifH*. The hybridization was performed under 10% formamide conditions, 0.1 SSC washing. Mi: *M. ivanovi,* Mv: *M. voltae.* Kp: the *K. pneumoniae nifH* probe was pPC1201, An: the *Anabaena nifH* probe was pAn154.3. Band sizes are expressed in kbp.

Table 2.   Size of HindIII Restriction Fragments of DNA from Methanogens Hybridizing with K. pneumoniae and Anabaena nifH Probes

Probes	M. ivanovi	M. voltae	M. barkeri	M. thermoautograophicum
pPCl201[a]	1.3 1.8[b]	3.3	4.4	3.6[b]
pAnl54.3	7.0 4.3	3.3	4.3	ND

Band sizes are expressed in kilobasepairs (kbp)

[a] or 0.4 kbp KpnI-AvaII fragment (intro-nifH)

[b] These bands were detected only when the blots were washed with a 6 SSC containing solution instead of 0.1 SSC

ND: No band detected

DISCUSSION

A new taxonomic treatment based on 16S RNA sequences allowed the construction of phylogenetic trees and the characterization of methanogens as a group distinct from other bacteria (Balch et al., 1979). Using DNA/DNA hybridization, we have shown that organisms as distant as *B. subtilis* and *M. ivanovi* or *M. voltae* possess homology at the rRNA gene level. This is a preliminary step in the characterization of these genes in methanogens.

Some sequences were detected, in the four strains tested, that hybridized with *nifH* which is the structural gene for the nitrogenase Fe protein. In *M. voltae* a single 3.3 kbp *Hin*dIII fragment hybridized with the *K. pneumoniae* and *Anabaena nifH* probes. In *M. barkeri* and *M. thermoautotrophicum*, only weak hybridizations were detected. This could be due to the fact that DNA was partially degraded in the case of *M. thermoautotrophicum*. In *M. ivanovi,* the situation is more complex since two regions of homology have been detected, one homologous to the *K. pneumoniae nifH* gene and the other to the *Anabaena nifH* gene. In the latter case, we did not use an intra-*nifH* probe, so we cannot completely exclude the possibility of sequence homology outside of the *nifH* coding sequence. Methanogens, like eucaryotes, are not known as nitrogen fixing organisms and no acetylene reducing activity has yet been detedted either in whole cells or in crude extracts (data not shown). Moreover, no biochemical complementation was observed between crude extracts of *M. ivanovi* or *M. barkeri* and purified *K. pneumoniae* Fe protein or Mo Fe protein (J.G. Zeikus, personal communication and this laboratory). Thus one may wonder what function *nif*-like genes (*nif*-like enzymes) would perform in methanogens. One hypothesis is that such proteins would play a role in electron transfer.

ACKNOWLEDGMENTS

The authors wish to thank Dr. C. Elmerich for initiating this work and for constant support, Drs. K. Bott, H.J. Evans, R. Haselkorn, G. Stanier and A. Ullmann for providing plasmids. This study was supported by a French Government Scholarship to L.B., by a research contract from La Société Lyonnaise des Eaux et de l'Eclairage, Le Pecq, France to the Institut Pasteur and by research funds from the University Paris 7.

REFERENCES

Balch, W.E., G.E. Fox, L.J. Magrum, C.R. Woese, and R.S. Wolfe, *Microbiol. Rev.*, *43*, 260-293 (1979).

Bhatnagar, L., M. Henriquet, and R. Longin, *Biotechnol. Lett.*, *5*, 39-42 (1983).

Bolivar, F., R.L. Rodriquez, P.J. Greene, M.C. Betlach, H.L. Heynecker, H.W. Boyer, J.H. Crosa, and S. Falkow, *Gene*, *2*, 95-113 (1977).

Bollschweiler, C. and A. Klein, *Zbl. Bakt. Hyg.*, *I. Abt. Orig.*, *C3*, 101-109 (1982).

Buchanan-Wollaston, V., M.C. Cannon, J.L. Beynon, and F.C. Cannon, *Nature*, *294*, 776-778 (1981).

Cannon, F.C., G.E. Riedel, and F.M. Ausubel, *Mol. Gen. Genet.*, *174*, 59-66 (1979).

Cantrell, M.A., R.A. Haugland, and H.J. Evans, *Proc. Natl. Acad. Sci. USA*, *80*, 181-185 (1983).

Chang, A.C.Y. and S.N. Cohen, *J. Bacteriol.*, *134*, 1141-1156 (1978).

Hamilton, P.T. and J.N. Reeve, *Microbial Chemoautotrophy*, (W.R. Strohl and O.H. Tuovinen, eds.), The Ohio State University Press, Columbus, USA, In press (1984).

Humphreys, G.O., G.A. Willshaw, and E.S. Anderson, *Biochim. Biophys. Acta*, *383*, 457-463 (1975).

Maniatis, T., E. Fritsch, and J. Sambrook, *Molecular Cloning, A Laboratory Manual*, Cold Spring Harbor Laboratory, New York (1982).

Mevarech, M., D. Rice, and R. Haselkorn, *Proc. Natl. Acad. Sci. USA*, *77*, 6476-6480 (1980).

O'Farrell, P.A., E. Kutter, and M. Nakanishi, *Mol. Gen. Genet.*, *179*, 421-435 (1980).

Reeve, J.N., N.M. Trun, and P.T. Hamilton, *Genetic Engineering of Microorganisms for Chemicals*, (A. Hollaender, R.D. De Moss, S. Kaplan, J. Konisky, D. Savage, and R.S. Wolfe, eds.), pp. 233-244, Plenum Publishing Corp., New York (1982).

Riedel, G.E., S.E. Brown, and F.M. Ausubel, *J. Bacteriol.*, *153*, 45-56 (1983).

Rigby, P.W.J., M. Dieckmann, D. Rhodes, and P. Berg, *J. Mol. Biol.*, *113*, 237-251 (1977).

Roy, A. and A. Danchin, *Mol. Gen. Genet.*, *188*, 465-471 (1982).

Southern, E.M., *J. Mol. Biol.*, *98*, 503-517 (1975).

Stewart, G.C., F.E. Wilson, and K.F. Bott, *Gene*, *19*, 153-162 (1982).

Tuli, R., R. Fisher, and R. Haselkorn, *Gene*, *19*, 109-116 (1982).

Whitman, W.B., E. Ankwanda, and R.W. Wolfe, *J. Bacteriol.*, *149*, 852-863 (1982).

Wood, A.G., A.H. Redborg, D.R. Cue, W.B. Whitman, and J. Konisky, *J. Bacteriol*, *156*, 19-29 (1983).

13

EXPRESSION OF A METHANOGEN GENE IN *ESCHERICHIA COLI* AND IN *BACILLUS SUBTILIS*

Christina J. Morris and John N. Reeve

Department of Microbiology
Ohio State University
Columbus, Ohio

ABSTRACT

A DNA sequence has been cloned from *Methanosarcina barkeri* which complements *arg*G mutations of *Escherichia coli* and an *arg*A mutation of *Bacillus subtilis*. Analysis of the polypeptides encoded by this DNA sequence, indicates that the arginine complementing activity is embodied in a 51,000 dalton polypeptide. This polypeptide is synthesized in both *E. coli* and *B. subtilis* minicells. Deletions and Tn5 insertions have been introduced into the cloned *M. barkeri* DNA to delineate the DNA sequence which encodes the polypeptide. The direction of transcription and translation of the cloned gene has been determined by analysis of polypeptides synthesized in minicells containing plasmids with deletions and insertions.

INTRODUCTION

Anaerobic digestion of biomass to produce methane is an established biotechnological process (Klass, 1984). The microbial communities employed in this process are very complex and only recently have attempts been made to identify the individual species involved and their interactions. The terminal step of methane biogenesis is catalyzed by a group of obligate anaerobes known collectively as methanogens (Balch, Fox, Magrum, Woese, and Wolfe, 1979). Laboratory studies of methanogens have been hampered by their extreme oxygen sensitivity. Techniques and facilities have, however, now been developed so that methanogens can be routinely grown in pure culture (Balch et al., 1979; Hook, Corder, Hamilton, Frea, and Reeve, 1984). It is now possible to apply established microbial handling

procedures to investigate the genetics, biochemistry and molecular biology of methanogens (Beckler, Hook, and Reeve, 1984).
Studies already completed have established that methanogens
are Archaebacteria and have biological characteristics which
distinguish them from both eubacteria and eukaryotes (Woese,
Magrum, and Fox, 1978; Balch et al., 1979). We have begun an
investigation of the genome organization and mechanisms of
regulation of gene expression in methanogens (Reeve, Trun, and
Hamilton, 1982). This work is intended as the initial step in
applying genetic engineering procedures to the process of
methane biogenesis and also should provide details of the
basic molecular biology of archaebacteria. We have established that DNA sequences, cloned from methanogens, may be
functionally expressed in eubacterial species (Hamilton and
Reeve, 1984; Morris and Reeve, 1984; Wood, Redborg, Cue,
Whitman, and Konisky, 1984). Expression has been identified
by complementation of auxotrophic mutations of eubacterial
species. In this report we provide additional details of the
expression of a DNA sequence cloned from *M. barkeri* which complements *arg*G mutations of *E. coli* and an *arg*A mutation of *B.
subtilis* (Morris and Reeve, 1984).

MATERIALS AND METHODS

Details of plasmid constructions, growth conditions,
electrophoretic techniques and the use of minicells to
characterize plasmid-encoded polypeptides have been described
previously (Hamilton and Reeve, 1984; Morris and Reeve, 1984;
Reeve, 1984).

Transposon mutagenesis of pET381

E. coli JA221 *hsr*$^-$, *hsm*$^+$, *trp*ΔE5, *leu*B, *rec*A containing
pET381 was infected (input multiplicity of infection of 0.1)
with λ467, *b*221, *rex*::Tn5, *c*Its857, *0am*29, *Pam*80 at 30°C.
Cultures were diluted tenfold, incubated at 30°C for 4h and
then diluted twenty-five fold into LB containing 50µg kanamycin (Kan) per ml and 30µg chloramphenicol (CAM) per ml.
Following overnight incubation at 37°C aliquots (0.1 ml) of
the cultures were plated on LB-agar plates containing 250µg
neomycin/ml plus 30µg CAM per ml. The plates were incubated
at 37°C for 24h. Cells were washed from the plates and resuspended in 25% w/v sucrose dissolved in 50 mM Tris.HCl (pH
8). Plasmid DNA was prepared from these cell suspensions and
used to transform *E. coli* X760 *ara*-1, *leu*-1, *Azi*^{r}, *ton*A, *lac*Y2,
*pro*C119, *tsx*, *pur*E1, *gal*k2, *trp*-3, *his*4, *arg*G36, *rps*L, *xyl*-1,
mtl-1, *ilv*A6, *thr*-1, *met*-12. Transformants were selected on
LB-agar plates containing 50µg Kan/ml and 30µg CAM/ml and

screened for their ability to grow in the absence of arginine by transfer to M9-glucose minimal agar plates containing all the growth supplements required by *E. coli* X760 except arginine. Plasmid DNA was prepared from clones incapable of growth in the absence of arginine. The locations of the Tn5 insertions in the plasmids so isolated were determined by digestion of the plasmid DNAs with appropriate restriction enzymes.

RESULTS

Complementation of E. coli argG and B. subtilis argA mutations

Figure 1 shows the structures and constructions of plasmids which contain a fragment of DNA cloned from *M. barkeri* (Morris and Reeve, 1984). The ability or inability of these plasmids to complement *arg*G mutations of *E. coli* is indicated. Plasmids pET376 and pET377 (see legend to Figure 1) complement both *arg*G mutations of *E. coli* and the *arg*A2 mutation of *B. subtilis* (Morris and Reeve, 1984). Plasmids pET371, pET372, pET374, pET375, pET380, and pET381 replicate in *E. coli,* plasmids pET376 and pET377 can replicate in both *E. coli* and *B. subtilis.*

Expression of M. barkeri DNA in minicells

Recombinant plasmids were introduced into the minicell producing strain *E. coli* DS410 (Reeve, 1984) and, where appropriate, into the minicell producing strain, *B. subtilis* CU403 *divIV*Bl (Reeve, Mendelson, Coyne, Hallock, and Cole, 1973). Minicells prepared from cultures of these strains were allowed to incorporate 35S-methionine. Radioactively-labeled polypeptides, synthesized in the plasmid-containing minicells, were separated by polyacrylamide gel electrophoresis and visualized by fluorography (Reeve, 1984). Polypeptides synthesized in *E. coli* minicells containing pET381, pET380, and pHE3 and in *B. subtilis* minicells containing pUBllO and pET377 are shown in Figure 2. Plasmid pET381 complements *E. coli* argG mutations and directs the synthesis of a polypeptide with an apparent molecular weight of 51,000. This polypeptide is not synthesized in *E. coli* minicells containing pET380, which does not complement *arg*G mutations, nor in minicells containing the vector plasmid pHE3 (see Figure 1). All three plasmids direct the synthesis of chloramphenicol acetyltransferase which confers CAM resistance to strains carrying these plasmids (Hennecke, Isolde, and Binder, 1982). Minicells from *B subtilis* containing pET377, which complements

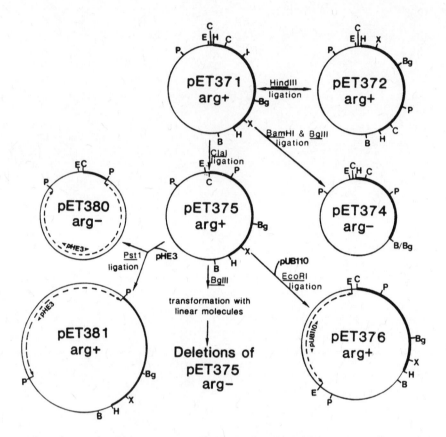

Figure 1.   Construction of plasmids containing DNA cloned from
M. bakeri.   Plasmids pET371 and pET372 contain the
same sequence of *M. barkeri* (heavy line) cloned in
opposite orientations in the *Hind*III (H) site of
pBR322 (Hamilton and Reeve, 1984).   Plasmids capable of complementing *argG* mutations of *E. coli* are
indicated as *argG+*.   Plasmids pET376 and pET377
(pET377 is not shown; it contains the same sequences
of pET375 and pUB110 as pET376 but ligated in opposite orientations) also complement *argA* mutations of
B. subtilis (Morris and Reeve, 1984).   Uses of plasmids pHE3 and pUB110 as cloning vectors are described by Hennecke et al., 1982, and Gryczan, 1982,
respectively.   Sites for restriction enzymes are designated as follows: *Bam*Hl, B; *Bgl*II, Bg; *Cla*I, C;
*Eco*RI, E; *Hind*III, H; *Pst*l, P and *Xba*I, X.

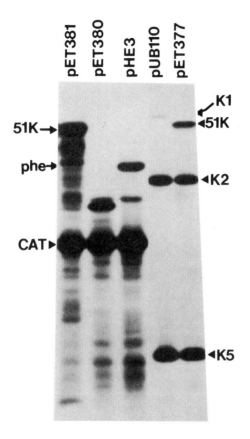

Figure 2.   Electrophorogram of the electrophoretic separation
            of radioactively-labeled polypeptides synthesized
            in plasmid containing minicells. Polypeptides en-
            coded by pET381, pET380 and pHE3 were synthesized
            in minicells of *E. coli,* polypeptides encoded by
            pET377 and pUBll0 were synthesized in minicells of
            B. subtilis. Plasmid pHE3 encodes phenylalanine
            tRNA synthetase *(phe)* which was inactivated in con-
            struction of pET380 and pET381. CAM resistance is
            conferred by pET380, pET381 and pHE3 and all three
            plasmids direct the synthesis of chloramphenicol
            transacetylase. The 51,000 dalton polypeptide
            whose activity complements *arg*G mutation of *E. coli*
            is indicated amongst the polypeptides encoded by
            pET381 and pET377. Polypeptides encoded by pUBll0
            designated Kl, K2 and K5 have been previously de-
            scribed (Shivakumar, Hahn, and Dubnau, 1979).
            Cloning into pUBll0 to produce pET377 destroyed the
            gene encoding polypeptide Kl.

*arg*G mutations of *E. coli* and the *arg*A2 mutation of *B. subtilis*
(see legend to Figure 1), synthesize a polypeptide with an
electrophoretic mobility identical to that of the 51,000 dalton polypeptide encoded by pET381 (Figure 2) synthesized in *E.
coli* minicells containing pET381. *E. coli* minicells containing pET377 also synthesize the 51,000 dalton polypeptide (results not shown). It appears therefore that complementation
of *arg*G mutations of *E. coli* and the *arg*A2 mutation of *B. subtilis,* by the cloned *M. barkeri* DNA sequence, is mediated in
both species by the synthesis of a 51,000 dalton polypeptide.
This strongly suggests that both eubacterial species recognize
and respond to the same sequences within the *M. barkeri* DNA
and that both species synthesize the same encoded polypeptide.

Mutagenesis of the M. barkeri DNA Sequence

Mutations have been introduced into the cloned *M. barkeri*
DNA to localize more precisely the DNA sequence encoding the
51,000 dalton polypeptide. Transformation of *E. coli* X760
with *Bgl*II linearlized pET375 DNA (Figure 1) resulted in the
isolation of a series of plasmids which had deletions located
around the *Bgl*II restriction site. These plasmids had lost
the ability to complement the arginine auxotrophy of *E. coli*
X760 (Figure 1; Morris and Reeve, 1984). The region of DNA in
which the deletions occurred is indicated in Figure 3. The
ability to complement arginine auxotrophy was also inactivated
by insertion of Tn5 into the *M. barkeri* DNA sequence in pET381.
The locations of these Tn5 insertions in pET381 are also shown
in Figure 3.

*Direction of expression of the M. barkeri gene which com-
plements arginine auxotrophy.* The locations of deletions and
transposon insertions which inactivate the arginine complementing activity delineate the *M. barkeri* DNA sequence which
encodes this activity but do not identify the direction of
gene expression from this sequence. Plasmids, containing
deletions or insertions, were therefore introduced in the *E.
coli* minicell producing strain to obtain this information.
The sizes of the polypeptides encoded by these plasmids and
synthesized in plasmid-containing minicells were correlated
with the locations of the deletions and insertions. Results
obtained with deletion mutants have been published (Morris and
Reeve, 1984). Figure 4 shows an example of the results obtained with Tn5 insertion mutants. As expected insertion of
Tn5, resulting in loss of arginine complementing activity,
also resulted in loss of the ability to synthesize the 51,000
dalton polypeptide. Smaller polypeptides are synthesized in
place of the 51,000 dalton polypeptide which presumably reflect
expression of the portion of the 51,000 dalton polypeptide

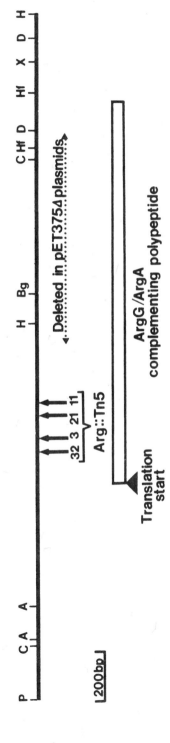

Figure 3. Map of the *M. barkeri* DNA in plasmid pET381. The upper line represents the physical map of the 2.4Kb fragment of *M. barkeri* DNA in pET381. Locations of Tn5 insertions which inactivate the complementation of arginine auxotrophy are shown by arrows below the physical map. The region around the *BglII* site which has been deleted in deletion derivatives of pET375 which are also incapable of complementing *argG* mutants of *E. Coli* is indicated. The box indicates the calculated location of the DNA encoding the 51,000 dalton polypeptide identified as complementing deficiencies in argininosuccinate synthetase in both *E. Coli* (*argG⁻*) and *B. subtilis* (*argA⁻*).

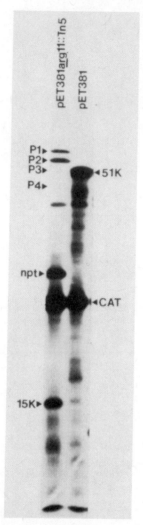

Figure 4.   Electrophorogram of the electrophoretic separation
of radioactively-labeled polypeptides synthesized
in plasmid-containing minicells.   The presence of
Tn5 in pET38*larg*11::Tn5 (see Fig. 3) prevents the
synthesis of the 51,000 dalton polypeptide encoded
by pET381.   A truncated polypeptide (apparent
molecular weight of 15,000) is synthesized in minicells containing pET38*larg*11::Tn5 which is not synthesized in pET381 containing minicells.   Polypeptides encoded by Tn5 are designated P1, P2, P3, and
npt (neomycin phosphotransferase) as previously
described (Johnson, Yin, and Reznikoff, 1982; see
also Klipp and Puhler, 1984).

encoded between the amino-terminus and the location of the
Tn5 insertion. Nonsense codons in the Tn5 sequence terminate
translation (Klipp and Puhler, 1984). The sizes of these
polypeptides correlated with the locations of the Tn5 insertions together with the results obtained with deletion mutants
(Morris and Reeve, 1984) indicate that the 51,000 dalton
polypeptide is transcribed and translated from left to right
as shown in Figure 3. Translation must begin approximately
700 bp to the left of the *Bgl*II site.

DISCUSSION

Results presented in this report and elsewhere (Wood et
al., 1983; Hamilton and Reeve, 1984; Morris and Reeve, 1984)
demonstrate that, although methanogens are archaebacteria, it
is possible to obtain functional expression of methanogen derived genes in *E. coli* and *B. subtilis*. Table 1 lists the
methanogens used as sources of DNA to date and eubacterial
genes for which successful complementation has been demonstrated. DNA sequences of cloned fragment of methanogen
derived DNA have now been obtained (Hamilton, P.T., Beckler,
G.S., and Reeve, J.N., unpublished results). The precise
sequences used to initiate transcription and translation are
being determined. It is most important to determine if the
sequences within cloned methanogen DNAs, recognized as signals
for initiation of transcription and translation in *E. coli*
and *B. subtilis*, are also used as such signals in the methanogenic cells from which the DNA was originally obtained. It
has been shown that introns exist in some archaebacterial
tRNA genes (Kaine, Gupta, and Woese, 1983) and it is possible
that introns may also occur in archaebacterial structural
genes. Such genes would presumably not have been isolated
by the procedures we have used to isolate genes which are
functionally active in *E. coli*, although the recent demonstration of an intron in a gene of coliphage T4 (Chu, Maley,
Maley, and Belfort, 1984) should not be forgotten in drawing
this conclusion.

As described in the Introduction, methanogens are obligate anaerobes. Many of the enzymes isolated from methanogens are inactivated by oxygen. Our studies have shown however, that some methanogen encoded enzymes can be synthesized
and function in aerobically grown eubacteria. It remains to
be determined if enzymes directly involved in methane biogenesis will be active if synthesized in aerobically grown
eubacteria under the direction of cloned genes. Fortunately,
E. coli grows well under the anaerobic conditions needed for
growth of methanogens (Hook et al., 1984) and therefore, if

Table 1. DNA cloned from methanogens which complements auxotrophic mutations of E. coli

Source of DNA	Mutation Complemented	Minimum size of cloned DNA[2]	Size of encoded polypeptide Mol. wt. x 10^{-3}	Reference
Methanobrevibacter smithii	proC	1.05	29	P.T. Hamilton; unpublished result
Methanobrevibacter smithii	purE	1.6	37	Hamilton and Reeve, 1984
Methanosarcina barkeri	argG[1]	2.4	51	Morris and Reeve, 1984
Methanococcus vannielii	argG	7	48	C.J. Morris, Unpublished result
Methanococcus voltae	argG	6.3	55	Wood et al., 1983
Methanococcus voltae	hisA	4	26	Wood et al., 1983
Methanococcus vannielii	hisA	1.2	26	Meile and Reeve, 1984

Footnote to Table 1.

1. Cloned DNA also complements argA2 mutation of B. subtilis (Morris and Reeve, 1984).

2. Current minimum size of the DNA fragment reported for the cloned DNA fragment capable of complementing the designated auxotrophic mutation.

necessary, it will be possible to assay for activity of cloned genes in anaerobically grown *E. coli*. DNA cloning procedures and construction of *E. coli* strains containing such recombinant DNA molecules may still, of course, be undertaken under aerobic conditions.

ACKNOWLEDGMENTS

This work was supported by Grant DE-AC02-81ER10945 from the Department of Energy. J.N.R. is the recipient of Research Cancer Development Award 1K04AG00108 from the National Institute on Aging.

REFERENCES

Balch, W.E., G.E. Fox, L.J. Magrum, C.R. Woese, and R.S. Wolfe, "Methanogens: Re-evaluation of a Unique Biological Group," *Microbiol. Rev., 43,* 260-296 (1979).

Beckler, G.S., L.A. Hook, and J.N. Reeve, "Chloramphenicol Acetyltransferase Should not Provide Methanogens With Resistance to Chloramphenicol," *Appl. Envir. Microbiol., 47,* 868-869 (1984).

Chu, F.K., G.F. Maley, F. Maley, and M. Belfort, "Intervening Sequence in the Thymidylate Synthetase Gene of Bacteriophage T4," *Proc. Nat'l. Acad. Sci. U.S.A., 81,* 3049-3053 (1984).

Gryczan, T.J., "Molecular Cloning in *Bacillus subtilis*," The Molecular Biology of Bacilli, (D.A. Dubnau, ed.), pp. 307-329, Academic Press, Inc., New York (1982).

Hamilton, P.T. and J.N. Reeve, "Cloning and Expression of Archaebacterial DNA from Methanogens in *Escherichia coli*," Microbial Chemoautotrophy, (W.R. Strohl, and O.H. Tuovinen, eds.), pp. 291-307, Ohio State University Press, Columbus, Ohio (1984).

Hennecke, H., G. Isolde, and F. Binder, "A Novel Cloning Vector for the Direct Selection of Recombinant DNA in *E. coli*," *Gene, 19,* 231-234 (1982).

Hook, L.A., R.E. Corder, P.T. Hamilton, J.I. Frea, and J.N. Reeve, "Development of a Plating System for Genetic Exchange Studies in Methanogens Using a Modified Ultra-Low Oxygen Chamber," Microbial Chemoautotrophy, (W.R. Strohl, and O.H. Tuovinen, eds.), pp. 275-289, Ohio State University Press, Columbus, Ohio (1984).

Johnson, R.C., J.C.P. Yin, and Reznikoff, "Control of Tn5 Transposition in *Escherichia coli* is Mediated by Protein from the Right Repeat," *Cell, 30,* 873-882 (1982).

Kaine, B.P., R. Gupta, and C.R. Woese, "Putative Introns in tRNA Genes of Prokaryotes," *Proc. Nat'l. Acad. Sci. U.S.A.*, *80*; 3309-3312 '(1983).

Klass, D.L., "Methane from Anaerobic Fermentation," *Science*, *223*, 1021-1028 (1984).

Klipp, W. and Puhler, A., "Gene Product Analysis of Wild-Type and Tn5 Mutagenized pACYC184 Plasmids in Minicells of *E. coli*: Determination of the Chloramphenicol Acetyl-transferase Coding Region," Advanced Molecular Genetics, (A. Puhler and K.N. Timmis, eds.), , pp. 230-235, Springer-Verlag, Berlin, W. Germany (1984).

Meile, L. and J.N. Reeve, "Potential Shuttle Vectors Based on the Methanogen Plasmid pME2001," *Biotechnology*, submitted for publication (1984).

Morris, C.J. and J.N. Reeve, "Functional Expression of an Archaebacterial Gene from the Methanogen *Methanosarcina barkeri* in *Escherichia coli* and *Bacillus subtilis*," Microbial Growth on C1 Compounds, (R.L. Crawford and R.S. Hanson, eds.) pp. 205-209, A.S.M., Washington, DC (1984).

Reeve, J.N., "Synthesis of Bacteriophage and Plasmid-encoded Polypeptides in Minicells," Advanced Molecular Genetics, (A. Puhler and K.N. Timmis, eds.), pp. 212-223, Springer-Verlag, Berlin, W. Germany (1984).

Reeve, J.N., N.H. Mendelson, S.I. Coyne, L.L. Hallock, and R.M. Cole, "Minicells of *Bacillus subtilis*," *J. Bacteriol.*, *114*, 860-873 (1973).

Reeve, J.N., N.J. Trun, and P.T. Hamilton, "Beginning Genetics with Methanogens," Genetic Engineering Micro-organisms for Chemicals, (A. Hollaender, R.D. DeMoss, S. Kaplin, J. Konisky, D. Savage, and R.S. Wolfe, eds.), pp. 233-244, Plenum Press, New York (1982).

Shivakumar, A.G., J. Hahn, and D.A. Dubnau, "Studies on the Synthesis of Plasmid Coded Proteins and Their Control in *Bacillus subtilis* Minicells," *Plasmid*, *2*, 279-289 (1979).

Woese, C.R., L.J. Magrum, and G.E. Fox, "Archaebacteria," *J. Mol. Evol.*, *11*, 245-252 (1978).

Wood, A.G., A.H. Redborg, D.R. Cue, W.B. Whitman, and J. Konisky, "Complementation of *arg*G and *his*A mutations of *Escherichia coli* by DNA Cloned from the Archaebacterium *Methanococcus voltae*," *J. Bacteriol.*, *156*, 19-29 (1983).

14

ANTIBIOTIC G418 FOR SELECTION AND MAINTENANCE OF PRO- AND EUKARYOTIC TRANSFORMANTS AND ITS POSSIBLE APPLICATION FOR METHANOGEN TRANSFORMATION

Daniel Feldman and Nancy W.Y. Ho

Laboratory of Renewable Resources Engineering
Purdue University
West Lafayette, Indiana

ABSTRACT

Antibiotic G418, an aminoglycoside antibiotic, is structurally related to kanamycin, gentamycin, and neomycin. The antibiotic is known to block protein synthesis in both eukaryotic and prokaryotic cells, and a wide variety of organisms including yeasts and plants are known to be sensitive or moderately sensitive to this antibiotic. Furthermore, G418, similar to kanamycin and other related antibiotics, can be inactivated by bacterial phosphotransferases encoded by the kanamycin and neomycin resistant genes. To date, the km^R and neo^R genes have been used as genetic markers for the celection of G418-resistant transformants from yeast, plant, and mammalian cells.

This paper reports that antibiotic G418 can also inhibit the growth of a number of methanogenic bacteria, particularly *Methanobacterium thermoautotrophicum* and *Methanococcus mazei*. Hence, antibiotic G418 might also be useful as a selective agent for the development of a transformation system for those species of methanogens that are sensitive to this antibiotic. Experiments are currently in progress in our laboratory to explore such a possibility.

INTRODUCTION

The establishment of a plasmid-mediated transformation system for a particular organism provides a powerful tool to facilitate the characterization of its various molecular mechanisms, particularly those involved in gene expression. It also provides a powerful tool for improving the capability of the organism via gene cloning. Up to now, no transformation system has been developed for any of the methanogen bacteria yet.

The construction of a suitable plasmid that can transform the host cell is essential in the development of a cloning system for a particular host. A plasmid that can be used to transform a specific host must contain two elements: an origin of DNA replication recognizable by the host DNA polymerase, and a functional gene that can serve as a genetic marker for the selection of the transformants. These two elements will enable the plasmid to be self-replicating and stably maintained in the host cell.

Recently, the existence of cryptic plasmid DNA in methanogenic species has been reported (1,2). These cryptic plasmids or their simplified derivatives could be the ideal plasmids for methanogen transformation, if suitable genetic markers can be inserted into these cryptic plasmids allowing them to be used for the selection of transformants from non-transformants.

A growing number of reports have shown that the bacterial kanamycin resistant genes, Km^R, which inactivate a number of antibiotics including antibiotic G418, can be used as genetic markers for the development of transformation systems for a variety of organisms including bacteria, yeasts and higher eukaryotes (3-7). Hence, we considered that Km^R might also possibly be used as a genetic marker to establish transformation systems for the methanogens, provided that methanogens are sensitive to one of the antibiotics inactivated by the enzyme product of Km^R.

Several structurally related aminoglycoside antiobiotics such as gentamicin, neomycin, kanamycin, and G418 (geneticin) can all be inactivated by the protein products of the Km^R genes. However, among them, only antibiotic G418 has a broad spectrum on inhibitory effect against yeast, fungi, algae, plant, and animal cells in addition to bacterial cells (3-9). It has been demonstrated that antibiotic G418 inhibits protein synthesis in both pro- and eukaryotic cells. Methanogens are known to be inhibited by some of the antibiotics such as chloramphenicol (10), which inhibit protein synthesis in microorganisms. Hence, G418 should be the antibiotic most likely able to inhibit the growth of the methanogens. In junction with Km^R, it can also possibly be used as a selective agent for the development of a DNA-mediated transformation system for the methanogens. In this paper, we report our preliminary results on the study of the effects of antibiotic G418 on the growth of several methanogenic species.

MATERIALS AND METHODS

Chemicals and Gases

Antibiotic G418 sulfate (geneticin) was purchased from Gibco Laboratories, Madison, WI. All other chemicals were of reagent grade. Gases were obtained from Matheson Scientific, Inc., Joliet, IL.

Organisms and Cultivation

All methanogen cells were cultivated in 250 ml serum bottles with anaerobic techniques described by Balch, et al. (11). *Methanobacterium thermoautrotrophicum* strain ΔH was cultured in a mineral salts medium at 65°C as described by Zeikus and Wolfe (12). *Methanococcus voltae* strain PS was cultured according to Whitman, et al. (14).

RESULTS AND DISCUSSION

The effect of antibiotic G418 on *M. thermoautotrophicum*, *M. mazei*, and *M. voltae* was studied by culturing these strains in liquid medium in the presence of various concentrations of G418 sulfate. The effect of antibiotic G418 on the growth of *M. mazei* is shown in Figures 1 and 2. Identical experiments were carried out for *M. thermoautotrophicum* and *M. voltae*. Among these three strains of methanogens, *M. thermoautotrophicum* is most sensitive to antibiotic G418 and its growth can be totally inhibited by the presence of less than 50 μg/ml G418 in the culture medium. *M. voltae* is least sensitive to antibiotic G418; its growth will not be totally inhibited even in the presence of 300 μg/ml G418 in the culture medium.

Figure 1. Antibiotic G418 inhibition of the growth of *M. mazei*. 250 ml serum bottles containing 50 ml identical medium but with different concentrations of G418 sulfate were used to culture *M. mazei*. 0.5 ml actively grown culture was innoculated into each bottle. The experiment was repeated three times and identical results were obtained.

right:	without G418
middle:	100 μg G418/ml
left:	200 μg G418/ml

Figure 2. Antibiotic G418 inhibition of the growth of *M. mazei*. Experimental conditions were identical to those described in Fig. 1. This experiment was also repeated three times and identical results were obtained.

right: without G418
left: with 300 µg G418/ml

Preliminary results also indicated that antibiotic G418 has inhibitory effects against most methanogen species. Experiments are now in progress to determine whether the protein products of the Km^R genes can inactivate antibiotic G418 and render it to be ineffective in the inhibition of the growth of the methanogen strains which are sensitive to intact G418.

As described above, the bacterial Km^R gene has been expressed in various bacteria, yeasts, and other lower eukaryotes, it is highly possible that the gene will also be expressed in methanogens.

If the inhibitory effect of antibiotic G418 towards methanogen can be abolished by treating the antibiotic with the protein products of Km^R genes, it is most likely that antibiotic G418 in junction with the Km^R gene can be used as a selection system for the development of transformation systems for those methanogens which are sensitive to antibiotic G418.

The antibiotic $G418/Km^R$ system provides a convenient way for the development of transformation systems for those organisms such as the methanogens, which have difficulty obtaining auxotrophic markers. It also provides a convenient system for examining the transformability of organisms or strains without requiring the introduction of auxotrophic mutations into the recipient cells and the isolation of the wild-type genes for complementation.

Research is now in progress in attempting to insert the Km^R gene into a methanogen plasmid. Such a plasmid will be used as a vector for the transformation of *M. thermoautotrophicum, M. mazei, M. voltae,* and possibly other methanogenic bacteria as well.

ACKNOWLEDGMENT

This work was supported by U.S. Department of Energy (ANL Contract No. 40462401).

REFERENCES

1.   Thomm, M., J. Altenbuchner, and K.O. Stetter, "Evidence for a Plasmid in Methanogenic Bacterium," *J. Bacteriol., 153,* 1060-1062 (1983).

---

2. Meile, L., A. Kiener, and T. Leisinger, "A Plasmid in the Archaebacterium *Methanobacterium thermoautotrophicum*," *Mol. Gen. Genet.*, *191*, 480-484 (1983).

3. Davies, J. and D.I. Smith, "Plasmid Determined Resistance to Antimicrobial Agents," *Annu. Rev. Microbiol.*, *32*, 469-518 (1978).

4. Berg, D., R. Jorgensen, and J. Davies, "Transposable Kanamycin-neomycin Resistance Determinants," in Microbiology (ed., D. Schlesinger), A.S.M., Washington, DC, pp. 13-15 (1978).

5. Jimenez, A. and J. Davies, "Expression of a Transposable Antibiotic Resistance Element in *Saccharomyces*," *Nature*, *287*, 869-871 (1980).

6. Southern, P.J. and P. Berg, "Transformation of Mammalian Cells to Antibiotic Resistance with a Bacterial Gene Under Control of the SV40 Early Regional Promoter," *J. Mol. and Appl. Genet.*, *1*, 327-341 (1982).

7. Ho, N.W.Y., H.C. Gao, J.J. Huang, P.E. Stevis, S.F. Chang, and G.T. Tsao, "The Development of a Cloning System for *Candida* Species," *Biotech. Bioeng. Symp.*, In press (1984).

8. Hirth, K., C.A. Edwards, and R.A. Firtel, "A DNA-mediated Transformation System for *Dictyostelium discoideum*," *Proc. Nat'l. Acad. Sci.*, *79*, 7356-7360 (1982).

9. Ursic, D., J.D. Kemp, and J.P. Helgeson, "A New Antibiotic with Known Resistance Factors, G418, Inhibits Plant Cells," *Biochem. Biophys. Res. Comm.*, *101*, 1031-1037 (1981).

10. Jones, J.B., B. Bowers, and T.C. Stadtman, "*Methanococcus vannielii:* Ultrastructure and Sensitivity to Detergents and Antibiotics," *J. Bacteriol.*, *130*, 1357-1363 (1977).

11. Balch, W.E., G.E. Fox, L.J. Magrum, C.R. Woese, and R.S. Wolfe, "Methanogens: Re-evaluation of a Unique Biological Group," *Microb. Rev.*, *43*, 260-296 (1979).

12. Zeikus, J.G. and R.S. Wolfe, "*Methanobacterium thermoautotrophicum* Sp. n., and Aerobic, Autotrophic, Extreme Thermophile," *J. Bacteriol.*, *109*, 707-713 (1972).

13. Mah, R.A, "Isolation and Characterization of *Methanococcus mazei*," *Current Microbiology*, *3*, 321-326 (1980).

14. Whitman, W.B., E. Ankwanda, and R.S. Wolfe, "Nutrition and Carbon Metabolism of *M. voltae*," *J. Bacteriol.*, *149*, 852-863 (1982).

15

NUTRIENT REQUIREMENTS FOR ANAEROBIC DIGESTION

Richard E. Speece and Gene F. Parkin

Drexel University
Philadelphia, Pennsylvania

ABSTRACT

The anaerobic digestion process was initially engineered for domestic wastewater sludges. These sludges generally have been assumed to inherently provide all possible nutrients required for the growth of all members of the microbial ecosystem responsible for anaerobic digestion. For the loading rates and solids retention times in common practice, this assumption appears to be validated in most cases by the fact that the process operates satisfactorily without any additional supplementation of nutrients. However, even with domestic sludges, malfunctioning of the digestion process has been reported. In addition, there is evidence that present design and operational practices may have been forged by existing nutrient bioavailability limitations within domestic wastewater sludges. As a case in point, stimulation of the acetate and propionate utilization rate by up to 100% over a control was observed in eight out of thirty municipal sludge digester samples supplemented with one or more of the trace metals iron, cobalt, and/or nickel. A second case in point is that iron supplementation to a municipal sludge digested having a chronically high volatile acids concentration of about 4,000 mg/l, resulted in a reduction to 400 mg/l within ten days.

Municipal solid wastes are less nutritionally balanced than domestic wastewater sludges. Therefore, even more consideration must be given to assure that nutrient bioavailability is not limiting the overall process of anaerobic

digestion of municipal wastes. Major emphasis will be given
to bioavailability of trace metals and sulfide in the anaerobic environment and the tight ecological niche which allows
mutual bioavailability of mutually precipitating ions.

INTRODUCTION

All living things have specific nutrient requirements
and it should not be tacitly assumed that they are inherently
satisfied, e.g., limes for the British sailors, zinc for
pecan trees, selenium for marsh disease in sheep, and iron
and nickel for methanogens. Some of these nutrient requirements are incidentally satisfied by the feedstock. Others
have to be specifically supplemented. The economic value
of the process product determines the extent to which nutrient supplementation can be practiced. Unfortunately, with
anaerobic digestion, the process product -- methane -- has
relatively low value -- about $0.10/lb. The cost different-
ial between anaerobic digestion and alternative processes
must also be taken into consideration and adds an additional
economic benefit.

The cost effectiveness of nutrient supplementation is
related to the unit cost of a nutrient, the concentration
required, and the relative increase in process rate which
results from nutrient supplementation. At one end of the
scale, iron has a low unit cost and is required in low concentration and yet can result in a significant increase in
process rate. At the other extreme of the scale vitamin
B_{12} supplementation may be quite expensive and yet may not
increase the process rate appreciably. Therefore, a cost/
benefit analysis must be determined.

The unit rate of activity of a microbial reaction is
not only related to the adequacy of the required nutrients
but also related to the ease of metabolism of the feedstock.
Additional factors such as temperature, toxicity, mixing
and biomass retention also affect the reaction rate. The
relatively low value of methane gas - about $0.10/lb - dic-
tates that feedstocks be of even lower cost. Consequently,
waste organics, fodder crops, straw, perhaps giant kelp and
a few other organics have been primarily considered as feedstocks. Generally, these materials are difficult to degrade
and often the rate limiting step in the overall process is
the conversion of these complex organics to organic acids.
This slow rate necessitates prolonged substrate retention

times within the reactor and this manifests itself in relatively large reactor volume requirements which is a
problem in itself.

However, in this paper, we wish to address the nutritional requirements. Methane fermentation involves a consortia of microorganisms, each manifesting peculiar nutrient
requirements. The absence of particular nutrients can pre-
clude the presence of those microorganisms which manifest
obligate requirements or result in a possible reduced rate
of activity for those microorganisms for which the nutrient
is merely stimulatory. In both cases, the overall methane
production rate is reduced, necessitating increased reactor
volume. In our research, there is evidence that *Methanobacterium söehngenii,* which has a relatively low specific
utilization rate, predominates in a mixed culture containing minimal nutrients. However, when proper nutrient supplementation is provided, a sarcina (possible *Methanosarcina
mazei* or *M. barkeri*) predominates, which has a specific
utilization rate 4 to 8 times greater.

Again, due to the relatively low value of the methane
gas, there is a severe economic restraint on the options
available to provide optimal nutrient concentrations for
methane production in most applications. Reduced unit rates
of activity must be accepted because complete nutrient supplementation costs are prohibitive. Whereas in common commercial and pharmaceutical fermentations, great care and
expense are invested in determination of nutrient requirements and their supplementation because the value of the
final product can absorb such costs.

Conventionally, alkalinity, nitrogen, and phosphorous
are the only chemicals considered for supplementation in
applications of biomass conversion to methane. It may even
be assumed by some that these are the only chemicals required for supplementation, but this assumption is not based
upon rational evidence. The nutrient requirements of the
methanogens and hydrogen producing acetogens are not well
documented and even less is understood of the hydrolyzing
and acid forming group of microorganisms which, often are
the rate-limiting step in the overall process as they mediate
the conversion of complex organic substrates such as municipal wastes, straw, fodder, or manure to soluble intermediates. Most of these soluble intermediates are not routinely
monitored, with the exception of gross measurement of volatile acids concentrations. Therefore, stimulation of their
rate of formation or utilization as a function of nutrient

supplementation is rarely assayed. On the contrary, since volatile acids concentrations are commonly determined, there are numerous cases reporting chronic or sporadic elevations in anaerobic digestion, indicating that their rate of utilization is out of balance with their rate of formation. This could be due to kinetic and/or nutrient limitations.

We are not proposing that all nutrient requirements have to be finally satisfied or even should be satisfied. Indeed, methanogenesis in the natural ecosystem occurs under sub-optimal conditions. But our point is that adverse trade-offs must be accepted when nutrient requirements of the microorganisms are not provided. This may mean that a more kinetically favorable microorganism may be precluded from the system or that the kinetics of the predominant, existing, microorganisms may be adversely affected by nutrient limitations. In either case, the biological safety factor is reduced for a given installation, or the installation size must be increased to maintain the same biological safety factor when nutrient limitations exist. Tangible economical penalties must be paid for lack of adequate nutrients and/or proper concentrations. The engineering decision is to determine which is more economical: (1) pay for the resultant increased capital cost of the installation due to nutrient limitations? or (2) pay for the added operational cost of supplementing nutrients?

Feedstocks differ markedly in their intrinsic ability to simultaneously satisfy the nutrient requirements of the microorganisms of the anaerobic digestion process. Heat treatment liquor from waste activated sludge is reputed to be "better than yeast extract in providing nutrients" for the anaerobic digestion process. On the other extreme may be the nominally "de-ionized" wastewaters from some industrial operations which are exceptionally limited in inorganic ions. There is some evidence that even domestic sludges do not always satisfy the nutrient requirements of the anaerobic digestion process. Examples will be given of this later.

In this paper, we would like to discuss some of the known nutrient requirements of the anaerobic digestion process. Since relatively more is known of the nutrient requirements for the methanogens than for the hydrolyzing or acid-forming bacteria, of necessity, this will be emphasized. However, we are not inferring that this is more important than the nutrient requirements of the non-methanogens.

Half Saturation Concentration - K_S

In microbial systems, Monod kinetics commonly describe the system. A rate limiting nutrient can thus be modeled. Often this is tacitly assumed to be the organic carbon source. However, in some studies it is noted that the unit rate of activity is quite low compared with prototype installations and this logically raises the question as to the possibility that some nutrient other than the carbon source is actually contributing to the limitation. Such was the case with the principal author in his doctoral research. When domestic sludge was fed a defined medium and a pure substrate, the gas production gradually declined. When a source of iron was injected, an abrupt increase in gas production was noted the following day. McCarty observed that supplementation of nickel to a digester stimulated gas production. Kugelman also observed that digestion of raw, domestic sludge improved when iron and yeast extract were supplemented. Jeris noted that iron supplementation, up to a point, stimulated digestion of a paper waste. Owen observed that iron supplementation to a domestic sludge digester caused a rapid drop in the volatile acids concentration.

The half saturation constnat, K_S, has not been reported for the inorganic nutrients known to be required by the various methanogens. The literature may state that a given concentration of a required nutrient was stimulatory but the entire response curve as a function of a particular nutrient concentration has not been reported to our knowledge. Figure 1 conceptually indicates how the overall digestion process is inter-related to nutrient concentrations.

Microbial Generation Time

The microbial generation time is a function of the nutrients present. Ideal nutrient concentrations are not essential, but a process compensation must be made in either lower loading rates of lower resulting treatment efficiency as a trade-off for non-ideality of nutrients present. In addition, it is difficult to say that one nutrient is more important than another because all of the required nutrients are important and must be supplied. Nutrient limitations at best prolong the minimum generation time of methanogens and at worst can cause generation to eventually cease altogether. The rates of substrate metabolism are also limited by nutrient limitations. Specific substrate utilization rates can be increased severalfold when all required nutrients are in excess. There is a possibility that some of the

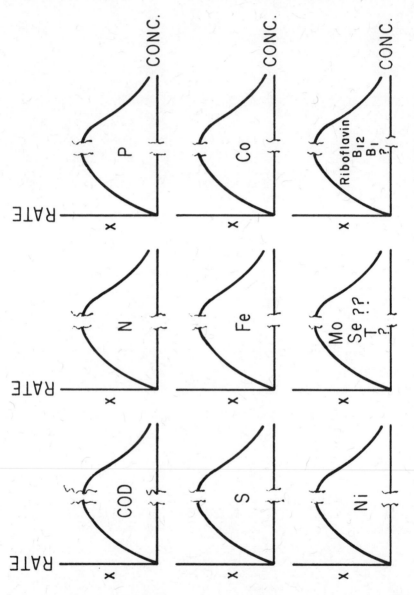

Figure 1.   Nutrients affecting the rate of methanogenesis.

published kinetic coefficients may reflect nutrient-limiting conditions under which the data were taken, e.g., nickel and iron defficiency. Also, toxicity response is compounded by nutrient limitations.

The biological safety factor is a measure of the sta-bility of the process and is calculated by dividing the actual substrate-solids retention time (SRT) by the microbial gene-ration times of the rate limiting organisms. If the acetate to methane conversion step is considered to be rate-limiting, then the generation time of the acetate-converting methano-gens is used. The sensitivity of the generation time of these organisms to nutrient sufficiency will therefore have a direct bearing on the biological safety factor.

One means of determining microbial generation time is to operate systems with a range of SRT and observe where washout occurs. Replicate systems were operated with one system receiving a defined chemical make-up feed and the other sys-tem receiving "boiled refed", i.e., the waste liquor was boilled to kill and lyse the methanogens and then refed as the make-up solution. The acetate concentration was stoi-chiometrically restored to 100 mg/l at the beginning of each day, based upon the previous day's production.

When both systems were operated at a 15 day hydraulic retention time (HRT) and SRT, the "boiled refed" system pro-duced more gas than the control. When the HRT and SRT were reduced to 3 days, the control system ceased gas production within 8 days due to washout of the biomass. However, the system receiving "boiled refed" continued gas production at a reduced, stable rate for 13 days (over 4 SRT) when the experiment was concluded (see Figure 2). The conclusion drawn from this is that the "boiled refed" contained nutri-ents which enabled the methanogens to have a reduced gener-ation time.

A similar experiment was conducted with various concen-trations of yeast extract supplemented to the make-up feed. At 12.5 day HRT/SRT, gas production declined but was measurable for 3 HRT/SRT for the control and all concentra-tions of yeast extract (Figure 3). At 10 day HRT/SRT, the control and the 10 mg/l yeast extract systems failed in 4.5 detention times (Figure 4). At 5 day HRT/SRT only the system receiving 1000 mg/l yeast extract in the make-up feed produced stable gas after 5 detention times

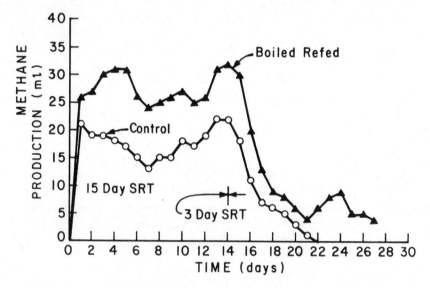

Figure 2. Effect of boiled refed on methane production
           - 3 day SRT.

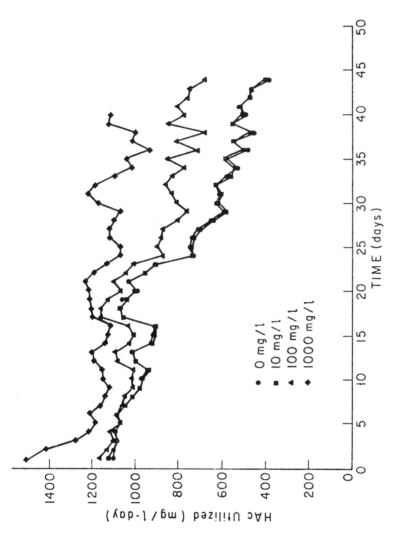

Figure 3. Effect of yeast extract on growth—12.5 day SRT.

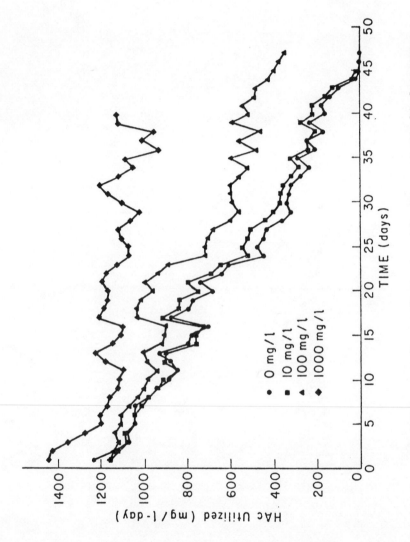

Figure 4. Effect of yeast extract on growth–10 day SRT.

(Figure 5). This again confirmed that nutrient sufficiency has a direct effect on microbial generation times and thus affects the biological safety factor of the process.

Economics of Nutrient Supplementation

Economics will dictate which nutrients can best be added to achieve a stimulation in activity. In the conversion of acetate to methane, approximately 3-6 kg of nitrogen is required per 1000 kg of acetate utilized depending on the SRT (Speece and McCarty, 1964). For a completely nitrogen deficient fatty acid wastewater, if supplemental nitrogen costs $0.40 per kg (as N) and methane has a value of $4.50 per 10^6 BTU, the nitrogen supplementation costs would be approximately 2% of the value of the methane produced. The phosphorous requirement is approximately 0.5 kg per 1000 kg of acetate utilized (Speece and McCarty, 1964).

For carbohydrate wastewaters the requirement for nitrogen, phosphorous, and sulfur may be as much as six times greater than for a fatty acid wastewater, due to the increased synthesis of fermentative bacteria. This has significant impact on some nitrogen deficient biomass feedstocks.

The heavy metals are relatively inexpensive to supplement. Due to the presence (and requirement) of sulfide, most of the heavy metals are precipitated from solution and it is difficult to determine their actual requirement. Generally for methanogens, iron is added at approximately 10 mg/l, cobalt at 5 mg/l and nickel, molybdenum, and selenium at 0.1 mg/l.

The purified form of riboflavin and vitamin B_{12} would not be supplemented, due to the high cost and because both vitamins are synthesized by many bacteria present in most mixed systems. However, addition of excess sludge heat-treatment-liquor could conceivably supply such vitamins in restricted cases. Presently, supplementation of these two vitamins is only of academic interest to achieve maximum stimulation rates. Recent studies in our laboratory have demonstrated that rates of acetate fermentation to methane can be stimulated ten-fold by proper inclusion of trace metals and yeast extract. The controls (a completely stirred tank reactor-CSTR day SRT) operated at about 3300 mg l^{-1} day^{-1} of acetate conversion to methane. Rates of 50,000 mg l^{-1} day^{-1} were achieved (Speece et al., 1983) at an SRT of 20 days when proper trace elements were included.

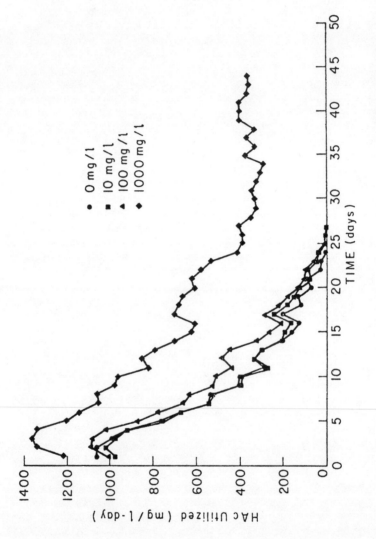

Figure 5.   Effect of yeast extract on growth—5 day SRT.

The search for possible stimulants is very dependent upon the nutrient media that the researcher is using and the situation is further complicated by the ecological inter-actions which must be taken into account. A compound which is stimulatory to a pure species of bacteria may have no effect on a mixed culture such as found in an anaerobic digester. Mah et al. (1976), found the rate of methane pro-duction differed in pure culture vs. an enriched culture and state:

"The interactions demonstrated between methanogenic and non-methanogenic species show that methane production in the mineral acetate enrichment is not the function solely of the organism catalyzing the split of acetate to methane and CO_2. It is a function of a community of organisms, each contributing nutrients to the common environment, and withdrawing others.... Decomposition of acetate to methane occurs far more rapidly in the community of mixed species than with the sarcina alone. This increase in rate through the interaction of many microbes is the most significant feature of these studies insofar as practical problems of methanogenesis are concerned. Complex organic growth factors are re-leased into the environment common to all the cells, hastening their growth and the rate of substrate dis-appearance. If provision of increased complex nutri-ents is the explanation for the increased methanogene-sis by the consortium, provision of the appropriate complex nutrients substrates to the enrichment ecosys-tem should increase the rate of methane formation. The greater rate of methanogenesis by the sarcina in the complex medium as compared to acetate alone is consis-tent with this view."

The organic growth factors which in some cases are re-quired for the activity of methanogens include: coenzyme M, Factor 420, acetate, 2-methyl butyric acid, vitamins, N-acetyl glucosamine, riboflavin, B_{12}, and possibly some as yet unidentified compounds. Tanner (1982) developed a defined medium for *Methanomicrobium mobile*. This organism was found to require acetate, isobutyrate, 2-methylbutyrate, isovalerate, tryptophan, para-aminobenzoic acid, pyridoxine, thiamine, biotin, and vitamin B_{12} for its growth. ATP is also required during the methane formation process (Roberton and Wolfe, 1969; Mountfort, 1980). Several of these com-pounds are only required by particular species. CoM and

F_{420} have been discussed above. Trace amounts of nickel are stimulatory to an enriched culture of acetate utilizing methanogens. Diekert et al. (1980), found nickel to be a component of F_{430}.

Taylor (1982) has given a good review of nutritional interactions of the methanogens. He suggests the stimulation of the growth rate of the primary anaerobes which occur in batch co-culture with a methanogen may in part be due to cross-feeding of growth factors or other nutrients. Excretion of lytic products of the various bacteria can support the growth of other bacteria in the mixed culture. An acetate-enrichment culture of *Methanosarcina* was shown to contain several types of non-methanogenic bacteria that had a varying degree of dependence on the methanogen (Baresi et al., 1978; Ward, Mah and Kaplan, 1978). Vitamin B_{12} was used to replace a requirement for the autoclaved, supernatant medium from a pure culture of *Methanosarcina*.

Growth of methanogens in pure culture can be stimulated by the presence of amino-acids (Bryant et al., 1971; Ferry and Wolfe, 1977) particularly cysteine (Zeikus and Wolfe, 1972; Zeikus and Henning, 1975). Cysteine is an essential amino-acid for the growth of *M. arboriphilus* A_z (Zehnder and Wuhrmann, 1977; Wellinger and Wuhrmann, 1977). The heterotrophic bacteria can stimulate the growth of the methanogens, presumably by the excretion of growth factors (Taylor and Pirt, 1977; Zhilina et al., 1973).

A. *Nitrogen*. Generally, nitrogen is required as the major nutrient other than an energy source for microbial systems. The empirical formula for the anaerobic microbial cells was found to be $C_5H_7O_3N$ which has a N/cell ratio of 11% (w/w). At a 5% net cell synthesis ratio, the nitrogen requirement would be 6 kg N/1000 kg of COD or 1 kg of N/60 m^3 of methane produced.

It was noted that microbial synthesis, and thus the nitrogen requirement of carbohydrate digestion, is about six times greater than for proteins and fatty acids respectively. This is shown in Figure 6. Therefore, the anaerobic digestion of high carbohydrate feedstocks to methane gas deserves special consideration because of the relatively high microbial synthesis to substrate ratio.

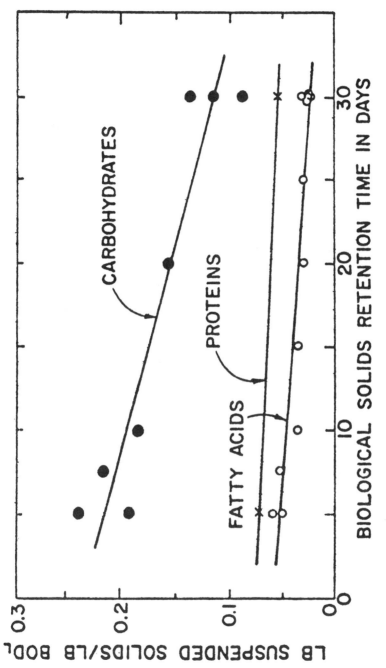

Figure 6. Biological solids production resulting from methane fermentation.

   B.   *Phosphorus.*.   The microbial uptake of phosphorus
in anaerobic digestion has been reported to be approximately
1/7 of that for nitrogen (Speece and McCarty, 1964).   This
would yield an empirical formulation of $C_5H_7O_2NP_{.06}$.

   C.   *Sulfur.*   The sulfur requirement for methanogens is
not extensively documented.   The limited references in the
literature make reference to only two of the many methanogenic
pure cultures and one mixed culture, which had cellulose as
the substrate   (Mountfort and Asher, 1979; Ronnow and Gunnarsson,   1981, 1982; Gunnarsson and Ronnow, 1982; Patel et
al, 1978; Khan and Trotter, 1978; Wellinger and Wuhrman,
1977.)   The sulfur requirement is part of a complex picture.
On the one hand, the presence of sulfate may limit methanogenesis because the sulfate reducers have a lower half saturation coefficient than the methanogens for substrates
on the production of sulfide for growth.   It is known that
methanogenic bacteria require fully reduced sulfur as a
sulfur source (Bryant et al., 1971; Zehnder and Wuhrmann,
1977; Wolfe, 1971).

   Scherer and Sahm (1981b) reported that optimal growth
of *M. barkeri* occurred on a defined medium containing methanol when 2.5 to 4.0 mM sodium sulfide was added.   This
resulted in a measured total soluble sulfide concentration
of 0.04-0.06 mM.   They also reported that iron sulfide,
zinc sulfide, or L-methionine could also act as sulfur
sources.   However, addition of sodium sulfate to a sulfide
depleted media failed to restore growth.

   Mountfort and Asher (1979) found that at least part
of the sulfide requirement for growth is used as a precursor
of HS-CoM which is 40% sulfide by weight.   Ronnow and Gunnarsson (1981) state that 2.6% of the cell mass of *M. thermoautotrophicum* is sulfur.   According to their data this would
give an empirical cell composition of: $C_5H_1O_2 \ NP_{.06}S_{0.1}$.
At sulfide levels below 0.1 mM, growth of *M. thermoautotrophicum* was poor and the methane production rate decreased.
Only frequent injection of sulfide could sustain rapid growth
and continuous methane production.   After injection of
sulfide, a large increase in methane production was noted
within 30 minutes.

   Ronnow and Gunnarsson (1982) later observed that
when sulfide was added in increments of 20 mg/l, the
methane production rate increased until sulfide was completely depleted in the medium (as determined by sulfide

analysis) and that the methane production rate then decreased. This observation provides a further indication of a sulfur compound requirement for high-rate methane formation rates.

The sulfide requirement can be easily overlooked in anaerobic digestion because there are many sinks for sulfide:

1. H_2S in the off-gas,
2. total sulfides in the effluent,
3. microbial synthesis of sulfide (150% of P synthesis), and
4. precipitation of sulfides by heavy metals.

One could reasonably hypothesize from these references that about 10 mg/l (corresponds to 1/2% H_2S in gas) of unionized H_2S is required for optimum growth of methanogens. If it was also assumed for convenience that the digester were operated at a pH of 6.85, equal to the pK of H_2S, then the total sulfide requirement for anaerobic digestion could be calculated as shown in Figure 7.

D. Trace metals. Nickel is generally not essential for growth of bacteria (Diekert et al., 1980b). However, in the last few years it has become evident that the methanogenic bacteria are clearly a unique group in this respect. Nickel is also an essential component of bacterial urease and in bacterial enzymes, which convert H_2-CO_2 to acetate.

In view of the apparently universal occurrence of F_{430} in methanogens and the inherent nickel requirement of F_{430}, the question arises as to the source of nickel, since it has not generally been included in defined media. The medium described by Balch et al. (1979), for growth of most methanogenic bacteria contains 0.2% yeast extract and 0.2% trypticase, but it is not supplemented with nickel salts. Diekert et al. (1981), found that the methanogenic bacteria tested grew readily on this medium. However, if the concentration of yeast extract was significantly lowered, good growth was only observed upon addition of nickel. They found that no other transition metal could substitute for nickel.

Other trace metal requirements have been demonstrated with methanogenic bacteria. Iron, cobalt, molybdenum, selenium, and tungsten have been shown to be stimulatory to methanogens (Jones and Stadtman, 1977; Schonheit et al., 1979; Speece and McCarty, 1964; Taylor and Pirt, 1977; Scherer and Sahm, 1981a).

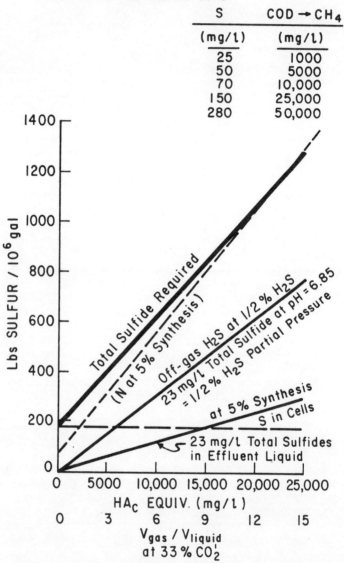

Figure 7. Sulfur requirement for anaerobic digestion.

McCarty and Vath (1962) have reported that very high
acetate utilization rates were observed when dried supernatant solids from a municipal digester were supplemented
to their digesters receiving a defined medium of acetate
and mineral salts. Conventionally, high-rate digestion is
considered to be 3.3 g l^{-1} day^{-1}, yet McCarty and Vath (1962)
observed acetate utilization rates as high as 21.9 g l^{-1}day^{-1}
when the digesters were supplemented with dried supernatant
solids. McCarty and Vath also stated that previous studies
on the digestion of pure volatile acids alone have indicated
that only relatively low rates of fermentation were possible.
They believed the low rates were not due to an inability
of the methane organisms to carry out much higher rates,
but rather to improper environmental conditions. They concluded that there appeared to be no practical limit to the
possible rate of volatile acid fermentation under proper
environmental conditions and process control.

Speece and McCarty (1964) attempted to define the
stimulatory components of the dried digester supernatant
solids and identified the requirement of the methanogens
for iron and cobalt. In addition, some organic compounds
such as proline, benzimidazole, and thiamine were also
shown to be stimulatory, resulting in acetate utilization
rates in excess of 20 g l^{-1} day^{-1}. By including iron and
cobalt in the medium, it was possible to steadily maintain
acetate utilization rates of approximately 2 to 3 g l^{-1} day^{-1}.

Hoban and van den Berg (1979) reported that addition
of iron to a methanogenic culture utilizing acetic acid
markedly increased conversion of acetate to methane. The
optimum soluble iron concentration was between 0.2 and 2 mM.
Conversion of acetic acid to methane in liquid from municipal
sewage digesters and from laboratory food processing waste
digesters was also increased markedly by addition of iron.
They concluded that optimization of the conversion of acetic
acid to methane in methanogenic fermentations requires
soluble iron concentrations which are many times higher
than those often required for maximum growth and activity
in microbial cultures (Jack et al., 1976; Lankford, 1973;
Patel et al., 1978). Also they stated that it is conceivable
that in practice some of the difficulties often encountered
in controlling the digestion process are caused by variations
in soluble iron content.

In our research, we observed that in the absence of
nickel, specific acetate-utilization rates were in the ranges
of 2 to 4 day^{-1} and did not exceed 4.6 day^{-1}. In the presence

of nickel (with no yeast extract), specific acetate-utilization rates of 10 day^{-1} were observed. When yeast extract was supplemented along with nickel, specific acetate-utilization rates as high as 12 to 15 day^{-1} were observed (Speece et al., 1983).

The methanogens require seemingly contradictory growth conditions. For good growth, a sulfide concentration of 10 to 27 mg/l is required (Mountford and Asher, 1979; Ronnow and Gunnarsson, 1981). But the sulfide may remove metal ions essential for growth, such as nickel, iron, and cobalt (Taylor and Pirt, 1977; Winfrey and Zeikus, 1977). Chelators must play a strong role. Likewise high rates of gas production tend to strip the essential sulfide from solution (Lawrence and McCarty, 1966).

It also appears that nickel, iron, and/or cobalt must be supplemented to achieve volatile suspended solids (VSS) concentrations (Speece et al., 1983). With yeast extract supplementation only, the highest VSS observed was 1800 mg/l. When nickel, iron, and cobalt only were supplemented, the highest VSS observed was 3000 mg/l. When nickel and yeast extract both were supplemented, the highest VSS observed was 7000 mg/l. The SRT in all cases was 20 days.

The inherent capability of achieving such high acetate-utilization rates at 20 days SRT (61 g l^{-1} day^{-1}, 20 vol. CH_4 vol. of culture^{-1} day^{-1}) and associated high specific acetate-utilization rates (15 day^{-1}) appears to be another bit of evidence indicating that the methanogens are indeed a unique biological group. With proper process control and adequate nutrient supplementation, there does not appear to be a practical limit to the possible rates of acetate fermentation.

Nutrition of Fermentative Bacteria. Much of our understanding of the anaerobic digestion process as it occurs in digesters is derived from studies of other anaerobic ecosystems, especially the rumen (Iannotti et al., 1981; Bryant, 1979; Mah et al., 1977; Kirsch and Sykes, 1971). In addition, much of the work on the acid phase of the domestic anaerobic digestion processes has been done with partially purified wastes, one segment of the population, or with minimally described cultures (Iannotti et al., 1982; Chynoweth and Mah, 1977; Mah and Sussman, 1968; Toerien, 1970). The acetogenic bacteria, except for a few cases, have not been isolated and characterized (Iannotti et al., 1982;

Boone and Bryant, 1980; Bryant, 1979). However, the methan-
gens have been studied in greatest detail (Iannotti et al.,
1982; Balch et al., 1979; Mah et al., 1977; Wolfe, 1979;
Zeikus, 1977).

The nutritional requirements of the fermentative bacteria are generally not well defined. Initially, rumen
fluid and digester fluid were added to the medium to supply
nutritional requirements. Caldwell and Bryant (1966) then
developed a more defined medium in which rumen fluid was replaced by trypticase, yeast extract, hemin, and volatile
fatty acids. The viable count per total microscopic count
with the above media has been about 10 to 30% (Iannotti et
al., 1978). The best recovery had been obtained from human
feces, in which the bacteria are apparently less fastidious
than those of other ecosystems (Eller et al., 1971). Mah
and Sussman (1968) recovered approximately 1 to 10% of the
bacteria from a domestic digester with a medium containing
glucose, digester fluid, minerals, a reducing agent, and
buffer. Kirsch (1969) recovered 8.6% of the bacteria using
Caldwell and Bryant's medium to which 16% digester fluid
was added.

Iannotti et al. (1978), developed a habitat-simulating medium
for enumeration and isolation of bacteria from a swine waste
digester. A roll tube medium with growth factors for strict
anaerobes from previously studied anaerobic ecosystems was
tried to evaluate the effects of deletion, addition or level
of digester fluid, digester fluid treated with acid or base,
rumen fluid, fecal extract, anaerobic pit extract, tissue
extract, carbohydrates, peptones, short-chain fatty acids,
minerals, vitamins, nitrogen and phosphorous sources, reducing and solidifying agents, buffers, and gases on colony
counts. With a medium containing digester fluid, peptones,
minerals, cysteine, sodium carbonate, and agar, colony counts
were 60% of the microscopic count. These yields represent
2.5 to 20 times greater improvements over previous media
compositions (Iannotti et al., 1978).

The level of growth of some of the isolates is still
low. The data are indicative of complex interactions within
the microbial population. Organisms that initiate the hydrolysis of complex molecules, such as cellulose, supply
partial breakdown products to bacteria with less hydrolytic
capabilities. The latter organisms supply growth factors
to the former specialists. Growth factors are also supplied
as the plant material is degraded(Iannotti et al., 1981).

Iannotti et al. (1981), have isolated and characterized 130 strains of bacteria from a swine digester. The isolates were divided into 11 groups that included organisms identified as *Peptostreptococcus, Peptococcus, Eubacterium, Lactobacillus, Bacterioides,* and unidentified genera plus miscellaneous facultative and strict anaerobes. The organisms were shown to require mixtures of known factors for growth plus unknown factors in crude extracts such as from digester fluid, swine manure extracts, and rumen fluid. The requirement for both known and unknown factors indicates that fermentative bacteria have complex requirements, and this limits the ability to define optimum conditions.

Most strains of fermentative anaerobes require one or more B-vitamins for growth (Bryant and Robinson, 1962). Sulfide often serves as the main sulfur source, although methionine and cysteine are sometimes required, and at least a few species carry out assimilatory or dissimilatory sulfate reductions (McInerney and Bryant, 1981; Bryant, 1973; Bryant, 1974; Emery et al., 1957). Nutritional interactions among bacterial species are very important in these ecosystems (McInerney and Bryant, 1981).

SUMMARY

Determination of nutrient requirements for anaerobic digestion should be an integral part of process design. The feedstock may incidentally satisfy the nutrient requirements in some cases. In other cases, certain nutrients may have to be specifically supplemented. In this paper, major emphasis has been placed upon insuring the adequacy of nitrogen, phosphorous, sulfide, and trace metals. If nutrients are not adequately provided, a reduction in the anaerobic digestion process rate and/or efficiency results. An economic analysis needs to be made to determine whether the increased capital costs required for increased reactor volume is more cost effective than increased operating costs for nutrient supplementation.

ACKNOWLEDGMENT

Some of this research was funded by the Solar Energy Research Institute.

REFERENCES

Balch, W.E., G.E. Fox, L.J. Magrum, G.R. Woese, and R.S.
    Wolfe, "Methanogens: Re-evaluation of a Unique Bio-
    logical Group," *Microbiol. Rev., 43,* 260-296 (1979).

Balch, W.E. and R.S. Wolfe, "Specificity and Biological
    Distribution of Coenzyme M (2-mercaptoethanesulfonic
    acid)," *J. Bacteriol., 137,* 256-263 (1979a).

Baresi, L., R.A. Mah, D.M. Ware, and I.R. Kaplan, "Methano-
    genesis from Acetate: Enrichment Studies," *Appl.
    Environ. Microbiol., 36,* 186-197 (1978).

Boone, D.R. and M.P. Bryant, "Propionate-degrading Bacterium,
    Syntrophobacter wolinii sp. nov. gen. nov. from Methano-
    genic Ecosystems," *Appl. Environ. Microbiol., 40,*
    626-632 (1980).

Bryant, M.P., "Nutritional Requirements of the Predominant
    Rumen Cellulolytic Bacteria," *Fed. Proc. Fed. Am. Soc.
    Exp. Biol., 32,* 1809 (1973).

Bryant, M.P., "Nutritional Features and Ecology of Predom-
    inant Anaerobic Bacteria of the Intestinal Tract,"
    Am. J. Clin. Nutr., 22, 1313 (1974).

Bryant, M.P., "Microbial Methane Production-Theoretical
    Aspects," *J. Anim. Sci., 48,* 193-201 (1979).

Bryant, M.P. and I.M. Robinson, "Some Nutritional Character-
    istics of Predominate Culturable Ruminal Bacteria,"
    J. Bacteriol., 84, 605 (1962).

Bryant, M.P., S.F. Tseng, I.M. Robinson, and A.E. Joyner,
    "Nutrient Requirements of Methanogenic Bacteria,"
    Adv. Chem. Ser., 105, 23-40 (1971).

Caldwell, D.R. and M.P. Bryant, "Medium Without Rumen Fluid
    for Non-selective Enumeration and Isolation of Rumen
    Bacteria," *Appl. Microbiol., 14,* 794-801 (1966).

Chynoweth, D.P. and R.A. Mah, "Bacterial Populations and
    End Products During Anaerobic Sludge Fermentation
    of Glucose," *J. Wat. Poll. Control Fed., 49,* 405-
    412 (1977).

Diekert, G., B. Klee, and R. Thauer, "Nickel, a Component
    of Factor F_{430}," *Arch. Microbiol., 124,* 103-106 (1980).

Diekert, G., R. Jaenchen, and R.K. Thauer, "Biosynthetic
    Evidence for Nickel Tetrapyrrole Structure of Factor
    F_{430} From *Methanobacterium thermoautotrophicum*," *FEBS
    Lett., 119,* 118-120 (1980).

Diekert, G., U. Konheiser, K. Peichulla, and R.K. Thauer,
    "Nickel Requirement and Factor F_{430} content of Methano-
    genic Bacteria," *J. Bact., 148,* 459-464 (1981).

Eller, C., M.R. Crabill, and M.P. Bryant, "Anaerobic Roll Tube Media for Nonselective Enumeration and Isolation of Bacteria in Human Feces," *Appl. Microbiol., 22,* 522-529 (1971).

Emery, R.S., C.K. Smith, and L. Fai To, "Utilization of Inorganic Sulfate by Rumen Microorganisms. II. The Ability of Single Strains of Rumen Bacteria to Utilize Inorganic Sulfate," *Appl. Microbiol., 5,* 363 (1957).

Ferry, G. and R. Wolfe, "Nutritional and Biochemical Characteristics of *Methanospirillum hungateii,*" *Appl. and Environ. Microbiol., 34,* 371-376 (1977).

Gunnarsson, L.A.H. and P.H. Ronnow, "Variation of the ATP Pool in Thermophilic Methanogenic Bacteria During Nitrogen or Sulfur Starvation," *FEMS Microbiol. Lett., 14,* 317-320 (1982).

Hoban, D.J. and L. van den Berg, "Effect of Iron on Conversion of Acetic Acid to Methane During Methanogenic Fermentation," *J. Appl. Bacteriol., 47,* 153-159 (1979).

Iannotti, E.L., J.R. Fischer, and D.M. Sievers, "Medium for the Enumeration and Isolation of Bacteria from a Swine Waste Digester," *Appl. Environ. Microbiol., 36,* 555-566 (1978).

Iannotti, E.L., J.R. Fischer, and D.M. Sievers, "Medium for Enhanced Growth of Bacteria from a Swine Manure Digester," *Appl. Environ. Microbiol., 43,* 247-249 (1982).

Iannotti, E.L., M.K. Wulfers, J.R. Fischer, and D.M. Sievers, "The Effect of Digester Fluids, Swine Manure Extract and Rumen Fluid on the Growth of Bacteria from an Aerobic Swine Manure Digester," Chapt. 49, *Developments in Industrial Microbiology,* Vol. 22, 565-576 (1981).

Jack, M.E., G.J. Farquhar, and G.M. Cornwall, "Anaerobic Digestion of Primary Sludge Containing Iron Phosphate," *Water Poll. Res. in Canada, 8,* 91-109 (1976).

Jones, J.B. and T.C. Stadtman, "*Methanococcus vannielii:* Culture and Effects of Selenium and Tungsten on Growth," *J. Bact., 130,* 1404-1406 (1977).

Khan, A.W. and T.M. Trottier, "Effects of Sulfur-containing Compounds on Anaerobic Degradation of Cellulose to Methane by Mixed Cultures Obtained from Sewage Sludge," *Appl. Env. Microbiol., 35,* 1027-1034 (1978).

Kirsch, E.J., "Studies on the Enumeration and Isolation of Obligate Bacteria from Digesting Sewage Sludge," *Dev. Ind. Microbiol., 10,* 170-176 (1969).

Kirsch, E.J. and R.M. Sykes, "Anaerobic Digestion in Biological Waste Treatment," *Progr. Ind. Microbiol., 9,* 155-239 (1971).

Lankford, C.E., "Bacterial Assimilation of Iron," *CRC Critical Reviews of Microbiol.*, 273-331 (1973).

Lawrence, A.E. and P.L. McCarty, "The Effects of Sulfides on Anaerobic Treatment," *Int. J. Air and Wat. Poll.*, *10*, 207-221 (1966).

Mah, R.A. and C. Sussman, "Microbiology of Anaerobic Sludge Fermentation. I. Enumeration of the Nonmethanogenic Anaerobic Bacteria," *Appl. Microbiol.*, *16*, 358-361 (1968).

Mah, R.A., D.M. Ward, L. Baresi, and T.C. Glass, "Biogenesis of Methane," *Ann. Rev. Microbiol.*, *31*, 309-341 (1977).

Mah, R., R. Hungate, and K. Ohwaki, "Acetate, a Key Intermediate in Methanogenesis," *Seminar on Microbial Energy Conversion*, (H.G. Schlegel, ed.), E. Goltz KG, Gottingen, Germany, 97-106 (1976).

McCarty, P.L. and C.A. Vath, "Volatile Acid Digestion at High Loading Rates," *Int. J. of Air and Wat. Poll.*, *6*, 65-73 (1962).

McInerney, M.J. and M.P. Bryant, "Review of Methane Fermentation Fundamentals," *CRC Press* (1981).

Mountfort, D.O., "Effect of Adenosine 5'-monophosphate on Adenosine 5'-triphosphate Activation of Methyl Coenzyme M. Methylreductase in Cell Extracts of *Methanosarcina barkeri*," *J. Bacteriology*, *143*, 1039-1041 (1980).

Mountfort, D.O. and R.A. Asher, "Effect of Inorganic Sulfide on the Growth and Metabolism of *M. barkeri*, strain DM," *Appl. Envir. Microbiol.*, *37*, 670-675 (1979).

Patel, G.B., A.W. Khan, and L.A. Roth, "Optimum Levels of Sulfate and Iron for the Cultivation of Pure Cultures of Methanogens in Synthetic Media," *J. Appl. Bacteriol.*, *45*, 347-356 (1978).

Roberton, A.M. and R.S. Wolfe, "ATP Requirement for Methanogenesis in Cell Extracts of *Methanobacterium strain M.o.H.*," *Biochem. Biophys. Acta*, 420-429 (1969).

Ronnow, P.H. and L.A.H. Gunnarsson, "Sulfide Dependent Methane Production and Growth of Thermophilic Methanogenic Bacterium," *Appl. Envir. Microbiol.*, *42*, 580-584 (1981).

Ronnow, P.H. and L.A.H. Gunnarsson, "Response of Growth and Methane Production to Limiting Amounts of Sulfide and Ammonia in Two Thermophilic Methanogenic Bacteria," *FEMS Microbiol. Lett.*, *14*, 311-315 (1982).

Scherer, P. and H. Sahm, "Effect of Trace Elements and Vitamins on the Growth of *Methanosarcina barkeri*," *Acta Biotechnologica*, *1*, 57-65 (1981a).

Scherer, P. and H. Sahm, "Influence of Sulfur Containing Compounds on the Growth of *Methanosarcina barkeri* in a Defined Medium," *European J. Applied Microbiol. Biotechnol., 12,* 28-35 (1981b).

Schonheit, P., J. Moll, and R.K. Thauer, "Nickel, Cobalt and Molybdenum Requirement for Growth of *Methanobacterium thermoautotrophicum*," Arch. *Microbiol., 123,* 105-107 (1979).

Speece, R.E., G.F. Parkin, and D. Gallagher, "Nickel Stimulation of Anaerobic Digestion," *Wat. Res., 17,* 677-683 (1983).

Speece, R.E. and Perry L. McCarty, "Nutrient Requirements and Biological Solids Accumulation in Anaerobic Digestion," *Adv. in Wat. Poll. Res., 2,* 305-322 (1964).

Tanner, R.S., "Novel Compounds from Methanogens: Characterization of Component B of the Methylreductase System and Mobile Factor," Ph.D. Dissertation, University of Illinois-Urbana (1982).

Taylor, G.T., "The Methanogenic Bacteria," *Progress in Industrial Microbiology,* (M.J. Bull, ed.), Elsevier (1982).

Taylor, G.T. and S.J. Pirt, "Nutrition and Factors Limiting the Growth of a Methanogenic Bacterium (*Methanobacterium thermoautotrophicum)*," Arch. *Microbiol., 113,* 17-22 (1977).

Torien, D.F., "Population Description of Nonmethanogenic Phase of Anaerobic Digestion. I. Isolation, Characterization and Identification of Numerically Important Bacteria," *Water Res., 4,* 129-148 (1970).

Ward, D.M., R.A. Mah, and I.R. Kaplan, Methanogenesis from Acetate: A Nonmethanogenic Bacterium from any Anaerobic Acetate Enrichment, *Appl. Environ. Microbiol., 35,* 1185-1192 (1978).

Wellinger, A. and K. Wuhrman, "Influence of Sulfide Compounds on Metabolism of *Methanobacterium* Strain AZ," *Arch. Microbiol., 115,* 13-17 (1977).

Winfrey, M.R. and J.G. Zeikus, "Effect of Sulfate on Carbon and Electron Flow During Microbial Methanogenesis in Freshwater Sediments," *Appl. and Environ. Microbiol., 33,* 275-281 (1977).

Wolfe, R.S., "Microbial Formation of Methane," *Adv. Microbial Physiol.,* (A.H. Rose and J.F. Wilkinson, eds.), Vol. 6, Academic Press, New York-London, 107-146 (1971).

Zehnder, A.J.B. and K. Wuhrman, "Physiology of a *Methanobacterium* Strain AZ," *Arch. Microbiol., 111,* 199-205 (1977).

Zeikus, J.G., "The Biology of Methanogenic Bacteria," *Bact. Rev., 41,* 514-541 (1977).

Zeikus, G. and D.L. Henning, "*Methanobacterium arbophilicum* sp. nov. An Obligate Anaerobe Isolated from Wet wood of Living Trees," *Ant. v. Leeu., 41,* 543-552 (1975).

Zeikus, J.G. and R.S. Wolfe, "*Methanobacterium thermoautotrophicum* sp. n., An Aerobic, Autotrophic Extreme Thermophile," *J. Bacteriol., 109,* 707-713 (1972).

Zhilina, T.N. and G.A. Zavarzin, "Cyst Formation by *Methanosarcina*," *Mikrobiologiya, 48,* 451-456 (1979).

16
PROPIONATE CONVERSION TO BUTYRATE IN AN ANAEROBIC DIGESTER

E. Samain, R. Moletta, H.C. Dubourguier, and G. Albagnac

Station de Technologie Alimentaire
Institut National de la Recherche Agronomique
Villeneuve.D'Ascq, France

ABSTRACT

During batch degradation of propionate by anaerobic sludge or by a highly purified culture, a transient accumulation of acetate and butyrate was observed. Labelling experiments and inhibition of methanogenesis by bromoethane sulfonic acid evidenced that butyrate is a facultative electron sink during acetogenesis rather than an end-product of acetate fermentation by butyric bacteria.

INTRODUCTION

Since pioneer work of Stadtman and Barker (1951), propionate degradation in digesters is regarded as an oxidative decarboxylation. Syntrophic associations between hydrogenophilic methanogens and obligate hydrogen producing acetogens (OHPA) such as *Syntrophobacter wolinii* are reported to degrade propionate to acetate and reductive methane (Boone and Bryant, 1980). In digesters, sulfate reducing bacteria such as *Desulfobulbus propionicus* (Widdel and Pfennig, 1982) or *D. elongatus* (Samain, Dubourguier, and Albagnac, 1984) play a limited role in propionate degradation. Recently, using ^{14}C labelled propionate, Koch et al. (1983), gave some evidences that batch degradation of this organic acid by enriched cultures proceeds through a randomizing pathway.

In this paper, we report the synthesis of butyrate and acetate during batch degradation of propionate, either by anaerobic sludge or by an highly purified culture.

MATERIALS AND METHODS

Sludges were sampled anaerobically from a 5000 m^3 digester treating vegetable canning wastewaters (Verrier, Moletta, and Albagnac, 1983). The anaerobic technique of Hungate modified by Miller and Wolin (1972) was used throughout this study. CBBM medium (Zeikus and Wolfe, 1972) modified according to Samain et al. (1982), served for enumeration and culture under low sulfate concentration. Enumeration of methanogens and propionate oxidizers was carried out by the three tubes Most Probable Number (MPN) method on the following substrates: H_2/CO_2, acetate, propionate, propionate plus sulfate. From the last tubes showing degradation of propionate, an enriched culture was established by repeating three times the MPN procedure.

Volatile fatty acids (VFA) and gases were determined by gas chromatography. Labelled VFA were individually counted after HPLC separation with a continuous flow scintillation detector (flow one hp). All standardizations were made by the channel ratio on a Intertechnique SL 4000 scintillation counter.

RESULTS

Enumeration

In the anaerobic sludge, 2.3 x 10^7 cells per ml degrading acetate into methane were counted after 12 weeks of incubation. The prevailing organisms in the last positive tubes were long filamentous non-fluorescent rods similar to *Methanothrix soehngenii*. Enumeration of hydrogenophilic methanogens led to 4 x 10^7 cells per ml of straight non motile fluorescent rods similar to *Methanobacterium formicicum*. Per ml of sludge, 10^7 and 2.3 x 10^6 cells of syntrophs degrading respectively propionate and butyrate were counted. On propionate, small amounts of butyrate were observed beside acetate in the liquid phase. Sulfate reducers oxidizing propionate were in low numbers, i.e., less than 10^5 cells/ml.

Batch Degradation of Propionate by Anaerobic Sludge

During degradation of $(1-^{14}C)$ propionate (Fig. 1), a transient accumulation of labelled acetate and butyrate was evidenced. The specific activity (SA) of propionate remained constant during the course of the experiment. Within the

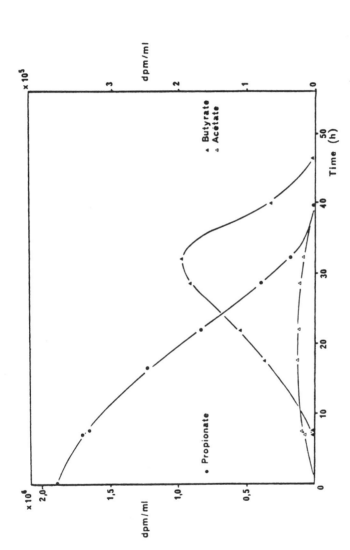

Figure 1. Batch degradation of (1- ^{14}C) propionate by digester sludge. Experiments were done in duplicate in 125 ml-anaerobic serum bottles containing 10 ml of sludge under a N_2/CO_2 (85/15) atmosphere. Incubation temperature was 35°C. Reaction was initiated by injecting 1 ml of labelled propionate (about 25 µCi/mM).

fifteen first hours of cultivation, the SA of butyrate reached a value very close to the SA of propionate. At the opposite, the final SA of acetate was twice lower.

This transient accumulation of butyrate and acetate was not due to an inhibition of their catabolism. In fact, upon addition of trace amounts of $(1-{}^{14}C)$ acetate during degradation of unlabelled propionate (Fig. 2), labelled acetate was rapidly metabolized to gases (CH_4 and CO_2) and radioactivity was also incorporated into propionate and then into butyrate. In the same way, labelled acetate and pro- pionate were formed after addition of trace amounts of $(1-{}^{14}C)$ butyrate (Fig. 3).

Batch Degradation of Propionate by the Purified Culture

The purified culture formed large wooly aggregates. Optical microscopy revealed a network of *Methanothrix*-like (Fig. 4) non fluorescent filaments associated with fluorescent non motile rods similar to *Methanobacterium*. In addition, a number of thick short rods (Fig. 5) were entrapped in flocs. These last organisms are assumed to be the propionate- and the butyrate-degrading organisms.

Propionate was degraded to methane and CO_2 by this purified culture but a significant transient accumulation of acetate and butyrate was observed (Fig. 6). The latter VFA was only degraded when complete consumption of propionate and acetate was achieved. Addition of bromoethane sulfonic acid (BESA) up to 0.5 mmol/l strongly inhibited methanogenesis and butyrate degradation (Table 1). In this case, 70% of propionate was still degraded within 40 days and 1 mole of this substrate splitted to 1 mol of acetate and 0.24 of butyrate, no hydrogen being detected in the gas phase.

CONCLUSION

Obviously, the purified culture on propionate still con- tains butyrate degraders. This may be explained by the syn- thesis of butyrate during batch degradation of propionate which allows maintenance of butyrate-degrading organisms which are present in high numbers in the digester micro- flora.

The results of labelling experiments with the digester sludge strongly suggest that some bacterial species or con- sortia are able to synthesize butyrate from propionate. This is also sustained by numerations which never evidenced

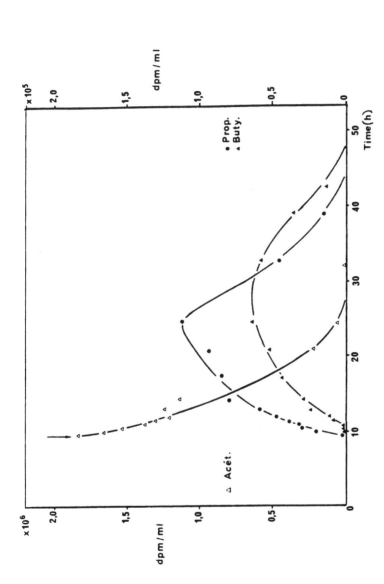

Figure 2. Acetate degradation during batch methanization of propionate. Experiments were done under the same conditions as in Fig. 1, except unlabelled propionic acid was used and 10 μCi of (1-^{14}C) acetate (57 mCi/m mole) were added to 9.5 hr after the beginning of propionate degradation.

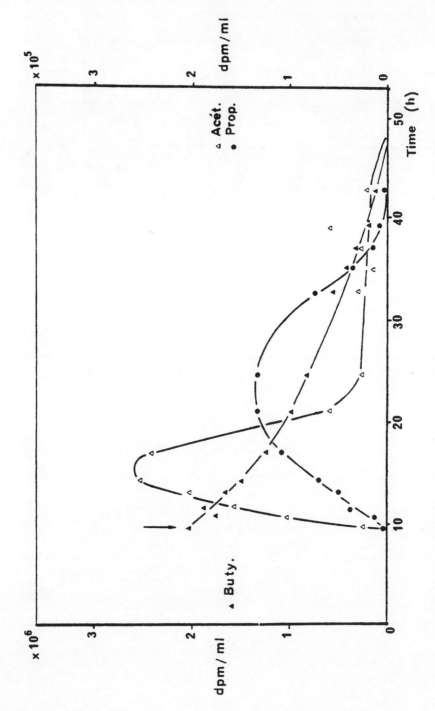

Figure 3.   Butyrate degradation during batch methanization of propionate.

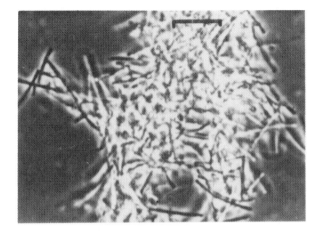

Figure 4. Wooly aggregates of the purified culture dis-
          mutating propionate. Bar represents 10 μm.

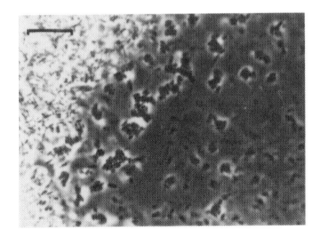

Figure 5. Thick short rods present in the wooly purified
          culture aggregates. Bar represents 10 μm.

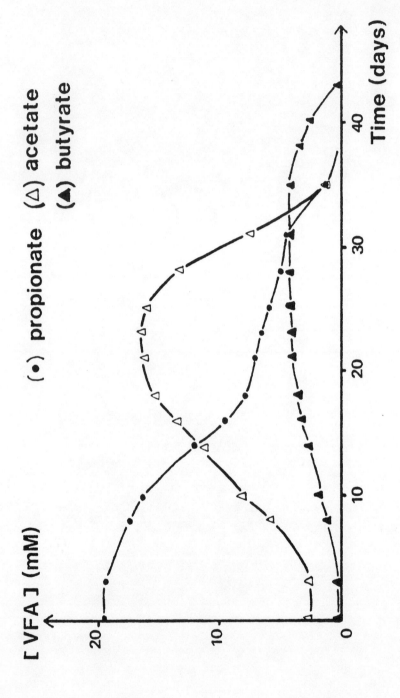

Figure 6.   Propionate degradation by the purified culture.

Table 1.
Effect of Bromo Ethane Sulfonic Acid on End-Products of Propionate

B.E.S.A. (mM)	Propionate Degraded (%)	End-Products (moles) per 10 moles of propionate degraded			
		Acetate	Butyrate	Methane	
0	100	0	0	14.0	
0.1	73.5	10.34	2.08	0.71	
0.5	69.5	10.30	2.35	0.60	
1.0	30.8	10.10	2.41	0.44	

BESA was added to propionate (20 mM) medium before inoculation with the highly purified culture (Incubation time = 40 days).

high numbers of acetate-fermenting butyrogens (less than 10^6 cells/ml) and by the dismutation of propionate in presence of BESA. Thus, propionate dismutation is possible in the absence of sulfate-reducers or hydrogenophilic methanogens. Fermentation balance can be expressed as follows:

$$1 \text{ propionate} + \frac{1-2x}{2} \text{ CO}_2 + \frac{1-2x}{2} \text{ H}_2\text{O} \longrightarrow$$

$$\frac{7-10x}{4} \text{ acetate} + x \text{ butyrate} + \frac{3-6x}{4} \text{ H}^+$$

in which (3 - 2x) acetate are provided through CO_2 reduction.

Experimental data gave x + 0.24.

But from our experiments, metabolic pathways remain unclear. In fact, acetate and butyrate may originate from a single four-carbon intermediate or butyrate may be synthesized from acetyl-CoA as in *Clostridia*.

Finally, complete acetogenesis from propionate in some digesters proceed not only through an oxidative decarboxylation as described by Koch et al. (1983), and Boone and Bryant (1980), but also through β-oxidation of butyrate by organisms similar to *Syntrophomonas wolfei* (McInerney et al., 1981).

REFERENCES

Boone, D.R. and M.P. Bryant, "Propionate Degrading Bacterium, *Syntrophobacter wolinii* sp. nov. gen. nov. from Methanogenic Ecosystems," *Appl. Environ. Microbiol., 40,* 626-632 (1980).

Koch, M., J. Dolfing, K. Wuhrmann, and A.T.B. Zehnder, "Pathways of Propionate Degradation by Enriched Methanogenic Cultures," *Appl. Environ. Microbiol., 45,* 1411-1414 (1983).

McInerney, M.J., M.P. Bryant, R.B. Hespell, and J.M. Costerton, "*Syntrophomonas wolfei* gen. nov. sp. nov., An Anaerobic Syntrophic Fatty Acid Oxidizing Bacterium," *Appl. Environ. Microbiol., 41,* 1029-1039 (1981).

Miller, T.L. and M.J. Wolin, "A Serum Bottle Modification of the Hungate Technique for Cultivation of Obligate Anaerobes," *Appl. Microbiol., 27,* 985-987 (1983).

Samain, E., G. Albagnac, H.C. Dubourguier, and J.P. Touzel, "Characterization of a New Propionic Acid Bacterium thatFerments Ethanol and Displays a Growth Factor-Dependent Association with a Gram Negative Homoacetogen," *FEMS Microbiol. Lett., 15,* 69-74 (1982).

Samain, E., H.C. Dubourguier, and G. Albagnac, "Isolation and Characterization of *Desulfobulbus elongatus* sp. nov. from a Mesophilic Industrial Digestor," *System. Appl. Microbiol., 219,* In press (1984).

Verrier, D., R. Moletta, and G. Albagnac, "Anaerobic Digestion of Vegetable Canning Wastewaters by the Anaerobic Contact Process: Operational Experience," 3rd Int. Symp. Anaerobic Digestion, Boston (U.S.A.) (1983).

Widdel, F. and N. Pfennig, "Studies on Dissimilatory Sulfate-reducing Bacteria That Decompose Fatty Acids: II. Incomplete Oxidation of Propionate by *Desulfobulbus propionicus* ge. nov. sp. nov.," *Arch. Microbiol., 131,* 360-365 (1982).

Zeikus, J.G. and R.S. Wolfe, "*Methanobacterium thermoautotrophicum* sp. nov., An Anaerobic Autotrophic Extreme Thermophile," *J. Bact., 109,* 707-715 (1972).

17

EFFECTS OF THE ADDITION OF PROPIONATE AND BUTYRATE TO THERMOPHILIC METHANE-PRODUCING DIGESTERS

J. Michael Henson, F.M. Bordeaux, and Paul H. Smith

Department of Microbiology and Cell Science
University of Florida
Gainesville, Florida

ABSTRACT

The effects of increased levels of propionate and butyrate on thermophilic bench-top digesters were studied by
continuous infusion of these fatty acids, in addition to
the regular daily feed, into 3.6 liter digesters. When
butyrate was added at the rate of 10 µmol/ml sludge per day,
it did not accumulate. The production of biogas (methane
and carbon dioxide) increased by 150%. Percent methane in
the biogas increased from 56% to 63%. Acetate and propionate
concentrations in the sludge increased from 4 to 35 µmol/ml
and 0.5 to 3.3 µmol/ml, respectively. When the addition
of butyrate was increased to 15 µmol/ml sludge per day,
digestion remained stable with elevated concentrations of
acetate (150 µmol/ml sludge) and propionate (25 µmol/ml
sludge). Daily production of methane was 5.2 liters versus
3.2 liters in a control digester. Butyrate accumulated to
only 10% of the cumulative added butyrate. A second thermophilic digester was infused with propionate at the rate of
10 µmol/ml sludge per day. Propionate accumulated in this
digester parallel to the total propionate added for 40 days
(two hydraulic detention times). Propionate concentrations
then decreased until they reached about 30% of the total
added propionate at day 100. Biogas production and methane
production was only slightly higher than control digesters.

INTRODUCTION

The microbiological formation of methane from agricultural, forestry, and domestic wastes provides two useful services. Treatment and removal of these wastes are the first of these services while the second is the production of a potentially valuable fuel. Agriculture and forestry wastes in Florida exceed 105 million metric tons per year (8). This waste material, in addition to waste material from other agricultural states, represents a substantial amount of organic matter which could be biologically recycled. The anaerobic microbiological conversion of organic matter results in the end product methane which is easily recovered. Methane is a valuable fuel source because it retains about 90% of the energy found initially in the organic matter (5).

The conversion of organic matter to methane is a microbiological process that involves at least three distinct groups of bacteria (2,5,7). The metabolic scheme of these three groups is shown in Fig. 1. Fermentative bacteria hydrolyze insoluble polymers, such as cellulose, to soluble carbohydrates which are then fermented to produce acetate, propionate, butyrate, hydrogen, and carbon dioxide. The intermediate group, the hydrogenic bacteria, utilize propionate and butyrate to produce acetate, hydrogen, and carbon dioxide (1,6). The methanogenic bacteria, the terminal group, produce methane from acetate and hydrogen/carbon dioxide.

The metabolism of propionate and butyrate may be the rate-restricting step in the methanogenic process (3,4). The successful conversion of organic matter from agricultural, forestry or domestic wastes to methane will involve organic loading rates higher than in anaerobic digestion systems such as sewage treatment facilities. Higher organic loading rates will increase the levels of the intermediate fermentation products propionate and butyrate. This study was undertaken to determine the levels of propionate and butyrate that could be infused into thermophilic digesters while maintaining a stable conversion of organic matter to methane.

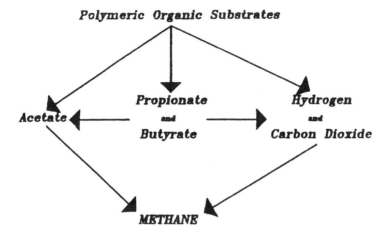

Figure 1.   Metabolic scheme for the anaerobic utilization
            of polymers such as cellulose. Fermentative
            bacteria hydrolyze the polymers, obligate hydro-
            genic bacteria utilize propionate and butyrate,
            and methanogenic bacteria produce methane from
            acetate and hydrogen-carbon dioxide.

MATERIALS AND METHODS

Digesters

    Digesters were constructed by wiring black rubber
stoppers into 4 liter aspirator bottles. Access ports were
drilled through the stoppers for introduction of daily feed,
exit of biogas, and passage of a stirring rod. The daily
feed consisted of Bermuda grass-cattle feed (Seminole Brands)
in a 3 to 1 mixture. The organic loading rate was 3.6 g
volatile solids per liter. The volume in the digesters was
3.6 liters. The retention time was 20 days. Temperature
in the digesters was maintained constant by placing the
digesters in water baths held at 55°C. Biogas was collected and measured by displacement of water in calibrated
glass vessels.

Infusion of Volatile Fatty Acids

    Concentrated stock solutions of sodium n-butyrate and
sodium propionate (Pfaltz and Bauer, Inc., Stamford, CT)
were prepared and frozen until used. These solutions were
continuously infused into separate digesters by syringe

pumps (Harvard Apparatus). Fresh solutions were placed
daily into 30-ml plastic syringes which were connected via
1mm (I.D.) Tygon tubing to 7.6cm, 18 ga. needles that pierced
the black rubber stoppers. Control digesters received only
the daily feed; whereas, the amended digesters received
daily feed plus the infusion of the volatile fatty acid.

Gas Chromatography

A Hewlett-Packard model 5880A gas chromatograph was
used. Gases were separated in a 1.8-m by 1.0-mm stainless
steel column packed with Carbosphere mesh 80/100 (Alltech
Associates, Inc., Deerfield, IL) and were measured with a
thermal conductivity detector. Helium was the carrier gas.
Column and detector temperatures were maintained at 130
and 145°C, respectively. Methane concentrations were determined by comparison to standards (Ultra-High-Purity Methane,
Matheson, Morrow, GA).

Volatile fatty acids (VFAs) were separated in a 1.8-m
by 1.0-mm glass column packed with 8% SP-1000, 2% SP-1200,
and 1.5% phosphoric acid on 80/100 mesh chromosorb W AW 8100
(Supelco, Bellefonte, PA), and were measured with a flame
ionization detector. Helium was the carrier gas. Injector,
oven, and detector temperatures were 145, 130, and 175°C,
respectively. Samples were mixed with an equal volume of
4% o-phosphoric acid, centrifuged at 12,800 x g for 2 minutes (22°C) and the supernatant was frozen until VFA determinations were made. Each VFA concentration was determined
by comparison to standards.

RESULTS

Characterization of Control Digesters

Volatile fatty acids (VFAs) were found in low concentrations in a thermophilic digester receiving only the daily
feed. When measured just prior to daily feeding, the concentrations, in μmol/ml sludge, were: acetate, 3 to 4;
propionate, 1 to 3; and butyrate, 0.1. Biogas (methane and
carbon dioxide) production was ca. 6 liters per 24h of which
ca. 3.4 liters was methane.

Infusion of n-butyrate into a Thermophilic Digester

A stock solution of sodium n-butyrate was prepared so
that it could be pumped into the digester at the rate of 10
μmol/ml of sludge per day. Fig. 2 shows the theoretical

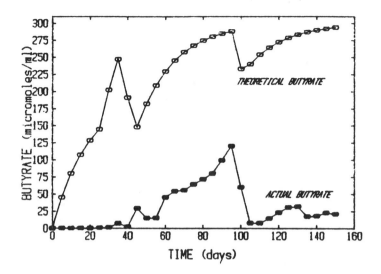

Figure 2.   The theoretical and actual concentrations of
butyrate in a thermophilic digester amended with
sodium n-butyrate. The rates of addition, in
μmol/ml sludge per day, were as follows: days
0 to 25, 10; days 25 to 35, 20; days 35 to 45, 0;
days 45 to 150, 15. The digester was moved on
day 55 and pump failure occurred on day 95.

concentrations of butyrate if not utilized by the digester
and the actual concentrations measured. Butyrate did not
accumulate in the digester. The concentrations of acetate
and propionate are shown in Fig. 3. Acetate concentration
increased from 3 to 35 μmol/ml of sludge of day 23. Propionate concentration increased to 3.3 μmol/ml of sludge
by day 23. The amount of biogas produced by the butyrate-
amended digester was ca. 150% of the control digester (Fig.
4). The percentage of methane in the gas phase increased
from 58% to 63%. The pH increased from 7.3 to 7.8, where
it remained stable (Table 1).

The butyrate amended digester appeared stable; therefore, the concentration of sodium n-butyrate infused into
the digester was doubled to 20 μmol/ml sludge per day. The
levels of all VFAs began to increase and by day 36 the concentrations, in μmol/ml, were: acetate, 167; propionate, 16.9 ·
16.9; and butyrate, 8.9. During the initial increase of

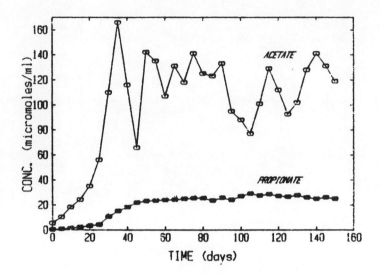

Figure 3.   The concentrations of acetate and propionate in a thermophilic digester receiving sodium n-butyrate. The rates of addition are given in Fig. 2.

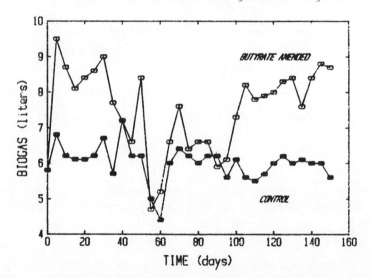

Figure 4.   Comparison of total biogas (methane and carbon dioxide) produced by a thermophilic digester amended with sodium n-butyrate versus one that was not amended. The rates of addition are given in Fig. 2.

Table 1. *The average methane concentration and pH in thermo-*
          philic bench-top digesters amended with sodium
          *propionate and sodium n-butyrate.**

	Percent Methane	pH
Control	56 ± 1.0**	7.3 ± 0.13
Propionate amended	60 ± 1.7	7.7 ± 0.16
Butyrate amended	62 ± 0.6	7.8 ± 0.14

*Rates of amendment: Sodium propionate, 10 μmol/ml sludge
 per day; Sodium n-butyrate, 15 μmol/ml per day.*
**Mean ± Standard deviation.*

VFAs, the biogas production remained near 150% of the control
digester but dropped to 135% of the control digester after
about 10 days. Because of the rapid increase in VFAs and
decrease in biogas production, the addition of butyrate was
stopped for several days.

Addition of sodium n-butyrate was restarted at 15 μmol/ml
of sludge per day on day 45. The digesters, however, were
moved on day 55 and it took 2 retention times (day 55 to
day 95) for the butyrate-amended digester to completely
recover. During this period the production of biogas
averaged 107%; slightly above the control digester. After
day 95 the concentration of butyrate in the digester decreased while the production of biogas averaged 138% of the
control digester. The addition of butyrate was continued
for 3 retention times. The parameters measured during this
period are given in Table 2.

Infusion of Propionate into a Thermophilic Digester

A stock solution of sodium propionate was prepared so
that it could be pumped into the digester at the rate of
10 μmol/ml sludge per day. Fig. 5 shows the theoretical
concentrations of propionate if not utilized by the digester
and the actual concentrations measured. Propionate did not
accumulate for the first 15 days. On day 20 the concentration increased from 3.46 to 49.7 μmol/ml and continued to
increase in parallel with the propionate added until day 50

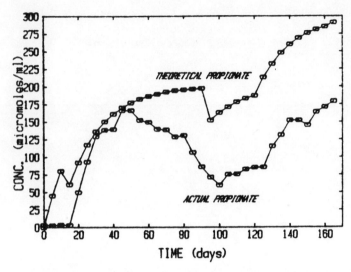

Figure 5.   The theoretical and actual concentrations of propionate in a thermophilic digester amended with
sodium propionate. The rates of addition, in
µmol/ml sludge per day, were as follows: days 0
to 120, 10; days 125 to 165, 15.   The digester
was moved on day 15 and pump failure occurred
on day 95.

Table 2.   *Parameters measured in a thermophilic digester continuously infused with sodium n-butyrate.   The rate
of addition was 15 µmol/ml sludge per day for 3
retention times, days 150 to 210.*

Parameter Measured	Concentration (µmol/ml
Acetate	121 ± 38 (113 to 156*
Propionate	25.2 ± 2.1 (21.5 to 28.3)
n-Butyrate	
Actual	22.7 ± 9.4 (9.3 to 33.3)
Theoretical	300
Biogas**	136 ± 9.9 (117 to 151)

*Mean ± Standard deviation (range).
**Percent of control.

when it decreased until day 125. The concentrations of acetate and butyrate are shown in Fig. 6. Acetate increased from day 0 to day 15, then decreased to the initial concentration where it remained stable until day 50. It then increased to a maximum of 35.8 μmol/ml on day 90 when the concentration decreased to 18.3 μmol/ml on day 125.

Biogas production is shown in Fig. 7. Initially the volume of biogas in the propionate amended digester was greater than in the control digester. Between day 15 and 50 the digesters produced similar volumes of biogas. From day 50 to day 125 volume of biogas produced by the propionate amended digester averaged 116% of the control digester.

Between day 15 and day 50 the digester was unstable because of relocation on day 15. The digester required about 2 retention times to restabilize. From day 50 to day 120 the digester appeared to be stable with an infusion of propionate at the rate of 10 μmol/ml sludge per day.

On day 125, the rate of infusion was increased to 15 μmol/ml of sludge per day. Concentrations of propionate (Fig. 5) and acetate (Fig. 6) increased. Biogas production averaged 110% of the control digester (Fig. 7).

DISCUSSION

The microbiological formation of methane from organic wastes removes these wastes and produces the fuel product methane. The rate-restricting step in the methanogenic fermentation of organic matter may be the conversion of propionate and butyrate (3,4). This study shows the potential of a thermophilic digester receiving daily organic input to maintain a stable digestion when infused with propionate and butyrate at concentrations greater than found in a control digester.

When infused with sodium n-butyrate, in addition to receiving the daily feed, a thermophilic digester exhibited increased production of biogas, an elevated percentage of methane, an increased concentration of acetate, and no accumulation of butyrate. The rates of infusion were 10 and 15 μmol/ml of sludge per day. These results show that the thermophilic digester has the capacity for metabolizing concentrations of butyrate much higher than found in a control digester. The additional butyrate contributes to an increased production of methane.

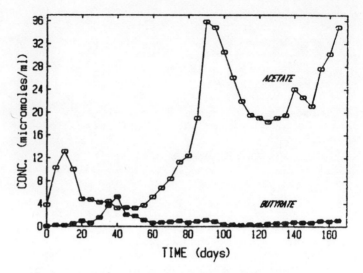

Figure 6.   The concentrations of acetate and butyrate in a thermophilic digester receiving sodium propionate. The rates of addition are given in Fig. 5.

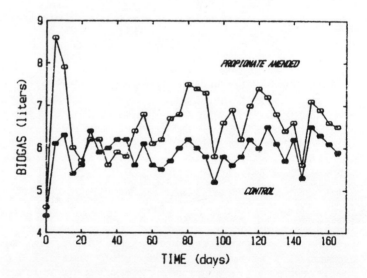

Figure 7.   Comparison of total biogas produced by a thermophilic digester amended with sodium propionate versus one that was not amended. The rates of addition are given in Fig. 5.

The infusion of sodium propionate into a thermophilic digester did not destabilize the operation of the digester receiving of the daily feed. The rates of infusion were 10 and 15 μmol/ml of sludge per day. However, the addition of propionate did not seem to stimulate production of methane to the same level that the addition of butyrate did.

Both digesters were adversely affected by being relocated. This result may show the sensitivity of digestion of organic matter at thermophilic temperatures. After relocation, the digesters required a period of two retention times in order to exceed metabolic activities of the control digesters receiving only the daily feed.

CONCLUSIONS

Two thermophilic digesters were infused with sodium propionate and sodium n-butyrate in addition to receiving daily feed. The addition of each of the organic acids and at rates of 10 and 15 μmol/ml of sludge per day did not result in digester failure. Therefore, it appears that increasing organic loading of thermophilic digestion systems with a concomitant increase in the organic acids propionate and butyrate, may result in a stable digestion process.

ACKNOWLEDGMENTS

This paper reports results from a project that contributes to a cooperative program between the Institute of Food and Agricultural Sciences of the University of Florida and the Gas Research Institute, entitled "Methane from Biomass and Waste."

The authors thank E.D.H. for excellent typing of the manuscript.

REFERENCES

1. Boone, D.R. and M.P. Bryant, "Propionate-degrading Bacterium, *Syntrophobacter wolinii* sp. nov. gen. nov., From Methanogenic Ecosystems," *Appl. Environ. Microbiol.*, *40*, 626-632 (1980).
2. Bryant, M.P., "Microbial Methane Production-Theoretical Aspects," *J. Anim. Sci.*, *48*, 193-201 (1979).

3. Mackie, R.I. and M.P. Bryant, "Metabolic Activity of Fatty-acid Oxiding Bacteria and the Contribution of Acetate, Propionate, Butyrate, and CO_2 to Methanogenesis in Cattle Waste at 40 and 60°C," *Appl. Environ. Microbiol.*, *41*, 1363-1373 (1981).
4. McCarty, P.L., "Energetics and Kinetics of Anaerobic Treatment," *Anaerobic Biological Treatment Processes, Advances in Chemistry Series 105*, (R.F. Gould, ed.), pp. 91-107, American Chemical Society, Washington, DC (1971).
5. McInerney, M.J. and M.P. Bryant, "Basic Principles of Bioconversions in Anaerobic Digestion and Methanogenesis," *Biomass Conversion Processes for Energy and Fuels*, (S.S. Sofer and O.R. Zaborsky, eds.), pp. 277-296, Plenum Publishing Corp. (1981).
6. McInerney, M.J., M.P. Bryant, R.B. Hespell, and J.W. Costerton, "*Syntrophomonas wolfei* gen. nov. sp. nov., An Anaerobic, Syntrophic, Fatty Acid-oxidizing Bacterium," *Appl. Environ. Microbiol.*, *41*, 1029-1039 (1981).
7. Smith, P.H., "Studies of Methanogenic Bacteria in Sludge," EPA-600/2-80-093, p. 112, Government Printing Office, Washington, DC (1980).
8. Smith, W.H. and M.L. Dowd, "Biomass Production in Florida," *J. Forestry, 79*, 508-511 (1981).

18
COMBINED METABOLISM OF BUTYRIC ACID AND WASTE ACTIVATED SLUDGE AT THERMOPHILIC TEMPERATURES

Ross E. McKinney and Reza Shamskorzani

Department of Chemical Engineering
University of Kansas
Lawrence, Kansas

ABSTRACT

A laboratory study was carried out using a 1.5 liter, fixed media anaerobic reactor with intermittent mixing and once daily feeding of Lawrence, Kansas, WWTP waste activated sludge and butyric acid. The research was divided into three phases. Phase 1 was a study of decreasing the hydraulic retention time 10 days to 2 days while keeping the temperature constant at 50°C and the butyric acid COD constant at 3 g/d. Once good operations were established at 2 days retention, the butyric acid COD was increased to 18 g/d at a constant waste activated sludge loading. The digester became saturated with solids at 6 g/d COD and remained saturated for the remainder of the study. The system became overloaded when the total BCOD fed rose from 13.7 g/d to 16.7 g/d. The third phase consisted of feeding waste activated sludge only and increasing the temperature from 50°C to 70°C. Good operations occurred at 50°C, 55°C and 60°C. The methane bacteria began to be adversely affected as the temperature increased. These results indicated that the biodegradable fraction of waste activated sludge was 92% metabolized with 3 days hydraulic retention and 11 days solids retention time at 50°C. The fixed media system gave good anaerobic digestion of both the waste activated sludge and the soluble butyric acid even when the system was saturated with solids.

INTRODUCTION

Anaerobic digestion of waste activated sludge has been a major problem in municipal wastewater treatment plants utilizing the activated sludge process. It was found by experience that waste activated sludge did not thicken as well as primary sludge and the waste activated sludge was not as digestible, based on volatile suspended solids (VSS) reductions. The net result has been the design of large anaerobic digestors with long retention times when waste activated sludge was included with the primary sludge. The large volumes of waste activated being produced has raised a question about the future of the activated sludge process. There is no doubt that a more rapid anaerobic digestion process is needed if waste activated sludge is to be successfully processed anaerobically. Fair and Moore (1937) studied the batch digestion of sewage sludge at various temperatures from 15°C to 60°C and found that optimum digestion occurred near 55°C. Unfortunately, the problems in heating and mixing anaerobic digestors at this time were sufficiently large to discourage consideration of thermophilic anaerobic digestion.

In Los Angeles, Garber (1954) examined anaerobic digestion of primary and waste activated sludges at the Hyperion WWTP at 30°C, 38°C and 49°C and found no differences in VSS destruction; but there was a difference in the filtering characteristics of the sludge. The thermophilic sludge showed better filtering characteristics than the mesophilic sludges. Later, Garber (1975) took another look at thermophilic digestion at Hyperion to confirm the better sludge dewatering characteristics. He found that energy costs were greater for thermophilic digestion and dewatering costs were less. Los Angeles did not adopt thermophilic digestion; but Popova and Bolotina (1964) reported that Moscow adopted thermophilic digestion of primary and waste activated sludges in 1958 at 51°C. They were able to reduce the retention period in the digestors from 18 days to 9 days with only a slight decrease in gas production. In spite of this application, few other plants have employed thermophilic digestion of sewage sludges on a large scale.

Thermophilic digestion has had a little more interest among researchers. Golueke (1958), Malina (1962, 1964), Pohland and Bloodgood (1963), and Maly and Fadrus (1971) all examined thermophilic digestion and found both positive and

negative results with sewage sludges. Other studies have
focused on synthetic wastes or industrial wastes, providing
little insight into the treatment of municipal sludges, especially, waste activated sludge.

In an effort to produce a better understanding of both the
anaerobic digestion of waste activated sludge and thermophilic
temperatures, this research project was established and carried out over a two year period. Waste activated sludge was
collected from the Lawrence, Kansas, WWTP to provide a field
activated sludge from a municipal plant; but rather than using
primary sludge which would introduce another major variable,
it was decided to use butyric acid which gave a better organic
substrate for evaluation of the data. In this way, the waste
activated sludge was digested under similar conditions as
normal municipal sludges to determine if the other organic
matter made a difference in the total digestibility of the
waste activated sludge.

MATERIALS AND METHODS

The anaerobic digestor was constructed from a wide mouth,
two liter, glass bottle with an initial liquid capacity of
1.5 liters. The digestor was filled with short lengths, 1/2",
of a small diameter, 1/2", aluminum tubing in a random fashion.
The addition of the aluminum media displaced some liquid; but
the 1500 ml liquid volume was retained within the two liter
bottle. As shown in Figure 1, there were three glass tubes in
the digestor. One tube was for introducing the feed. A small
funnel on the end of the feed tube aided in introducing the
substrate. The opposite end of the feeding tube discharged
near the bottom of the digestor. The second tube was the discharge tube and it was set so that the inside end was just below half submergence, permitting withdrawal of half the liquid.
The other end had a rubber tubing connected to the glass tube
to form a siphon to assist in effluent withdrawal. The third
tube was at the top of the digestor and was connected to a
three liter plastic bottle which was in series with a second
three liter plastic bottle. The two plastic bottles were gas
collection bottles with the acid, salt solution from the first
bottle being displaced into the second bottle which was open
to the atmosphere. A siphon connected the two plastic bottles
so that the production of gas in the digestor resulted in displacement of fluid from the first bottle to the second bottle.
By adjusting the liquid levels in the plastic bottles to the
same height when reading the gas volumes, the gas pressure was
essentially at atmospheric pressure. The plastic bottles were
calibrated to provide gas volume measurements. Two glass tees

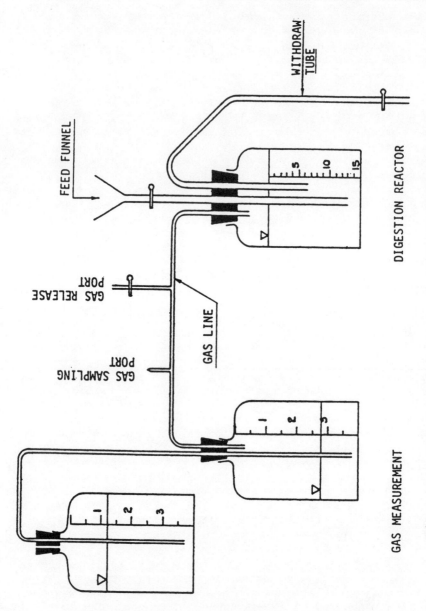

Figure 1.   Schematic diagram of anaerobic digestion system.

were located in the line between the digestor and the first
plastic bottle. One glass tee had a syringe cap to allow gas
collection for analysis. The second glass tee allowed the gas
to be wasted to the atmosphere and the gas measuring liquid
level returned to zero each day. The digestor was kept in a
thermostatically controlled water bath while the gas collection bottles were allowed to remain at room temperature. The
gas volume temperatures were measured daily for correction to
standard conditions along with pressure measurements.

The digestor was fed once daily on a batch feed basis.
The gas volume readings were collected and then the second
reservoir bottle was raised to create a positive pressure head
on the digestor. The effluent was forced from the center of
the digestor by the pressure and the siphon action. The fresh
substrate was added to return the liquid volume back to the
desired level and the excess gas was wasted to the atmosphere.
The system was shaken and the reaction measurements restarted.
The anaerobic digestor was shaken by hand 4 or 5 times each
day and allowed to remain quiescent overnight prior to effluent drawoff. The solids accumulated in the media and were
removed only when the system began to accumulate excess solids.
It was noted that while solids accumulated around the media,
they did not become attached to the media. The media simply
allowed the solids to accumulate to their maximum concentration before being removed.

The basic substrate was waste activated sludge concentrated to approximately 1.0% with a definite COD as a mixture
of butyric acid and sodium butyrate. The combination of
butyric acid and sodium butyrate varied during the study in an
effort to keep the pH and the alkalinity at reasonable levels.
The pH was maintained between 7.2 and 7.4 throughout most of
the study and the alkalinity was kept between 2000 and 4000
mg/L as $CaCO_3$.

RESULTS

The experimental studies were divided into three phases.
The first phase was concerned with reducing the liquid retention time from 10 days to 2 days while maintaining a constant butyrate COD at 3 g/d and 50°C. Reducing the hydraulic
retention time resulted in increasing the waste activated
sludge VSS from 1.0 g/d at 10 days to 3.6 g/d at 2 days. The
system shifted easily from the 10 day retention time to the
2 day retention time. Solids accumulated in the fixed media
and increasing amounts of volatile solids were lost in the
effluent as the hydraulic retention time decreased.

Unfortunately, there was no way to accurately measure the accumulated solids in the digestor. The total gas production rose with increasing solids loading. Correcting the total gas production for the butyrate metabolized gave a total gas production of approximately 215 ml/d (STP) per g VS added. With 80% methane in the gas, 0.49 g COD was converted to gas, 35% destruction. The effluent volatile acids dropped from 440 mg/L at 10 days retention to 140 mg/L at 3 days retention. The gas production data and the volatile solids data are shown in Figure 2. It was noted that the carbon dioxide fraction of the gas varied as the ratio of butyric acid:sodium butyrate varied.

The second phase was concerned with increasing the butyrate COD while keeping the hydraulic retention time at 2 days. The butyrate COD was increased from 3 g/d to 18 g/d in increments while the temperature was kept constant at 50°C, as in the first phase. The gas production rose with the increased COD load as shown in Figure 3. The increased COD load not only produced more gas but also produced more volatile solids in the effluent. In fact, the effluent volatile solids began to approach the influent volatile solids. The volatile acids began to show a change between 12 g/d and 15 g/d COD loading, rising from 290 mg/L to 580 mg/L and reaching 1930 mg/L at 18 g/L. The volatile solids loading rate averaged about 3.4 g/d waste activated sludge. The total gas production from the volatile solids averaged slightly over 300 ml/g until the 12 g/d COD when it dropped to 165 ml/g based on volatile solids added. The 15 g/d COD showed 76 ml/g while the 18 g/d COD showed incomplete metabolism of the butyrate and essentially no degradation of the waste activated sludge. The 18 g/d COD operation was much too short to reach equilibrium.

The third phase was carried out using only waste activated sludge. This phase examined the effect of increasing temperature on the digestion process. The temperature of the digestor was increased in increments from 50°C to 70°C while the hydraulic retention time was kept at 3 days. A change in waste activated sludge characteristics resulted in greater volatile solids in the waste activated sludge, ranging from 4.6 g/d to 6.8 g/d overall. The gas production data shown in Figure 4 indicated that the digestion of waste activated sludge slowed between 60°C and 65°C. The volatile acid data, also plotted in Figure 4, began to increase slightly between 55°C and 60°C with a sharp increase after 60°C. When it became apparent that the system was in trouble at 70°C, the temperature was dropped to 65°C for two weeks and then dropped to 60°C. The system slowly recovered to normal gas production

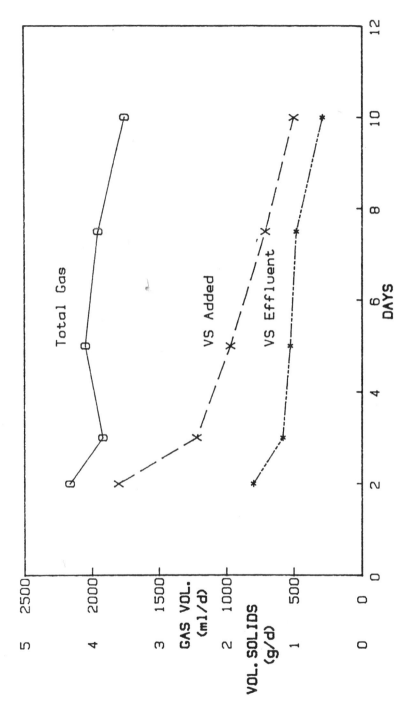

Figure 2. Gas production and volatile solids data for varying hydraulic retention times.

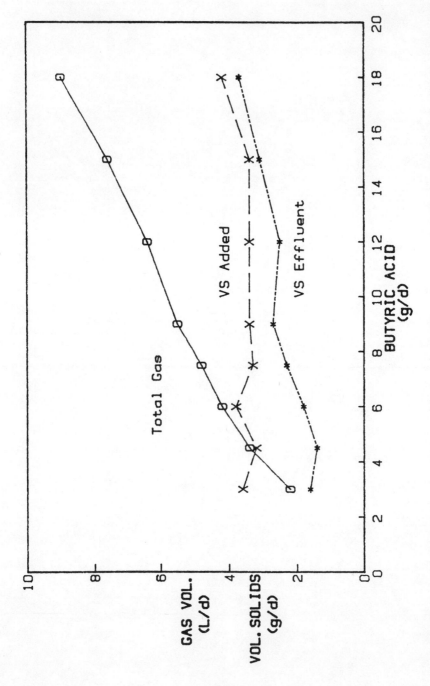

Figure 3. Gas production and volatile solids data for varying butyric acid loading rates.

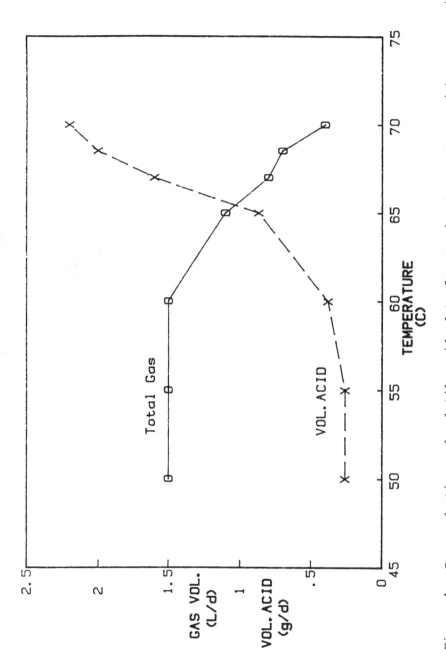

Figure 4.   Gas production and volatile acids data for varying temperatures with waste activated sludge feed.

and then the feed was stopped and the system allowed to undergo endogenous respiration for three weeks. Figure 5 shows the
gas production over the endogenous respiration period at the
end of the study. The final analysis of total solids indicated that the system contained 30.0 g/L with 19.5 g/L volatile
solids. Based on the rate of endogenous respiration the volatile solids destruction was only 0.05 g/d at the end of the
study, indicating that the majority of the solids were inert.

DISCUSSION

The keys to good anaerobic digestion are the retention of
the microbes in the reactor and their ability to obtain substrate for metabolism. The use of aluminum media assisted in
retention of microbial solids by providing surfaces for attachment and quiescent settling zones around the media. The
microbial attachment was very loose with most of the solids
easily shaken off by regular mixing. The microbes tended to
flocculate and settle rapidly once mixing was stopped. It
appeared that microbial flocculation followed a similar pattern as with activated sludge where flocculation occurs with
mixing once the organic substrate has been metabolized. The
long solids retention time helped retain the active microbial
mass with the inert solids. The inert solids in the waste
activated sludge fed to the digestor also provided settleable
mass that assisted in the retention of the active microbial
mass.

The metabolism by the microbes results in synthesis of
new cell mass and endogenous respiration of old cell mass.
The use of fixed media and waste activated sludge prevented
evaluation of the solids data directly. Data by Speece and
McCarty in 1964 indicated a yield factor (Y) of 0.054 based on
COD and an endogenous factor (Kd) of 0.038/d at 35°C for the
metabolism of butyric acid. The yield factor should not be
affected by temperature; while the endogenous respiration
factor should be greater at the warmer temperature. One would
expect that Kd might increase to 0.11/d at 50°C and to 0.22/d
at 60°C. Thus, in Phase 1, the metabolism of butyrate would
be expected to contribute only 77 mg/d volatile solids with a
10 day solids retention time. Since the solids retention time
was longer than the hydraulic retention time, the volatile
solids resulting from butyrate approached the expected minimum
of 32 mg/d. The waste activated sludge fed to the digestor
had 650 mg/d inert volatile solids at the 10 day retention
period and 2350 mg/d at 2 days retention. Needless to say,

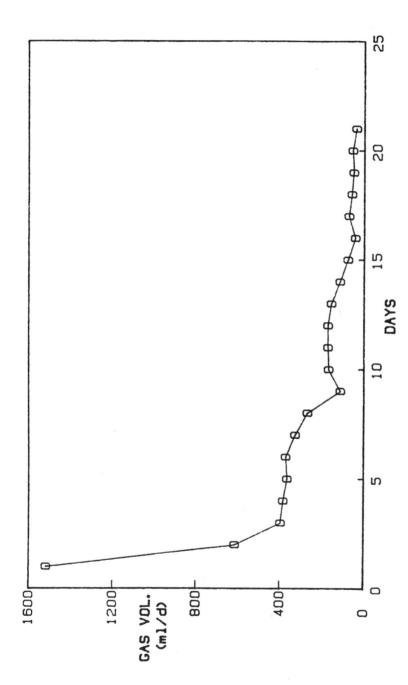

Figure 5.   Daily gas production data for endogenous respiration in the anaerobic digestor at 60°C.

the accumulation of inert volatile solids from the waste activated sludge controlled the accumulation of solids in the
digestor and their ultimate discharge in the effluent.   It
appeared that some solids were still accumulating in the digestor at the end of Phase 1.

Operations during Phase 2 resulted in the rapid accumulation of inert volatile solids in the digestor from the waste
activated sludge and in additional solids from the increased
butyrate metabolism. By the end of Phase 2 it appeared that
the digestor had reached its maximum capacity and was dis-
charging all excess solids, including unmetabolized solids, in
the effluent. The volatile acids in the effluent were the
keys to when the system became overloaded. When the total
BCOD rose from 13.7 g/d to 16.7 g/d, the digestor effluent
began to show an increase in volatile acids, indicating leakage of biodegradable organics. It appeared that the system
became saturated with solids during the 6 g/d butyrate feeding.
The effluent solids were equal to the influent solids less the
biodegradable fraction. The butyrate produced solids were
relatively small compared to the waste activated sludge volatile solids, contributing little to the total solids removed
each day.

The addition of waste activated sludge without any butyrate confirmed that the system was saturated with solids.
Normal metabolism appeared to occur with solids additions at
50°C, 55°C and 60°C. Gas production and volatile acids showed
that metabolism was slowing down at 65°C. Raising the temperature confirmed that problems were occurring with the methane
bacteria. The volatile solids showed an average reduction of
36% during Phase 3; while the gas data indicated 31% reduction. It appeared that the hydrolytic bacteria were not affected as much by the increased temperature as the methane
bacteria. The pH data confirmed this reaction, dropping from
7.1 at 60°C to 6.7 at 70°C.

The temperature was dropped back to 60°C and the system
was returned to normal operations before stopping operations
and allowing the system to undergo endogenous respiration.
The measured rate of endogenous respiration based on gas production averaged 0.15/d at 60°C in contrast to the 0.22/d expected from Speece and McCarty's data (10). Since this data were
obtained from dieoff data in contrast to the routine operational data of Speece and McCarty, the endogenous rate during
normal operations was probably close to that indicated by
Speece and McCarty as corrected to 60°C.

The anaerobic digestion of waste activated sludge is limited to the active microbial fraction which is related to the solids retention of the activated sludge in the aeration system prior to wasting. As the solids retention period increases, less active mass remains for digestion as it has already undergone aerobic digestion in the aeration tank. Since only about 80% of the active microbial mass in the waste activated sludge is biodegradable, one normally expects to find only 30-40% of the volatile solids in the waste activated sludge being metabolized. The 35% volatile solids destruction in Phase 1 and the 36% in Phase 3 appear to be typical for municipal activated sludge. Another study on mesophilic digestion of Lawrence waste activated sludge at 37°C by McKinney and van Rensburg (7) indicated 32% volatile solids destruction with 15 days total retention and 25% with 6 days retention.

All of the data confirmed that solids retention was the key to good destruction of waste activated sludge. At the end of Phase 3 the system contained 37 g volatile solids prior to the endogenous respiration test. It appeared that the solids were retained in the digestor for 10.8 days with a 3 day hydraulic retention at saturation. At 50°C it appeared that 92% of biodegradable volatile solids were digested.

CONCLUSIONS

Based on the results obtained in this study, the following conclusions were obtained.

1. Waste activated sludge can be digested anaerobically at 50°C with a short retention time, 2 days, provided the solids retention is adequate, 10 days or more.

2. Approximately, 92% of the biodegradable fraction of waste activated sludge can be converted to methane gas with a fixed media, anaerobic digestion system.

3. Fixed media allows the solids to accumulate in the anaerobic digestion system so that a short hydraulic retention time can be combined with a long solids retention time.

4. Inert activated sludge solids will accumulated in the fixed media to saturation and will reach equilibrium with good metabolism depending upon contact with the microbes.

5. Thermophilic digestion of waste activated sludge was best between 50°C and 60°C.

6. Soluble organics added together with the waste activated sludge gave good digestion up to 10 g/L-d BCOD at solids saturation in the fixed media digestion system.

REFERENCES

1. Fair, G.M. and E.W. Moore, *Sew. Works J.*, *9*, 3-5 (1937).
2. Garber, W.F., *Sew. & Ind. Wastes*, *26*, 1202-1216 (1954).
3. Golueke, C.G., *Sew. Works J.*, *30*, 1225-1232 (1958).
4. Malina, J.F., *Proc. 16th Ind. Waste Conf.*, Purdue University, 232-250 (1962).
5. Malina, J.F., *Wat. & Sew. Works*, *95*, 52 (1964).
6. Maly, J. and H. Fadus, *J. Wat. Poll. Cont. Fed.*, *43*, 641-650 (1971).
7. McKinney, R.E. and J.J. van Rensburg, *Trans. 34th Ann. San. Engr. Conf.*, University of Kansas, 1-9 (1984).
8. Pohland, F.G. and D.E. Bloodgood, *J. Wat. Poll. Cont. Fed.*, *35*, 11-42 (1963).
9. Popova, N.M. and O.T. Bolotina, *Adv. in Wat. Poll. Research*, *2*, 97-110 (1964).
10. Speece, R.E. and P.L. McCarty, *Adv. in Wat. Poll. Research*, *2*, 305-322 (1964).

19
ANAEROBIC DIGESTION OF HEMICELLULOSIC FEEDSTOCKS

C. Rivard, M. Himmel, and K. Grohmann

Biotechnology Branch
Solar Fuels Division
Solar Energy Research Institute
Golden, Colorado

ABSTRACT

Lignocellulosic biomass in its native form is highly resistant to microbial attack. Exceedingly long reaction times (10-90 days) are required for complete microbial conversion of even the more degradable feedstocks such as straw, corn stover and MSW to methane (Ramasamy and Verachtert, 1979). Controlled acid pretreatment of these feedstocks, besides enhancing the digestibility of the cellulosic residue, liberates soluble hemicellulosic polymers which are further degraded to free sugars by enzymatic hydrolysis within the anaerobic digestion system and ultimately converted to biogas. Pretreatment hydrolysate liquor expressed from straw (a model low value biomass feedstock) cooked at 120°C with dilute phosphoric acid was successfully used as feed in a mesophilic anaerobic digestion system. Stirred and packed bed anaerobic reactors were examined as systems for the conversion of biomass hydrolysate liquor to methane. Various column packing materials for packed bed reactors were also evaluated. After maturation of microbial attachment to the bed material, preliminary rates of 3.5 volumes of biogas (65% methane) per reactor volume per day were attained in the packed bed reactors. The effect on digestor performance of trace quantities of several compounds present in acid hydrolysates of biomass, such as furfural, 5-hydroxymethyl furfural, low molecular weight lignin and lignin precursors, was also investigated.

INTRODUCTION

Wheat straw is composed primarily of lignocellulosic material and has been previously shown (Clausen and Gaddy, 1983; Jewell et al., 1980; Wujcik and Jewel, 1980) to be difficult to digest anaerobically to methane with short reactor residence times. An increase in reaction rates and the extent to which lignocellulosic biomass is fermented to methane would make this process financially more attractive by permitting a reduction in reactor volume and thus capital cost. Several ways to increase the rate and extent of biomass digestibility include pretreatment through both chemical and physical methods. Chemical pretreatment with dilute mineral acids generates a solid lignocellulosic residue and hydrolysate liquor containing solubilized hemicellulosic sugars, soluble lignin components, acetic acid and other substances which can be utilized as substrate (Grohmann et al., 1984). However, several of these products may prove inhibitory to microorganisms in anaerobic digesters. Therefore, the use of straw pretreatment liquor as a feed for microbial methane generation under anaerobic conditions was studied.

Continuously stirred tank reactors (CSTR) and packed bed reactors were compared with respect to reduction of volatile solids and COD, rates of gas production and gas product conversion efficiency. Also, the effects of variation in dilution rate were examined with respect to digester performance. The effects of hydrolysis (pretreatment) byproducts such as furfural, 5-hydroxymethyl furfural, low molecular weight lignins, and lignin precursors on gas production performance by digester sludge was investigated. The possible effect of saccharinic acids, an alkaline sugar degradation product, was also studied.

The biomass particle size distribution for various milling methods was also examined since the sugars solubilize more rapidly for smaller particles, and weight loss during chemical pretreatment is faster. The results of acid pretreatment at 120°C with variation in phosphoric acid concentration and hydrolysis time were also examined.

Finally, extraction of the hydrolysate liquor from the lignocellulosic residue following pretreatment has been previously shown to be an energy intensive process if substantial liquor recovery is desired. Therefore, an analysis of the force required for expression of pretreatment liquor was undertaken.

MATERIALS AND METHODS

Straw Comminution

Wheat straw was obtained directly from farms on the eastern slope area around Denver, Colorado. The straw chosen was free from fungal rot and harvested in 1983. This straw was subjected to knife milling with a Wiley mill model No. 4 fitted with a 2 mm rejection screen. A Mitts and Merrill model 10/12 with a three knife open rotor and a 10 h.p. motor was fitted with either a 1/4" or a 1/8" round hole rejection screen for further studies. The milled straw was applied to a Tyler model RX-24 portable shaker. The sieves used were 7, 10, 12, 14, 16, 18, 20, 35, 60, 80, and 120 mesh (U.S. standard). Shaking duration was until the fractions showed little or no loss in weight (e.g., about 55 minutes).

Dewatering of Straw

Biomass expression studies were initiated by the design and construction of a piston-type expression chamber which is operated in a Bueler high pressure hydraulic press. Design of this device will be discussed in detail elsewhere. The milled straw was weighed after acid hydrolysis and a portion (10.0 grams) of the slurry (10-12% w/v) was loaded into the expression cell. The pressure was then increased and maintained at 100 psi increments as the liquor was fully expressed and collected. The pressed pellet was then removed for subsequent gravimetric analysis (wet and oven dried).

Acid Hydrolysis

Acid treatment of straw samples (from the Mitts & Merrill mill, without classification) was carried out in large scale using glass carboys (3 x 10L.) in a walk-in type autoclave programmed to maintain 121°C (Sybron-Castle model 3220). Acid samples were prepared with reagent grade phosphoric acid in v/v percentages from 0.1% to 1.0%.

Analysis of Straw Hydrolysate and Digester Feed

The molecular weight distribution of soluble polymers was determined by high performance size exclusion chromatography (HPSEC). The samples to be analyzed were prepared by neutralization with NaOH and followed by centrifugal clarification. This supernatant was then subjected to

batch ion exchange with an equal volume of Bio-Rex RG-501-X8
mixed bed resin. Finally, the samples were pelleted for
two minutes with an Eppendorf desk top centrifuge (model
5415) before injection. Chromatography was performed on a
Waters chromatographic system equipped with the following:
a model 6000A dual piston pump, a model 680 system con-
troller, a model 401 refractive index detector and a Wisp
autoinjector. Two Toyo Soda G2000 SW series columns were
used to obtain the molecular weight resolution of interest.
Specific problems and solutions to biomass component analysis
with HPSEC is discussed in detail by Himmel et al. (1983).

Volatile fatty acids and furfural were determined with
a Hewlett Packard model 5840A gas chromatograph equipped
with a flame ionization detector and a Carbopack C/ 0.3% CW
20M/ 0.1% H_3PO_4 packed column (6 ft. x 2 mm, Supelco). The
column injection port and detector temperatures were maintained at 125, 145, and 250°C, respectively. Nitrogen
served as the carrier gas. The column was calibrated with
C-2 to C-5 volatile fatty acids (0.1%, Supelco) and furfural (Aldrich).

Total solids were determined by evaporating 25 ml aliquots of SPF feed in preweighed crucibles in a 105°C oven to
constant weight. Volatile solids (VS) were determined from
total solids by ignition of the dried residue at 550°C in
a muffler furnace for 30 minutes. Chemical oxygen demand
(COD) was determined by the dichromate reflux method as
described by Greenberg et al. (1981a). In preparation for
the analysis of total carbohydrates, free ammonia, Kjeldahl
nitrogen, and phosphate, the samples were first clarified
by centrifugation for 30 min. at 7,000 rpm at 4°C using a
GSA rotor in a Sorvall RC-5B centrifuge. Total carbohydrates
were determined by the anthrone method as described by Ashwell
(1957). Free ammonia and Kjeldahl nitrogen were determined
by the phenate and Kjeldahl methods respectively as described by Greenberg et al. (1981b). Phosphate was determined by the vanadomolybdic acid method as described
by Greenberg et al. (1981c).

Feed Preparation

Feed for the packed bed reactors and fermenters was
produced by the addition of 10% solids (Mitts & Merrill
mill, 1/8" screen product) to 0.2% phosphoric acid followed
by cooking this slurry in an autoclave at 121°C at 18 psi
for 3 hours. This procedure is presented in the process
summary, Figure 1. After autoclaving, the slurry was cooled

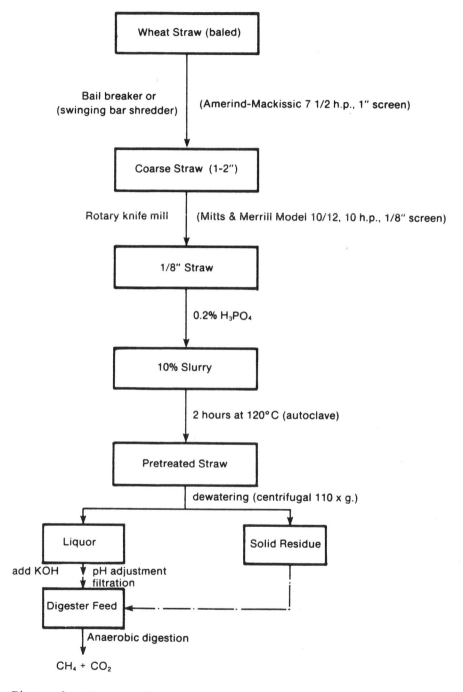

Figure 1. Process Summary.

in an ice bath and either filtered through cheese cloth or
centrifuged in a modified General Electric washing machine
lined with white linen cloth for 15 minutes. The straw
residue was discarded. The filtrate was adjusted to pH
9.2 and the precipitate which formed was removed by filtration with Whatman #1 filter paper. The feed was refriger-
ated at 4°C until use.

Enrichment Medium

Methanogenic enrichments were cultivated in 155 ml serum
bottles containing sterile prereduced growth medium with a
80-20% H_2-CO_2 gas phase. Medium was prepared by a modification of the Hungate (1950) technique as revised by Balch
and Wolfe (1976). The enrichment medium utilized was that
described by Balch et al. (1979), (medium #1) for the cultivation of fresh water methanogens. The medium contained
the following components in grams per liter deionized water:
yeast extract (2.0), trypticase (2.0), sodium acetate (2.5),
sodium formate (2.5), $NaHCO_3$ (5.0), $FeSO_4 \cdot 7H_2O$ (0.002),
L-cysteine HCl (0.5, trace mineral solution (10 ml), trace
vitamin solution (10 ml), mineral solution #2 (50 ml),
mineral solution #1 (50 ml). The pH was adjusted to 7.2 with
2*M* KOH prior to sterilization.

Fermenters and Packed Bed Reactors

Continuously stirred fermenters (CSTR's) were constructed from either 500 ml or 3 liter adjustable hanging
magnetic bar spinner flasks (Bellco). The flasks were
modified by the addition of a barbed nipple to the side
of the flask close to the bottom to allow fluid to be
drained from the system. The screw tops were attached with
a silicone adhesive to inhibit leakage. The side arms were
fitted with rubber stoppers. One side arm stopper was connected with 3/8" Tygon tubing to the water displacement
gas monitoring system.

Four packed bed reactors were constructed from standard
2" flanged glass process pipe, two 18 and two 12" x 2",
(Kimax) with custom machined teflon end caps compressed to
the glass pipe with O-ring (Viton[R]) seals. The Bed support
in one 18" and one 12" reactor consisted of foam-glass
(Corning Glass Co.) while ceramic insulator formed a packing for the other two reactors. Both packing types were
crushed to pieces ranging from 0.5 to 1.0" in diameter. The
effluent port was connected with tubing to a gas/liquid
separator which was made by modifying a 250 ml graduated

cylinder. The gas/liquid separator had three tubing con-
nection ports. The input port was located just above the
250 ml calibration mark, the liquid recirculation port was
at the bottom, and the gas venting port was at the top.
Temperature control for the entire system was accomplished
by operation at 37°C in a 12' x 12' controlled temperature
room (LabConco). The feed was continuously infused to both
the fermenter and packed bed reactors using a polystatic
four channel pump (Buchler). Connections were made with
intravenous-tubing on the packed bed reactors with injection
sites previous to the bed, after the bed and after the
gas/liquid separator.

Start-up of the Stirred and Packed Bed Fermenters

Initially, two 500 ml continuously stirred fermenters
were set up. The fermenters were flushed with oxygen free
N_2 gas and 100 ml of Balch et al. medium #1 was introduced,
followed by 100 ml of active culture. The fermenters were
fed daily with 25 ml of hydrolysate and allowed to accumulate to a liquid level of 500 ml. These two small fermenters
were subsequently used to seed both the 3 liter fermenter
and the packed bed reactors. The large fermenter was started
using effluent from the 500 ml fermenters and allowed to
reach a liquid level of 3 liters. The packed bed fermenters
were first flushed with oxygen-free N_2 gas and pumped with
prereduced water. Then 100 ml of fermenter liquid was injected into each column. This was followed by a continuous
feed infusion at a rate of 50 ml/day. A second peristaltic
pump was set up to recirculate the effluent collected in
the gas/liquid separators back into the columns, thus serv-
ing to recycle cells which had not attached previously.

Gas Analysis

Gas compositions were determined using a GOW-MAC model
550 thermal conductivity gas chromatography equipped with a
Porapack Q column (6ft. x 2 mm, Alltech Assoc.). The column
temperature was maintained at 30°C and nitrogen served as
the carrier gas. The gas composition was determined by
calibration with high purity methane (Matheson). Gas production in serum bottles was monitored using a pressure
transducer (Setra Systems, Inc.) to determine overpressure.
In packed bed reactors and continuously stirred fermenters,
gas production was monitored using calibrated 2 liter water
displacement reservoirs.

Microscopy

A Nikon Labophot microscope equipped for epifluorescence
was used for observation of wet mounts and fluorescence of
F-420 containing methanogenic bacteria (Cheeseman et al.,
1972). Light in the 420-nm wavelength range was provided by
a mercury light source (HBO 50w) and a Nikon V filter package.

INHIBITION EXPERIMENTS

Experiments conducted with sludge external to the CSTR
were performed in 150 ml serum bottles (Wheaton). Sludge
was removed from the effluent tubing of the stirred fermenter
in 50 ml aliquots using a 50 ml syringe and transferred to
serum bottles while flushing the bottles with an oxygen-
free N_2-CO_2 (80-20%) gas mixture. After flushing sufficiently
to remove oxygen, butyl rubber stoppers were inserted and
aluminum crimp caps attached. The serum bottles were incubated with shaking in a New Brunswick shaking incubator
(model G25) at 175 rpm and 37°C. One milliliter of SPF feed
was introduced to each serum bottle using a 1 ml syringe
with a 25 ga x 7/8" needle. The cultures were incubated
for 2 hours before the test compounds were added. Experiments on the effects of test compounds on sludge gas production were performed in duplicate. Furfural, 5-hydroxy-
methyl furfural, levulinic acid, acetovanillone, and p-
hydroxy cinnamic acid were obtained from Aldrich Chemical
Co. The sample of low molecular weight lignin was obtained
from aspen wood chips by ethanol extraction *in vacuo* and
were a gift from Helena Chum, Solar Energy Research Institute. Saccharinic acid was prepared from lactose by the
calcium hydroxide method as described by Whistler and Be-
Miller (1963).

Concentrated stock solutions or suspensions of test
compounds were made in deionized water and 0.5 ml was added
to each serum bottle to give a final concentration of 0.01,
0.05, 0.1, or 0.5% w/v. Controls received 0.5 ml deionized
water. After the test compounds were added, overpressure
within the serum bottles was exhausted to atmospheric
pressure using a 1 ml syringe with the plunger removed.
Gas production was measured as overpressure after various
time intervals using a Setra pressure transducer. Overpressure was released after every reading.

RESULTS

The first step in the two-stage size reduction of wheat straw employed an Amerind-Mackissic hammer shredder with a 1" rejection screen which produced pieces approximately 1-2" long and practically no fines. Figure 2 shows the product size characteristics from three milling configurations. The Mitts and Merrill mill with 1/4" screen produced the smallest quantity of fines (i.e., sizes finer than 80 mesh). Although the Wiley mill (Figure 2, C) produces the highest quantity of product in the 40 to 80 mesh range, the Mitts and Merrill 1/8" screen mill (Figure 2, B) was chosen for these studies because the product is intermediate in size to those tested and is also representative of a large scale mill product.

The relationship between phosphoric acid concentration and weight reduction at different hydrolysis times is shown in Figure 3. It is apparent that hydrolysis times from 2 to 5 hours produce a small difference in weight loss at acid concentrations of 0.2% and less. Also, an increase in phosphoric acid concentration from 0.2 to 0.5% results in at least a doubling of straw weight loss.

Figure 4 shows the molecular weight (MW) distribution of straw hydrolysis samples obtained from two hour pretreatments with 0.1 to 1.0% H_3PO_4 at 121°C. The large refractive index peak shown to elute at V_o (void volume), decreased almost entirely after deionization. This indicates that the salts or organic ions known to elute at this volume by ion exclusion mechanisms can be effectively removed with the highly crosslinked ion exchange resin. The distribution of soluble oligomers shown in Figure 4 represents the size distribution of the neutral oligosaccharides present in the hydrolysate. Figure 4 also shows the steady trend to lower average molecular weights in the intermediate oligomer fraction and the generation of monomeric sugars as the acid concentration is increased from 0.1% to 1.0% phosphoric acid.

Expression of hydrolysate liquor after pretreatment performed in a modified washing machine exerts approximately 15 psi on the slurry and results in a residue of 76% moisture (Table 1). Use of the hydraulic piston press indicates that greater pressures imparted to the sample result in a higher percentage of solids up to 42%, at which point greater pressures do not result in an increase in solids content.

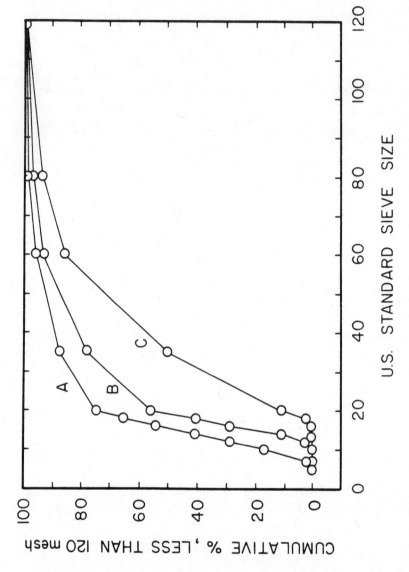

Figure 2.   Size distribution comparison of milled straw from (A) Wiley mill, (B) Mitts and Merrill mill (1/8" screen), and (C) Mitts and Merrill mill (1/4" screen).

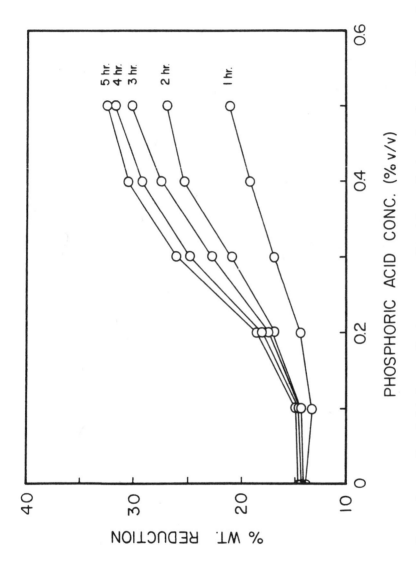

Figure 3. Comparison of weight loss of straw as a function of phosphoric acid concentration and hydrolysis time at 120°C.

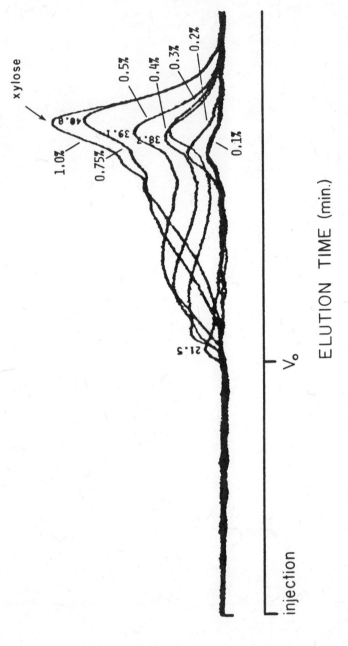

Figure 4.   High pressure size exclusion chromatography (HPSEC) of straw hydrolysis supernatants produced at various phosphoric acid concentrations after 2 hr hydrolysis at 120°C. The void volume (V_o) of the column system used was found from the elution of Dextran T10 (9700MW).

Table 1.
Dewatering Pressure Requirements of Pretreated Straw

	Final Required Pressure (PSIA)	% Solids	
		Start	Final
Hydraulic piston press[1]	20	12	17
	120	25	29
	220	32	34
	320	34	37
	620	37	42
	920	42	43
Perforated rotary drum[2]	15	10	24

[1] *Dewatering interval was 200 seconds (time required for reduction in expressate flow).*

[2] *Conditions were 110 x g for 15 minutes. $P = \rho\omega^2(r_2^2-r_1^2)/2g_cP_o$ (See McCabe and Smith, 1967).*

Analysis of the straw pretreatment filtrate (SPF) composition is shown in Table 2. Total sugars constitute slightly greater than 50% of the volatile solids present in SPF. The carbohydrate fraction was found to contain primarily xylose and xylodextrins by HPSEC (see Figure 4). The pretreatment by-product furfural was identified at a concentration of 0.01%. Although free ammonia was determined to be present in low levels, organic nitrogen resulted in a C:N ratio of 88.

Anaerobic cultures were collected from various sources to analyze the fermentability of the straw pretreatment filtrate. Numerous sediment samples from the Denver area fresh water ponds and streams, cow manure from a North Denver dairy farm, and sludge from the Denver municipal sewage treatment plant's anaerobic digester were analyzed for methane production from SPF in batch cultures in a defined, reduced medium. Cultures were analyzed over a three week period for gas production, methane composition, volatile fatty acid pools and populations of methanogenic bacteria. Sludge from the Denver municipal anaerobic digester was observed to have had the highest populations of methanogenic bacteria and exhibited the most rapid conversion of added

Table 2.
Analysis of Straw Pretreatment Filtrate (SPF) Feed

Total Solids	*2.08%*
Volatile solids	*14.5 g/l*
Total carbohydrates[1]	*7.95 g/l*
Volatile fatty acids (acetate)	*0.05%*
Furfural (2-furfuraldehyde)	*0.01%*
Phosphate	*0.22%*
Nitrogen:	
* Free ammonia*[2]	*0.00044 g/l*
* Total*[3]	*0.14 g/l*
COD	*12.3 g/l*
C:N Ratio	*88*

[1]*Data from anthrone assay.*

[2]*From phenate assay.*

[3]*From Kjeldahl assay.*

SPF to methane. Two 500 ml CSTR's were set up with continuous SPF feed introduction as previously described and
innoculated with SPF enrichment culture.

     The two small CSTR's were used to seed a 3 liter CSTR
and four packed bed reactors. After maturation of microbial
biomass on the packed bed media, effluent recycle was dis-
continued and gas production, gas production efficiency,
and % destruction of VS and COD of the SPF feed were assessed.
Effluents from the 3 liter CSTR and the 18" packed bed
reactors underwent a similar reduction in VS and COD (~80%),
while the 12" packed bed reactors lagged in volatile solid
reduction and to a greater extent in COD destruction
(Figure 5).

     Although the total gas production data shown in Figure
6 indicate the 3 liter CSTR (F) is on a similar slope with
the 18" packed bed reactors (B) and (D), increasing the
dilution rate to 0.4 results in CSTR digester failure.
Rates of total gas production in the 12" packed bed reactor
(A) and (C) lagged behind those of both the 18" packed bed
reactor and the CSTR.

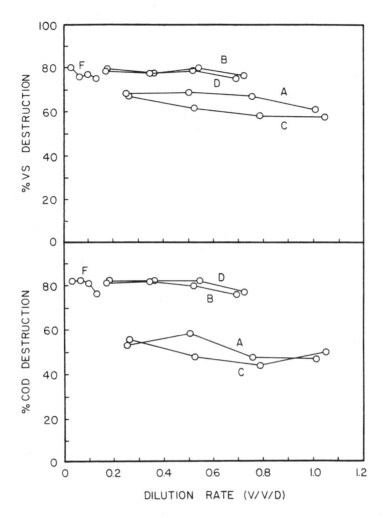

Figure 5.   Comparison of volatile solids (VS) and chemical
            oxygen demand (COD) destruction with variation
            in dilution rate. (A) Foam glass, 12 x 12",
            395 ml void volume. (B) Foam glass, 18 x 2",
            575 ml void volume. (C) Ceramic, 12 x 2", 380
            ml void volume. (D) Ceramic, 18 x 2", 550 ml
            void volume. (F) Continuously stirred tank
            reactor, 3,000 ml volume.

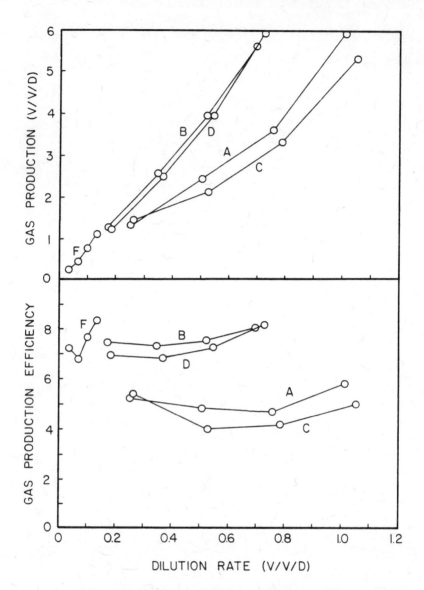

Figure 6.   Comparison of total gas production rates (volume
of gas produced/volume of reactor/day) and total
gas production efficiency (volume of gas pro-
duced/volume of feed introduced/day) with variation in dilution rate.   The reactors A through
F are identical with those described in Figure 5.

The 3 liter CSTR and 18" packed bed reactors were determined to have similar total gas production efficiency rates in comparison with the 12" packed bed reactors which exhibited depressed gas production efficiency (Figure 6).

The effect on gas production in digester sludge taken from the 3 liter CSTR upon addition of test compounds is shown in Table 3. Furfural, 5-hydroxymethyl furfural (5-HMF), levulinic acid, acetovanillone, and p-hydroxy cinnamic acid were all observed to cause inhibition of gas production at the 0.5% (w/v) addition level. Furfural and 5-HMF, although exhibiting initial inhibitory effects, resulted in gas over-production at longer incubation times when compared to the controls. Lignin and a saccharinic acid preparation did not inhibit gas production at the levels tested.

DISCUSSION

Wheat straw in its native state, the low value biomass model used in this study, has been shown to be difficult to digest anaerobically to methane in reactors with short retention times. Preliminary investigations of chemical and mechanical pretreatment have been carried out. Physical pretreatment using first a bailbreaker or swinging bar shredder followed by a rotary knife mill was used success-fully to reduce this biomass to the smallest particle size attainable with the commercial knife mill available. Also, use of a pneumatic transfer system following rotary knife milling dramatically increased throughput rates.

The particle size distribution data indicates that the Wiley mill equipped with a 2mm screen produces straw meal with the highest percentage of product in the 40 mesh and finer size. However, the Mitts & Merrill mill equipped with a 1/8" rejection screen represents a milling configura-tion attainable for a large scale process. It was con-sidered of interest, therefore, to characterize the product obtainable with this mill. A comparison of the product distribution indicates that the Mitts & Merrill 1/8" screen mill is a good compromise because it substantially reduces the quantity of coarse product obtained with the Mitts & Merrill 1/4" screen and approaches the fine meal content of the Wiley mill product.

Increasing the phosphoric acid concentration for chemi-cal pretreatment at 120°C results in increased sugar solu-bilization and weight loss with a final acid concentration optimum of 0-5%. Increasing hydrolysis (pretreatment) time results in increased weight loss and sugar solubilization

Table 3.
Effects of Test Compounds on Gas Production in Anaerobic Digester Sludge

Compound	Final Concentration (%w/v)	Analysis Time (hours)			
		12	24	48	72
Furfural	0.01	139	129	98	93
	0.05	146	176	149	80
	0.1	95	234	242	134
	0.5	53	72	106	137
5-hydroxymethyl furfural	0.01	99	105	104	102
	0.05	110	100	101	120
	0.1	99	103	93	124
	0.5	41	51	72	106
lignin[1]	0.01	98	101	100	102
	0.05	103	103	98	98
	0.1	106	101	96	95
Levulinic acid	0.01	99	101	102	103
	0.05	100	95	98	97
	0.1	101	96	94	88
	0.5	54	57	32	10
Acetovanillone	0.01	106	95	98	96
	0.05	121	116	93	84
	0.1	119	110	94	85
	0.5	60	71	48	37
p-hydroxy cinnamic acid	0.01	99	102	97	100
	0.05	102	102	94	99
	0.1	103	103	92	90
	0.5	67	73	46	19
Saccharinic acid[2]	0.01	106	106	106	120
	0.05	103	110	119	120
	0.1	105	107	186	127

[1] Insoluble lignin represented part of total added lignin.

[2] Saccharinic acid prep was considered crude and was not completely soluble.

up to 4-5 hours. The 0.2% acid concentration was chosen for
this study because of the high fraction of oligomeric sugars
produced during pretreatment and this amount of acid is
close to the smallest amount required for catalysis of the
reaction. This concentration of acid also results in a more
consistent feed with variation in pretreatment time for
anaerobic reactor performance studies.

Expression of pretreatment liquor has been shown to be
a problem for substantial hydrolysate recovery. Results
from the hydraulic piston press studies indicate that pretreated straw can be dewatered to 29% solids at 120 psi.
Increasing the pressure to 620 psi and above produces straw
with a solids content of no better than 42%. It is interesting to note that this is very similar to the moisture
content of native vegetable fiber.

The straw pretreatment expressed liquor, having a COD
of 12.3 g/l and C:N ratio of 88, was successfully utilized
in an anaerobic digestion system as the substrate for methane
production. Furthermore, the pretreatment liquor does not
require additives such as trace vitamins and minerals to sustain metabolic activity.

In light of capital costs required for large scale
anaerobic fermentation of the pretreatment liquor, two
digester designs were examined. Packed bed reactors constructed from conical-flanged glass process piping utilized
two different types of column media for microbial attachment:
crushed foam glass and ceramic tile. Both are commonly
used in insulation applications and are relatively inert
to microbial attack. While microbial attachment occurred at
a faster rate with the foam glass packing media, both media
served equally well to immobilize microbial mats after
several months.

The second digester type, the continuously stirred tank
reactor, employed a hanging magnetic bar stirrer for agitation. Although the volatile solids and COD destruction was
found to be comparable in effluent from packed bed reactors
and the CSTR, the CSTR was not competitive with respect to
dilution rates. In fact, for a preliminary study in which
the dilution rate was increased for the CSTR to 0.4 resulted
in impending digester failure. To document this observation,
lower dilution rates were established in the CSTR and smaller
dilution rate increments were used to assess the dilution

rate threshold at which microbial wash out and resultant digester failure occur. When considering dilution rates, however, the CSTR was not competitive with the packed bed reactors with respect to total gas production rates and efficiency.

Although packed bed reactors utilizing the foam glass media resulted in faster microbial attachment, the reaction rates after maturation were similar (Figure 6, B & D and A & C). Variation in packed bed reactor dimensions (length/ diameter ratio) had a significant effect on digester per- formance. The shorter, 12" packed bed reactors (380-395 ml) (Figures 5 and 6, A and C) resulted in depressed gas pro- duction rates and efficiency as well as lower destruction of volatile solids and COD.

Biomass pretreatment by-products such as furfural, 5- hydroxymethyl furfural and levulinic acid, as well as the lig- nin precursors acetovanillone, and p-hydroxy cinnamic acid, all resulted in inhibition of gas production in the anaerobic digestion system at levels of 0.5%. The levels of furfural, 5 HMF and levulinic acid required for inhibition of gas pro- duction in the sludge fed SPF was higher than previous re- ports (McCarty et al., 1983; Young and McCarty, 1981; Owen et al., 1979). Higher levels required for inhibition are most likely due to acclimation in the sludge to continuous addition of these acid pretreatment by-products in the SPF feed. Although inhibitory at first in our digestion system, 0.5% furfural and 5-HMF resulted in increased gas production over the controls upon longer incubation times, consistent with previous reports of fast adaptation and degradation to methane. Lignin and a crude saccharinic acid preparation were not observed to cause inhibition at the levels tested.

In summary, it is apparent that the physical/chemical pretreatment conditions for a model low value biomass such as wheat straw can be identified and optimized resulting in an expressed liquor which can be utilized in an anaerobic diges- tion system as a substrate for methane production without supplementation. Also, packed bed reactors can be utilized to immobilize the microbial population, allowing increased dilution rates relative to the CSTR and leading to decreased capital costs for fermentation. Finally, the possible by- products from dilute acid biomass pretreatment do not inhibit gas production in the anaerobic digestion system at levels below 0.5%; on the contrary, furfural and 5-HMF may be degraded to methane.

ACKNOWLEDGMENTS

The authors wish to express their appreciation to the Biomass Energy Technology Division of the U.S. Department of Energy for sponsoring the work under WPA #387.

REFERENCES

Ashwell, W.L., *Methods in Enzymology*, (S.P. Colowick and N.O. Kaplan, eds.), pp. 84-85, Academic Press, Inc. (1957).

Balch, W.E., G.E. Fox, L.J. Magrum, C.R. Woese, and R.S. Wolfe, *Microbiol. Rev.*, *43*, 260-296 (1979).

Balch, W.E. and R.S. Wolfe, *Appl. Environ. Microbiol.*, *36*, 781-791 (1976).

Cheeseman, P., A. Toms-Wood, and R.S. Wolfe, *J. Bacteriol.*, *112*, 527-531 (1972).

Clausen, E.C. and J.L. Gaddy, *Fuel Gas Systems*, (D.L. Wise, ed.), pp. 11-140, CRC Press, Boca Raton, FL (1983).

Greenberg, A.E., J.J. Connors, and D. Jenkins (eds.), *Standard Methods for the Examination of Water and Waste-water, 15th Ed.*, pp. 489-493, APHA-AWWA-WPCF (1981).

Greenberg, A.E., J.J. Connors, and D. Jenkins (eds.), *Standard Methods for the Examination of Water and Waste-water, 15th Ed.*, pp. 383-388, APHA-AWWA-WPCF (1981).

Greenberg, A.E., J.J. Connors, and D. Jenkins (eds.), *Standard Methods for the Examination of Water and Waste-water, 15th Ed.*, pp. 409-425, APHA-AWWA-WPCF (1981).

Grohmann, K., M. Himmel, C. Rivard, M. Tucker, J. Baker, R. Torget, and M. Graboski, *Biotech. Bioeng. Symp.*, No. 14, In press (1984).

Himmel, M., M. Tucker, and K. Oh, *Biotech. Bioeng. Symp.*, No. 13, 583-595 (1983).

Hungate, R., *Bacteriol. Rev.*, *14*, 1-49 (1950).

Jewell, W.J., S. Dell-Orto, K.J. Fanfoni, T.D. Hayes, A.P. Leuschner, and D.F. Sherman, "Anaerobic Fermentation of Agricultural Residue: Potential for Improvement and Implementation," Vol. 2, U.S. Department of Energy Report, DOE/ET 20051-TI (1980).

McCabe, W.L. and J.C. Smith, *Unit Operations of Chemical Engineering, 2nd Ed.*, pp. 36-40, McGraw-Hill, New York (1967).

McCarty, P.L., K. Baugh, A. Bachmann, W. Owen, and T. Everhart, *Fuel Gas Developments*, (D.L. Wise, ed.), pp. 49-72, CRC Press, Boca Raton, FL (1983).

Owen, W.D., J. Stuckey, J. Healy, Jr., L. Young, and P.L. McCarty, *Water Res.*, *13*, 485 (1979).

Ramasamy, K. and H. Verachtert, *Straw Decay and Its Effects on Disposal and Utilization,* (E. Grossband, ed.), pp. 155-163, John Wiley & Sons (1979).

Whistler, R.L. and J.N. BeMiller, *Methods in Carbohydrate Chemistry, Vol. 2,* (R.L. Whistler and M.L. Wolfrom, eds. pp. 477-479, Academic Press, Inc. (1983).

Wujcik, W.J. and W.J. Jewell, *Biotech. Bioeng. Symp., 10,* 43-65 (1980).

Young, L.Y. and P.L. McCarty, *Fuel Gas Production from Biomass, Vol. II,* (D.L. Wise, ed.), pp. 133-176, CRC Press, Boca Raton, FL (1981).

20
HIGH SOLIDS DIGESTION OF MSW TO METHANE BY ANAEROBIC DIGESTION

E.C. Clausen and J.L. Gaddy

Department of Chemical Engineering
University of Arkansas
Fayetteville, Arkansas

ABSTRACT

The anaerobic digestion of lignocellulosic wastes, such as MSW, is quite slow, with retention times of 20-60 days quite common. Because of these slow rates of digestion, reaction vessels for producing methane by anaerobic digestion are quite large. Preliminary process economics have shown capital costs to be quite high, with about two thirds of the costs required for large digesters. Research clearly needs to concentrate on reaction kinetics and decreasing reactor volume in commercial anaerobic digestion processes.

This paper examines the feeding of high solids concentrations as a means of directly reducing the required reactor volume. Feed solids concentrations of nearly 30% are possible by feeding liquids and solids separately. The reaction kinetics must not be adversely affected and fluidity of the reactor contents must be maintained. The results of a project to determine the reaction kinetics and maximum MSW concentrations possible are presented. Economic projections for a large-scale facility are developed.

INTRODUCTION

The United States presently imports about 40% of its crude oil. Imports are 3.1 billion barrels annually. The development of alternative fuels has been stimulated by the increase in price of foreign cruds from $3.00 per barrel in 1973 to about $30 per barrel in 1984.

One relatively simple energy alternative is the production of methane gas by anaerobic digestion. Anaerobic digestion is a very old process, used extensively for treating domestic sewage prior to 1930. More recently, the process has received attention as a method of producing energy in the form of gaseous fuel from lignocellulosic wastes.

The United States generates approximately 90 million tons of municipal solid waste (MSW) each year. This quantity could generate 900 billion cubic feet of methane per year, or nearly 5% of our natural gas requirements. The conversion of MSW to methane would not only help to alleviate our fuel shortage, but would also significantly reduce the problems of MSW disposal.

One of the major problems associated with anaerobic digestion is the high capital costs required for large-scale reactors. A preliminary economic analysis shows that a unit producing 500 million cubic feet of methane per day (enough methane for a city with a population of 1 million) would require a capital investment of nearly $90 million. The capital costs for the reactors represent almost 60% of the total capital costs (Clausen and Gaddy, 1982). Clearly, research should be undertaken to decrease reactor volume, thereby decreasing capital and operating costs.

There are a number of different types of reactors under study for the anaerobic digestion of biomass. Batch reactors, above ground or as a landfill, are inexpensive but afford slow reaction rates, low densities and, generally, low yields. Dry fermentation systems have been shown to be unstable and to require liquid circulation for good conversions. Fixed film reactors give good reaction rates for soluble substrates, but cannot be used for extended periods with solid substrates. Plug flow reactors also give good reaction rates with the proper inoculation, but are not appropriate for slurries, such as MSW. The traditional continuously stirred tank reactor (CSTR) offers the best possibility for successful

application to large-scale MSW digestion. Mixing is essential
for good conversion of solid substrates in a reasonable reaction time and good reaction rates are obtained in the CSTR.
The CSTR will be utilized in this study, and the following
discussions relate to this reactor system.

Maximum Solids Concentration

In order to speed up the digestion reaction, the size
of the biomass particles must be reduced to increase the
accessibility of the polymeric structures to the enzymes.
A high solids concentration is desirable since this concentration controls the size of the fermentation equipment.
The size of the particles also affects the fluidity of the
solids/acid slurry. It is desirable to maintain fluidity
of the slurry to promote mass transfer and to facilitate
pumping and mixing. Therefore, the particle size is an important variable in the biomass conversion process.

Table 1 gives the maximum solids concentration to maintain fluidity of a biomass slurry, as a function of particle
size. A maximum concentration of about 10% is possible with
particle sizes less than 40 mesh. Grinding to 20 mesh gives
a particle size distribution in which 90% of the material
is less than 40 mesh. Therefore, grinding the biomass to
pass 20 mesh gives the appropriate size and produces the
maximum possible slurry concentration. Laboratory studies
also show that grinding to smaller sizes does not improve the
reaction rates.

Table 1.
Maximum Solids Concentration for Fluid Slurry

MESH RANGE (Sieve #'S)	SOLIDS CONC. (wt. %)
0 - 1	4.6
12 - 20	4.6
20 - 30	8.4
30 - 40	8.2
40 - 45	10.2
45 - 70	10.4
70 - 100	10.6
100+	10.8

Factors Affecting Reactor Volume

To determine the factors that influence the reactor volume, a summary of the calculational procedure is presented. For a continuous flow reactor, the volume, V, is determined as the product of the volumetric flow through the reactor, F, and the retention time, θ:

$$V = F\theta \tag{1}$$

The flow rate is defined as the weight of the biomass introduced into the reactor, B, divided by the product of the solids fraction (by weight) in the biomass/water mixture, W_B, and the density of the mixture, d:

$$F = B/dW_B \tag{2}$$

The reaction rate is determined from a mass balance around the CSTR and the first-order kinetic relationship:

$$-r = \frac{C_i - C_o}{\theta} = kC_o \tag{3}$$

where

$-r$ = reaction rate;
C_i = inlet substrate concentration;
C_o = outlet substrate concentration; and,
k = reaction rate constant.

Substituting Eqs. (2) and (3) into Eq. (1) gives:

$$V = \frac{F(C_i - C_o)}{kC_o} = \frac{B}{dW_B} \frac{(C_i - C_o)}{kC_o} \tag{4}$$

The conversion in the reactor, X, is related to biomass substrate concentration by the relation:

$$X = 1 - \frac{C_o}{C_i} \tag{5}$$

Substitution of Eq. (5) and rearrangement of Eq. (4) gives:

$$V = \frac{B}{dW_B k} \left(\frac{X}{1-X}\right) \tag{6}$$

The volume of methane, M, produced from a given quantity of biomass, is given by:

$$M = G B X = 6 B X \qquad (7)$$

where G = gas conversion constant, depending upon the carbon content of the biomass and the methane concentration of the product gas. For a biomass that is 40% carbon and produces 50% methane, G = 360/12 x .4 x .5 = 6. Substituting Eq. (7) into Eq. (6) gives:

$$V = \frac{M}{6dW_B k} \frac{1}{1-X} \qquad (8)$$

Eq. (8) can be used to analyze the parameters that influence the reactor volume and, hence, the process economics. For a required volume of methane production, M, the four factors that would lead to a reduction in the volume are an increase in the density, solids fraction or reaction rate constant, and a decrease in the biomass conversion, X. Decreasing the conversion results in an increase in the consumption of biomass and an increase in the quantity of residue. Since one of the objectives of utilizing MSW is to reduce the volume of residue for disposal, reducing the conversion is an expensive and unattractive option. A savings can be achieved, however, by using a series of reactors to give the same total conversion.

One method for reducing the reactor volume is to increase the reaction rate constant. There are a number of possible ways to improve the reaction kinetics. One method, under investigation by others, is to pretreat the biomass to make it more readily digestible. Various pretreatment processes are under study, including the use of caustic, acid and high temperature (autocatalytic).

An examination of the traditional Monod equation shows that the reaction rate, and, hence, the reactor volume, are dependent upon the microorganism concentration and type of microorganisms. Other ways to improve the reaction rate constant would, therefore, involve increasing the microorganism concentration and improving the culture.

The reactor volume also varies in proportion to changes in the density. The density inside the reactor cannot be increased beyond the point of fluidity; thus the maximum density is close to that of water. However, the density can be reduced if the solids fraction is increased beyond the point of fluid mixture. Fluidity is not possible for mixtures exceeding about 10% (by weight) biomass in water.

Therefore, if the concentration in the reactor exceeds 10%, an increase in volume would result in proportion to the reduction in density.

Perhaps the most promising method of decreasing reactor volume is to increase the biomass fraction, W_B, into the reactor, which results in a proportionate decrease in reactor volume. In the past, the upper limit of the feed fraction has been considered to be 10%, such that the mixture could be pumped into the reactor. It is possible to feed more concentrated mixtures by introducing the solids and liquids separately. The concentration within the reactor must be maintained at 10% or less, and this places an upper limit on the feed concentration. For example, a 30% mixture could be fed to the reactor and a 10% reactor concentration maintained, if the conversion in the reactor were 66%. A 30% feed mixture would result in a reactor only one-third the original size and capital cost savings of about 60% for the reactors.

Purpose

The purpose of this paper is to present experimental data and preliminary economics for the anaerobic digestion of high solids concentrations of MSW to produce methane. At this point in the study, only preliminary data are available for high solids MSW digestion. These data parallel the digestion of corn stover, which has been studied extensively. Thus, high solids digestion data for corn stover will be presented along with preliminary economics for the digestion of MSW.

A COMPARISON OF MSW AND CORN STOVER AS FEEDSTOCKS FOR ANAEROBIC DIGESTION

The composition of municipal solid refuse has been well documented in the literature (Blum, 1976; Snyder, 1974). Technology now exists for removing metals and glass from MSW, either by gravity separation or steam treatment. The resulting material is a lignocellulosic mixture well suited for methane production by anaerobic digestion.

The typical material composition of MSW with glass and metals removed is shown in Table 2. As noted, the feedstock is rich in paper and food waste, with these two components comprising over 85% of the dry weight. The remaining 15%

Table 2.
Composition of MSW

Weight Percent, Dry Basis

	Weight Percent, Dry Basis
Paper	61.5
Food Waste	24.8
Plastic, rubber, leather	6.7
Textiles	2.6
Wood	4.5

of the MSW is plastic, rubber, leather, textiles and wood. A synthetic mixture of MSW has been prepared for use in anaerobic digestion studies having the composition shown in Table 2.

As previously mentioned, MSW and corn stover are similar feed materials for anaerobic digestion. A comparison of compositions and elemental analyses of stover and MSW is shown in Table 3. As noted, MSW is richer in cellulose and stover is richer in hemicellulose. The total carbohydrate content, however, is nearly the same, 60-65% hemicellulose or cellulose. The elemental analyses of both feedstocks is also nearly identical. Both feedstocks contain approximately 45% carbon, 6% hydrogen and 1% nitrogen.

Table 3.
Comparison of MSW and Corn Stover As
Anaerobic Digestion Feedstock

	Weight % MSW	Dry Basis Corn Stover
Cellulose	57.44	36.39
Hemicellulose	3.56	29.52
Lignin	12.26	8.42
Ash	1.66	1.44
Carbon	44.22	45.82
Hydrogen	6.78	5.20
Nitrogen	1.01	1.80

Based upon the data shown in Table 3, it might be expected that the digestion of these two feed materials would be quite similar. This is indeed the case as will be demonstrated in the following sections.

Before addressing the anaerobic digestion of MSW in continuous culture using high solids feed, it is instructive to trace the anaerobic digestion of corn stover in continuous and high solid culture. This work, carried out in the University of Arkansas laboratories, directly parallels the work outlined in this study. Techniques developed for the feeding of high solids corn stover are directly applicable to the feeding of MSW. Limitations found in the stover study will be instructive in dealing with potential limits in the MSW study.

THE ANAEROBIC DIGESTION OF CORN STOVER IN CONTINUOUS CULTURE

A standard culture has been developed at the University of Arkansas over the years with an inoculum of sewage sludge, animal rumin, and animal wastes. This culture has been demonstrated to be stable and capable of digesting lignocellulosic residues without nutrient addition or continuous pH control. This culture was used to establish a reliable kinetic model to serve as a standard for comparison for future improvements.

Experimental Procedure

The reactors employed in these studies are 3 liter continuously stirred tank reactors (CSTRs) maintained at 37°C in a constant temperature cabinet. The reactors are fed a 10% solids slurry of biomass in water once per day, and effluent is also removed at this time to maintain a constant volume. The pH, ammonia nitrogen, and phosphate concentration in the reactor are checked periodically, as needed. No nutrient additions are made to the reactors after the initial culture development is completed.

The gas from the reactors, containing approximately 50% methane and 50% carbon dioxide, is collected and measured once per day by liquid displacement. Gas composition is measured using gas-solid chromatography. Steady-state data for kinetic studies are determined after observing a steady gas output for 2-3 retention times.

Reaction Kinetics

An empirical relation used to describe many fermentation reactions is given by the Monod equation:

$$-r = \frac{k_1 C_A C_M}{k_2 + C_A} \tag{9}$$

where $-r$ = reaction rate
            C_A = concentration of substrate
            C_M = microorganism concentration
     k_1, k_2 = constants.

For mixed-culture, slurry digestion, the measurement of C_M is difficult or impossible. Also, for a complex substrate such as lignocellulosic materials, the determination of the substrate concentration, C_A, is complicated by the presence of cells and solubilized solids. For the anaerobic digestion of lignocellulosics in 10% feed concentrations, a first-order expression has been found applicable when using a mixed culture (Clausen and Gaddy, 1982):

$$-r_c = k \, C_c \tag{10}$$

where $-r_c$ = reaction rate based upon carbon concentration
             (g mol C/ℓ-day),
       k = reaction rate constant (days $^{-1}$) and
       C_c = substrate carbon concentration (g mol C/ℓ).

In Eq. (10), no term is included for the microorganism concentration, which is lumped with the reaction rate constant. The reaction rate is based upon the carbon concentration in the reactor, C_c. This equation has been found reliable for a variety of retention times using up to 10% solids feed.

Results

The data obtained for the semicontinuous digestion of corn stover and MSW are shown in Table 4. The retention time was varied from 20 to 100 days at feed concentrations of 8 and 10% solids. Gas production rates ranged from 2.8 to 6.4 ℓ/day at steady state. Reaction rates are calculated from the CSTR reaction equation previously given in Eq. (3).

These rates are expected to be low, and indeed range from 0.026 to 0.079 g moles/ℓ·day. Conversions in the reactor range from 39 to 71%, increasing with retention time. A plot of conversion as a function of the inverse retention

Table 4. Kinetics of Laboratory Biomass Digestion

Retention Time (Days)	Gas Production ℓ/day	Carbon Concentration gmoles/ℓ		Reaction Rate (gmoles/ℓ·day)	Conversion (percent)
		$Cin(C_{CO})$	$Cout\ (C_C)$		
Corn stover Digestion					
20	6.366	2.904	1.369	0.0568	39.1
20	8.815	3.630	2.059	0.0786	43.3
30	6.635	3.630	1.710	0.0642	52.9
40	6.052	3.630	1.472	0.0540	59.5
40	5.944	3.630	1.418	0.0553	60.9
40	3.760	3.630	1.390	0.0560	61.7
40	3.720	3.630	1.420	0.0553	60.9
60	2.660	3.630	1.250	0.0396	65.6
100	2.889	3.630	1.054	0.0258	71.0
MSW Digestion					
40	3.862	3.682	1.383	0.0575	62.4
40	3.903	3.682	1.359	0.0581	63.1

time yields a straight line as shown in Figure 1. This plot shows an ultimate conversion at infinite retention time of 78%, which is equal to the carbohydrate content of biomass. It should be noted that the same plot holds for both stover and MSW digestion, with MSW exhibiting a slightly higher conversion.

The first-order kinetic model suggests that a plot of reaction rate, $-r_c$, as a function of reactor carbon concentration, C_c, should yield a straight line passing through the origin. Such a plot is shown in Figure 2. As shown, linear behavior is obtained with a correlation coefficient of .95, yielding a slope, equal to the rate constant, k, of 0.045 days^{-1}. However, the plot does not pass through the origin, but near a point equal to the carbon concentration of lignin. Since the lignin cannot be converted to methane, the reaction rate would be expected to be zero at this concentration. This kinetic model provides a good data base for economic projections and comparison with high solids digestion. Again, approximately the same kinetic behavior is seen with MSW and stover, with MSW, perhaps, providing slightly higher reaction rates.

Process Design and Economic Considerations

Utilizing the kinetic model presented in Eq. (10) and a first-order rate constant of 0.045 days^{-1}, a preliminary process design and economic analysis may be made. Figure 3 shows a flow diagram of an industrial process to produce methane from lignocellulosic residues. The material is first ground and mixed with water to form a slurry that is 10% solids by weight. Storage facilities are provided for a one-day supply of solid residue. The slurry is fed to digesters operated in series and maintained at 35°C.

The undigested solids and nutrients from the reactor are available to be returned to the soil as an amendment. The product gas from the reactors, containing 50-55% methane with the balance carbon dioxide, is compressed to 15 psig. Following compression, the carbon dioxide is removed by scrubbing with diethanol amine. The remaining methane is dried in a glycol absorber to produce pipeline quality natural gas.

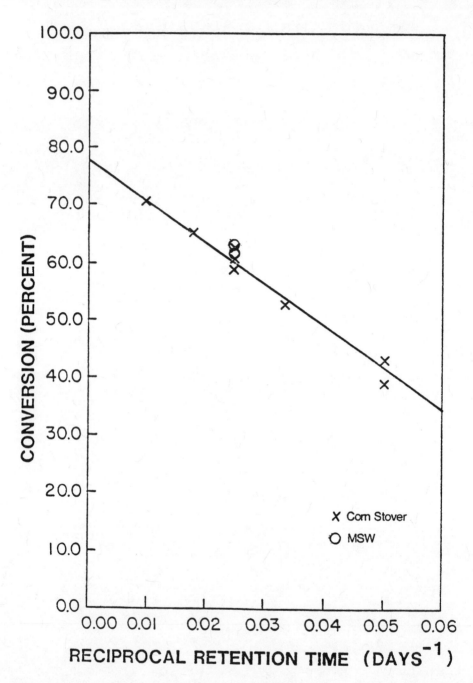

Figure 1.   Biomass conversion as a function of retention time.

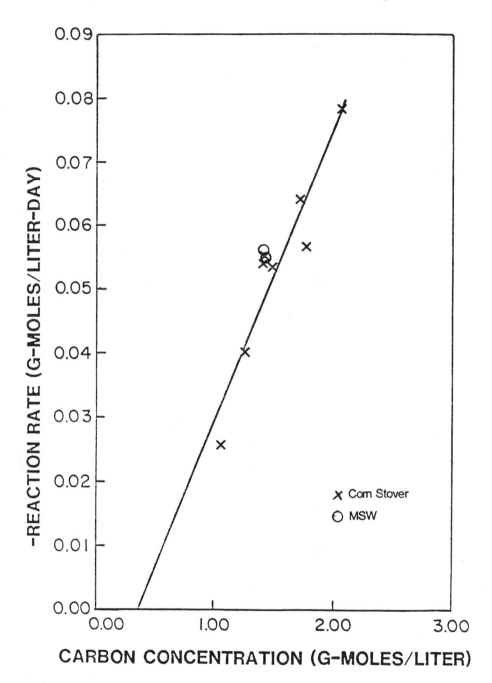

Figure 2. Test of first order kinetic model.

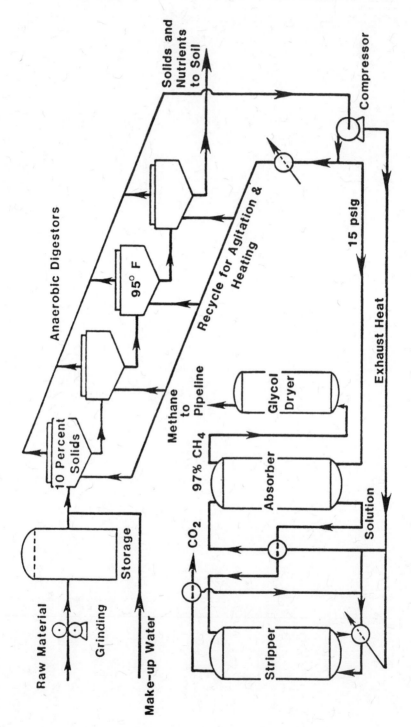

Figure 3.   Process for methane production from biomass.

Exhaust gases from the compressor are used as a source of heat to maintain the reaction temperature and to operate the adsorber and stripper. The energy balance of the process shows that only 7.5% of the methane product is consumed for compression and heat.

Table 5 presents the estimated capital and operating costs for the process presented in Figure 3 to produce 50 million SCF of methane per day. This quantity of methane would supply the energy needs of a city of one million population. The reactors were designed to give a conversion of 75% using four reactors in series. As is shown in Table 5, a total capital investment of $137 million is required, with $84 million for the reaction vessels. Since these are preliminary economics, a contingency of 30% has been included.

The revenue for the methane at $4/MCF comes to $73 million per year. Total operating costs including labor, maintenance, utilities, depreciation, taxes and insurance are $28.2 million. No raw material (MSW) costs are included. Costs for processing MSW to remove metals and glass offset the value of the recovered by-products.

The breakeven cost for the methane is $1.55 per MCF, and the return on investment at $4/MCF is 18.4%. Prices for "new" natural gas are as high as $8 per MCF. Thus, the process appears to be economically attractive at today's energy prices. However, the process requires a large capital investment because of the large reaction vessels, and the economics are sensitive to raw material cost.

The economics of methane production from MSW with 10% feed is only marginal. However, the potential improvement in these economics that can be achieved if high solids feed concentrations are employed is significant. If the feed concentration to the reactor is doubled at constant conversion, the required reactor volume can be halved. The large capital investment for reactors can be decreased 50%, with an attendent increase in the return on investment.

HIGH SOLIDS DIGESTION WITH BIOMASS

Experimental Procedure

The procedures required for experiments involving high feed solids concentrations are much the same as for the standard reactors. Two reactors were constructed with oversized motors and dual axial flow impellers. Since these

Table 5.
Capital Cost and Economics of Methane Production from MSW

Equipment		Capital Cost $ x 10^6
Reactors		84.0
Compressors		5.0
Absorber and stripper		5.0
Heat exchangers		2.3
Pumps and piping		9.0
Contingency (30%)		31.6
	Total	$136.9

Revenue (18250 MCF/yr. at $4/MCF) $73 x 10^6/yr

Operating Costs		Operating Cost $/yr x 10^6
Utilities		3.5
Labor and supervision		1.5
Maintenance (5%)		6.8
Depreciation (10%)		13.7
Taxes and Insurance (2%)		2.7
	Total	28.2

Break-even price ($/MCF)	1.55
Gross Profit ($/yr x 10^6)	44.8
Net Profit ($/yr x 10^6)	22.4
Return on Investment	18.4%

experiments required agitation of heavier, more viscous slurries, a motor with more power and a higher torque was necessary. Axial flow impellers are required to provide good mass transfer with the highly concentrated slurries. Larger polyethylene bags were also constructed to handle the increased gas volumes.

Since the performance of the higher feed solids reactors was unknown, caution was taken to avoid sudden upsets to the reaction system. After obtaining a standard culture representative of the data in Table 4, the solids concentration was gradually increased from 10 to 20% over a three month period. The reaction pH, gas production rate and load on the motor were monitored very closely. Once the 20% feed concentration was obtained at a 60-day retention time,

the reactor was fed at the same rate for 2-1/2 retention
times before sampling at steady state. Other data at other
retention times and solid concentrations were obtained
similarly.

Experimental Results

Data for 20% feed solids concentration using corn stover
as the feed are shown in Table 6. A constant retention time
of 60 days was used. Shorter retention times were not used
because of the resulting high solids concentrations in the
reactor. As the retention time is lowered at a constant feed
concentration, the conversion decreases, so that the solids
concentration in the reactor will be higher at shorter retention times.

Since the kinetics of MSW and corn stover digestion
are nearly identical, the data presented for the high solids
digestion of corn stover and MSW should be quite similar.
Preliminary data for the digestion of MSW using a 20% feed
concentration are indeed quite encouraging, approaching desired steady state production levels.

As shown in Table 6, the conversion for a stover feed
remained at the level achieved with a 10% feed when feeding
concentrations of 20%. The data in Tables 4 and 6 both show
a 66% conversion for 10 and 20% solids feeds. For a 20%
feed, the solids concentration in the reactor is about 6.8%.
The solids concentration in the reactor using a 10% feed is
about 3.5%. Since the maximum concentration for fluidity
is 10%, higher feed concentrations are possible.

The real significance of the high solids data is that
by employing this technique in a commercial facility, the
reactor volume can be cut in half for the same methane output,
at a 60-day retention and 20% solids. At higher solids concentrations the reactor volume requirements can be decreased
further. These same procedures are being applied to MSW
and produce similar improvements.

*The Effect of Increased Solids Concentrations on Process
Economics*

High solids studies have been successfully carried out
without a decrease in reaction rate or conversion at a 20%
solids concentration. It is interesting to re-evaluate
the process economics presented in Table 5 for a 20% solids

Table 6.

Kinetic Data for High Solids Feed Concentration Studies
(Corn Stover Feed)

Retention Time (days)	Percent Feeds Solids	Gas Production ℓ/day	C_i gmoles/ℓ	C_o gmoles/ℓ	Reaction Rate gmoles/ℓ·day	Conversion (percent)
60	20.0	5.38	7.26	2.456	0.0801	66.2
60	20.0	5.44	7.26	2.403	0.0810	66.9

feed. The process produces 18,250 million cubic feet of
methane per year, selling at $4/MCF. The analysis presented
in Table 5 yields a break-even methane price of $1.55 per MCR
and a return on investment of 18.4% at a methane price of
$4/MCF.

Table 7 shows that if 20% solids are used as the feed
for the reaction system, the capital costs for reactors is
cut in half (as is the volume) to $42 million. Other capital
costs remain the same, giving a total capital cost of $82.3
million, including a 30% contingency. The revenue at $4/MCF
is $73 million/year.

Operating costs decrease to $19.4 million/year. The
utilities increased slightly, because of a higher power demand
for agitation. Maintenance, depreciation, and taxes decreased
since they were calculated as a percentage of the capital
investment. The break-even price for methane is then $1.06/
MCF. The return on investment increases to 42.5%, more than
double the return using a 10% feed.

It is interesting to note that the return on investment
(ROI) increases significantly if even higher feed concentrations are possible. The ROI increases to 49% with a 25%
solids feed and to 54% with a 30% feed, if constant conversion
is assumed. The potential of high solids digestion is encouraging indeed.

SUMMARY

The composition of stover and MSW is quite similar,
so that is expected that digestion will proceed in a similar
fashion. Digestion studies at a 10% solids feed concentration
show nearly identical digestion results. The digestion of
stover at a feed concentration of 20% solids shows an identical conversion as with a 10% feed. Preliminary studies with
MSW employing high feed solids concentrations are again pro-
ceeding in nearly identical fashion.

The economics of methane production from MSW are quite
encouraging. The return on investment for a large-scale
methane production facility increases from 18.4% to 42.5% as
the feed solids concentration is increased from 10 to 20%.
The capital cost for a facility producing 50 million cubic
feet of methane per day is estimated at 82.3 million when
using a 20% solids feed.

Table 7.
*Capital Costs and Economics of Methane Production from MSW
(20% Solids Feed)*

		Capital Cost x 10^6
Equipment		
Reactors		$42.0
Compressors		5.0
Absorber and stripper		5.0
Heat exchangers		2.3
Pumps and piping		9.0
Contingency (20%)		19.0
	Total	$82.3

Revenue (18,250 MCF/yr at $4/MCR)	$73 x 10^6/yr

		Operating Cost $/yr x 10^6
Cost Item		
Utilities		4.0
Labor and supervision		1.5
Maintenance (5%)		4.1
Depreciation (10%)		8.2
Taxes and Insurance (2%)		1.6
	Total	$19.4

Break-even price ($/MCF)	1.06
Gross profit ($/yr x 10^{-6})	53.6
Net profit ($/yr x 10^{-6})	26.8
Return on investment	42.5%

REFERENCES

Blum, S.L., "Tapping Resources in Municipal Solid Waste,"
    Science, 191, Feb. (1976).
Clausen, E.C. and J.L. Gaddy, Methane Production by Fermentation of Agricultural Residues, Vol. 1, Alan R. Liss,
    Inc. (1982).
Snyder, N.W., "Energy Recovery and Resource Recycling," *Chem.
    Engr.,* Deskbook Issue, Oct. 21 (1974).

21

GASIFICATION OF CONCENTRATED PARTICULATE AND SOLID SUBSTRATES BY BIPHASIC ANAEROBIC DIGESTION

S. Ghosh

Institute of Gas Technology
Chicago, Illinois

ABSTRACT

This paper presents the development and performances of two innovative configurations of the two-phase anaerobic digestion process for gasification of concentrated organic slurries and high-solids feeds. Data are presented to show that the conventional single-stage digestion process is unsuitable for application to high-solids-content and concentrated feeds, and that efficient and rapid-rate gasification of these feeds can be better achieved by separated acidogenic and methanogenic fermentations. A process configuration utilizing upflow acid- and methane-phase digesters was demonstrated to be considerably superior to the conventional single-stage digestion process in terms of methane yield and production rate and net energy production. This process afforded a methane yield of 0.48 SCM/kg VS added (7.7 SCF/lb VS added) for a sewage sludge feed at a hydraulic residence time of 5.9 days; this performance was better than those reported in the literature for this type of feed. For low-moisture feeds such as MSW, a downflow leach-bed acid-phase fermenter was utilized to liquefy an organic bed of solids to generate a high volatile acids-content bioleachate product, which was sequentially gasified in a separate upflow methane digester. This system, operated under ambient conditions, provided for recycling of the indigenous nutrients, and did not require the addition of any external water; it afforded a methane yield which was 80% of the theoretical value.

INTRODUCTION

In municipal wastewater treatment, organic pollutants are removed from the contaminated water and collected in the form of primary and secondary sludges, the ultimate disposal of which is difficult and expensive because of their very dilute (0.5-3 wt % solids) nature and hetero- geneous physical and chemical character. Concentration of the dilute sludge by a series of unit operations and dis- posal of the high solids-content (8-30 wt %) end-product in lagoons, sludge drying beds, or landfills are common approaches to ultimate disposal. Thermal or biochemical oxidative conversion processes have been used to reduce the volume of the concentrated residue requiring ultimate disposal. However, these latter processes are capital and energy intensive and have other environmental con- sequences. None of the above sludge treatment processes can recover the material and energy resources of sludge.

The only proven sludge treatment process that can effect stabilization with simultaneous recovery of a fuel gas, and possibly chemicals, is anaerobic digestion. Be- cause municipal sludges are usually very dilute and con- tain 97 wt % or more of water, digester feeds are prepared by concentrating at least a part of the plant sludge stream. Total solids (TS) concentrations of 4-5 wt % are quite common for feed sludges for the conventional single-stage stirred-tank digesters, which are usually operated at hy- draulic residence times (HRT) of 15-20 days. More con- centrated feed sludges and shorter HRT's are desirable to reduce digestion-tank volume and sludge-heating require- ment, but are not used for single-stage mixed-tank digesters-- which are the only commercial processes available at this time--because these operational conditions could lead to an imbalance between the sequential acidogenic and methano- genic fermentation phases of anaerobic digestion.

The limitations of the conventional single-stage pro- cess are illustrated by the data presented in Table 1. As indicated by the methane yield and low residual volatile acids concentration, stable and balanced digestion of muni- cipal sludge could be obtained when a high HRT of 18 days and a low TS concentration of 4.6 wt % was used. However, volatile acids production and accumulation increased and methane production decreased indicating an imbalance be- tween the acidogenic and methanogenic steps when a 12.6 wt % TS-concentration sludge was fed to the digester.

Table 1. Limitations of Single-stage Stirred-tank Mesophilic (35°C) Sludge Digesters

Operating Conditions	Feed TS, wt %	HRT, days	Loading Rate kgVS/m³-day	Methane Content, mol %	Methane Yield, SCM/kg VS added	Effluent Volatile Acids, mg/L as acetic
A. Conventional Operation	4.7	18	1.44	70	0.31	80
B. Operation with Concentrated Feed	12.6	6.9	10.09	67	0.21	2500
C. Operation with Concentrated Feed and at Short HRT	6.1	4.3	9.13	51	0.05	3220

The digester exhibited even higher accumulation of volatile
acids when the HRT was reduced from 6.9 days to 4.3 days.
The accumulated VA inhibited methane fermentation and the
single-stage stirred-tank digester in effect became an
acid-phase digester (or a volatile acids producer) at low
HRT's and with concentrated feeds; methane fermentation,
which is tantamount to organics stabilization, was very
inefficient under these operating conditions because the
growth rate of the VA-utilizing methane bacteria is much
lower than that of the acid-forming organisms, and because
methanogenic metabolism is inhibited by the accumulated
acids. Overall, the result of digester operation at low
HRT's with high solids-content feed was separated acid-
phase fermentation. Clearly, conventional single-stage
complete-mix digesters do not seem to be suitable for
high-solids feeds.

BIPHASIC DIGESTION

     The biphasic or two-phase anaerobic digestion process
was developed to overcome the limitations of the conventional
single-stage process and to stabilize high solids-content
feeds at low HRT's and high loadings (2-5). Two-phase
anaerobic digestion takes advantage of the phase separation
phenomenon that tends to occur naturally as indicated above,
and provides for separate acidogenic digestion of concentrated feeds at low HRT's followed by separate methane
fermentation of the acidogenic end-products. Various
operating modes and digestion reactor designs can be envisioned within the broad framework of the two-phase fermentation concept. This paper presents the development
and performance of two two-phase process designs. The
first process design utilizes upflow digesters without
mechanical mixing and is suitable for concentrated organic
slurries such as municipal sludge. The second process
design is applicable to such high-solids feeds as municipal
solid waste (MSW), municipal sludge cakes, certain biomass
feeds, dry manures, etc., and provides for solid-phase
acidic fermentation in a leach-bed digester operated in
tandem with a separate methanogenic digester. The leach-
bed digester is operated to produce a low suspended solids-
content volatile acids-rich effluent which is then charged
to an upflow methane digester. The upflow digester could
be fully or partially packed with a selected medium, or it
could be of a sludge-blanket type.

UPFLOW TWO-PHASE DIGESTION OF CONCENTRATED PARTICULATE FEEDS

Materials and Methods

The upflow two-phase system consisted of a 7-L (6-L culture volume) upflow acid-phase digester operated in tandem with a 25-L (22-L culture volume) upflow methane digester. Both digesters were maintained at a mesophilic temperature of 35°C. The digesters were not mixed mechanically or by compressed gas. As shown in Figure 1, the raw sludge feed was introduced into the bottom of the acid-phase digester. The water fraction of the sludge traveled upward toward an overflow port at the culture surface while incoming sludge solids were deflected downward to effect high-SRT (Solids Residence Time) digestion of the retained solids. The overflowing supernatant and concentrated digester bottoms from the first-stage acid digester were fed to the bottom of the methane digester, which was similar in design and operation to the acid digester. However, there was no withdrawal of bottom sludge from the methane digester.

The upflow two-phase system was fed with raw primary sludge collected from the water pollution control plant of the South Essex Sewerage District (SESD), Salem, Massachusetts. The raw sludge had volatile solids (VS), total carbon, hydrogen, total nitrogen, total sulfur, and total phosphorus contents of about 61 wt%, 38.4 wt%, 5.9 wt%, 2.7 wt%, 0.88 wt%, and 2.94 wt% of TS, respectively. Since the SESD plant flow was constituted of 40-60 vol% industrial waste, the primary sludge contained a number of heavy metals including cadmium, copper, nickel, lead, zinc, chromium, mercury, and selenium in concentrations of 1-2.5 mg/L, 16-19 mg/L, 2-6 mg/L, 20-30 mg/L, 30-40 mg/L, 300-600 mg/L, 0.01 mg/L, and 0.5-1 mg/L, respectively. The sludge also contained high concentrations (3000-4000 mg/L) of calcium.

The elemental composition of the sludge was utilized to calculate stoichiometric biogas and methane yields of 0.97 SCM/kg VS reacted and 0.62 SCM/kg VS reacted, respectively, assuming that all the carbon in the feed was digestible and that 20% of the VS was utilized for cell biosynthesis. These theoretical yields serve as bench marks against which experimental yields were compared.

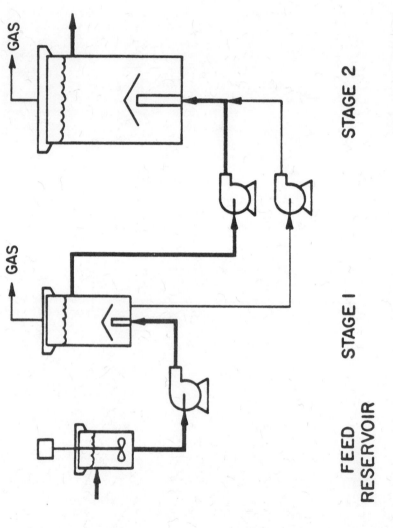

Figure 1. Upflow two-phase anaerobic digestion of concentrated sludge.

Results and Discussion

The upflow two-phase system was operated with the high metal-content concentrated primary sludge for about one year. Various operating conditions were used for the upflow acid- and methane-phase digesters. Research completed thus far indicates that for mesophilic conditions, acid and methane digester HRT's of about 1.5 days and 4.5 days, respectively, would afford high methane yields and production rates. Steady-state performance of the upflow two-phase system at an overall HRT of 5.9 days afforded a methane yield of 0.48 SCM/kg VS added which was 77% of the theoretical yield. Details of the steady-state performance data are reported in Table 2. The system methane yield of 0.48 SCM/kg VS added and the volumetric methane production rate of about 3 volumes of methane per day per unit culture volume were much higher than those reported in the literature for sewage sludge. The data in Table 2 indicate that about 74 wt% of the feed VS was recovered in the gas; the overall volatile solid reduction was thus higher than 74% which is about twice the VS reduction efficiency generally experienced in conventional single-stage mechanically mixed digesters. Since less than 80% of sewage sludge VS is biodegradable, the VS reduction exhibited by the upflow two-phase digestion process represents "complete" conversion of the biodegradable organic fraction of sludge.

Engineering Significance

The two-stage upflow digestion process has the notable advantage of producing a substantially larger quantity of methane with reduced fermenter volume (and reduced plant capital cost) relative to that of a conventional digestion process. As an example, for a 90.9 metric tons/day sludge solids (dry) load, a conventional single-stage digestion plant operated with a 3.6 wt% TS-content feed slurry would require 13,600 m^3 (480,000 ft^3) of fermenter volume (Table 3) for operation at a 5.5-day HRT. This process is a net energy consumer because the energy required for plant operation is larger than the energy value of the produced digester methane. By comparison, the two-stage upflow digestion process operated with a 6 wt% TS-content feed slurry was projected to require a total fermenter volume of 8270 m^3 (292,000 ft^3), and could exhibit a net energy production of 823 x 10^6 KJ/day (780 million Btu/day)- Thus, whereas there is no net energy production from conventional digestion, about 83% of the methane from the two-stage process is available as surplus.

Table 2. Performance of the Biphasic Two-stage Upflow Mesophilic (35°C) Digestion System at an HRT of 5.9 Days (1.3 Days for Acid Phase and 4.6 Days for Methane Phase) with a 5.8-Wt % TS-Content Feed

Operating Conditions/Performance	Acid Phase	Methane Phase	System
Run Duration, No. of HRT's	35	10	8
Loading Rate, kg/m³-day	28.84	7.85	6.25
Gas Production			
Methane Yield,			
SCM/kg VS added	0.06	0.42	0.48
Methane Content, mol %	59.2	70.1	68.4
Methane Production Rate, vol/vol-day	1.77	3.27	2.96
Effluent Characteristics			
pH	6.6	7.2	7.2
Volatile Acids, mg/L			
Acetic	643	77	77
Propionic	2251	48	48
Isobutyric	123	0	0
n-Butyric	141	0	0
Isovaleric	266	0	0
n-Valeric	79	0	0
Caproic	0	0	0
Total as Acetic	2827	118	118

Table 3. Comparison of Hypothetical Conventional and Two-stage Upflow Mesophilic (35°C) Digestion Systems to Stabilize and Gasify 90.9 Metric Tons/day (100 Tons/day) (Dry Solids Basis) of Sludge at an HRT of 5.5 Days

Operation and Performance	Conventional	Two-Stage Upflow
Feed VS, wt %	2.2	3.7
Loading Rate, kg/m^3-day	4.01	6.57
Methane Yield, SCM/kg VS added	0.125	0.481
Methane Production Rate, vol/culture vol-day	0.5	2.8
VS Reduction to Gas, %	24	75
Gross Methane Production, SCM/day	6797	26,621
Estimated Operating Energy Requirement, 10^6 kJ/day		
Feed Sludge Heating	256	161
Mixing	6.33	0
Pumping	2.11	3.2
Heating, Ventilation, Lighting, Other	8.44	5.3
Total	273	170
Net Energy Production, 10^6 kJ/day	-20	823
Digester Volume, 1000 m^3	13.6	8.3

Another important advantage of the two-stage upflow digestion process is the substantially higher solids reduction that is achieved relative to that of conventional digestion (Table 3). Thus, the cost of digested sludge disposal for the two-stage system could be one-third of that for the conventional process.

LEACH-BED TWO-PHASE DIGESTION OF HIGH-SOLIDS/DRY FEEDS

Methane fermentation of high-solids feeds occurs naturally in municipal landfills, manure piles, peat beds, and other solid-bed organic deposits, but these natural processes are uncontrolled, are slow to start, and require an unduly long time--frequently many years--to achieve completion of the methane fermentation process. Three important reasons for impeded fermentation within a high-solids bed are mass transport limitations within the bed, the occurrence of an imbalance between volatile fatty acids production by acidogenic bacteria and the conversion of acids to methane by methanogenic bacteria, and the absence of adequate amount of moisture in the solid bed.

An approach that has considerable potential of overcoming the above problems of high-solids feed digestion is to induce rapid bioleaching of the organic bed by application of an appropriate acidogenic culture, and to accelerate the leaching process by reactivating the percolating culture under controlled environment for reapplication on and recirculation through the bed (Figure 2) (1). As the bioleaching process is continued solubilized organics, volatile-acid precursors, and volatile fatty acids accumulate in the recirculating stream. Once the total fatty acids concentration reaches an inhibitory level, a part of the bioleachate is conveyed to an external methane-phase digester for the production of acetate--a substrate for methane fermentation--and the conversion of acetate to methane and carbon dioxide. The methane-phase effluent is recycled to the acid-phase digester and ultimately to the organic bed to conserve the nutrients indigenous to the solid waste, and thus to eliminate the need for addition of external water or nutrients. The overall process thus consists of a leach bed, and acid and methane-phase fermenters external to the bed.

The leach-bed two-phase anaerobic digestion process described above incorporates the basic principles of two-phase digestion, and is expected to be applicable to MSW deposits (landfills), manure and gob piles, industrial

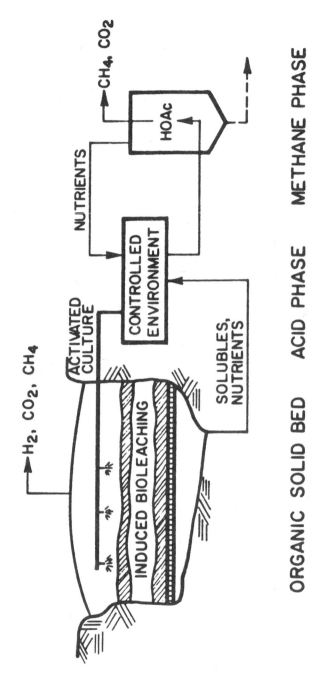

ORGANIC SOLID BED   ACID PHASE   METHANE PHASE

Figure 2.   Leach-bed two-phase anaerobic digestion of MSW.

solid wastes, agricultural and forestry residues, peat beds, and other heterogeneous organic deposits. This digestion process is expected to be superior to slurry-phase digestion and the so-called "dry" fermentation process because of the following potential advantages:

- Minimum feed processing (e.g., shredding and separation) is required
- No slurrification is necessary
- No mixing is needed
- Nutrients need not be added
- Accelerated decomposition of organic beds is promoted
- The process lends itself to *in-situ* gasification with ultimate disposal of dry wastes and biomass
- Fermentation is conducted in simple containment vessels
- Possibility of auto-heated digestion
- Recovery of methane at controlled and predictable rates
- Reduced operating energy requirement
- High net energy production
- Feasibility of energy storage in the form of fatty acids and other liquefaction products
- The substrate remains moist without addition of external water

Materials and Methods

The leach-bed digestion studies were conducted with RDF (refuse-derived feed) produced in a Midwestern refuse processing plant. Based on the elemental analyses, the MSW feed had a theoretical methane yield of 0.46 m^3/kg VS reacted (7.3 SCF/lb VS reacted) assuming that all carbon in the refuse is converted to bacterial mass. However, long-term anaerobic bioassay fermentation tests showed methane yields of 0.26 m^3/kg VS added (4.1 SCF/lb VS added) without external nutrients and 0.29 m^3/kg VS added (4.7 SCV/lb VS added) with external nutrients indicating refuse biodegradabilities of about 56% and 64% respectively.

The experimental work was conducted in a bench-scale system similar in design to that depicted in Figure 2. The system consisted of a 100-L culture-volume refuse bed, a 1.8-L facultative acid-phase digester, and a 12.5-L (gross volume) methane-phase upflow anaerobic filter. All units were hermetically sealed. Only the filter was heated to maintain a mesophilic temperature of 35°C. All units were constructed from Plexiglas.

The refuse bed contained 26.21 kg of shredded and separated Midwestern MSW having total solids, VS, and total carbon concentrations of 68.6 wt %, 59.4 wt % (of TS), and 33.8 wt % (of TS), respectively. The leach bed was inoculated with a carbohydrate-acclimated acidogenic culture. Solid-phase acidogenic fermentation of the refuse bed was started by percolating the applied culture and recirculating it around the bed.

System operation and performance were monitored by analyzing the influents and effluents for volatile acids, pH, and chemical oxygen demand (COD). Gas productions from the refuse bed and the anaerobic filter were measured daily or at suitable intervals, and the gases were analyzed for methane, carbon dioxide, and hydrogen concentrations.

Results and Discussion

As indicated above, operation of the leach-bed two-phase process was initiated by first starting solid-phase acidic fermentation of the dry organic substrate. Within about four days of solid-phase fermentation, significant liquefaction and acidification of the RDF occurred as indicated by increases in VA, soluble COD, and gas-phase hydrogen concentrations from 0 mg/L to about 5000 mg/L, 5000 mg/L to 12,000 mg/L, and 0 mol % to about 50 mol %, respectively. These observations indicated that accelerated liquefaction of the organic bed could be achieved rapidly under the fermentation conditions employed. Operation of an anaerobic filter methane-phase digester was started at this time with the high VA-content effluents from the solid-phase acid digester used as a feed. The methane digester had been operated earlier with refuse-derived acidic effluents; but it was not operational for several months before this run. Anaerobic filter operation was initiated at a high HRT and a low COD loading rate, but the HRT could be reduced to 8 hours and the loading rate could be increased to 2.4 kg COD/m^3-day after four hours of operation without affecting methane production. The methane digester effluent was recycled to the leach-bed digester via the acid-phase digester operated at an HRT of 1-2 hours. The ORP (Oxidation Reduction Potential) of the acid-phase digester contents was controlled between -200 mV and -300 mV by limited diffused-air aeration to eliminate the methanogens in the recycle stream, and thus to minimize volatile acids gasification in the bed.

The rate of liquefaction of the leach-bed solids increased to a peak value after about 20 days of operation as
indicated by the COD, volatile acids, and methane-concentra-
tion profiles in Figure 3. The liquefaction efficiency,
defined as the ratio of the sum of the bioleachate and methane
COD's to the raw-refuse COD, increased exponentially to about
28% during the first 20 days of operation, after which time
it only increased by about 10 percentage points during the
next 80 days when bed stabilization was virtually completed.
Liquefaction products were volatile acids and other soluble
compounds as detected by the filtrate-COD determinations.
Acetate was the major volatile acid produced during the
initial 15 days of operation, after which time propionic
acid predominated over the other acids. Peak concentrations of acetate and propionate were about 3300 mg/L and 8200
mg/L occurring after 7 days and 30 days into the run. A
peak VA concentration of about 1% was achieved in the bio-
leachates. If these high VA concentrations could be maintained and exceeded, then it could be worthwhile to recover
these valuable chemicals.

The refuse bed exhibited little gas production during
the initial 15 days of system operation. Gas production
started after this time and leveled out after about 65 days
of bed operation. A gas yield of 0.27 std m^3/kg VS added
(4.3 SCF/lb VS added) was observed after about 97 days of
operation.

Considerable consolidation of the solid refuse bed took
place as a result of gasification. The bulk density of the
hand-packed bed material increased from 268 kg/m^3 (3.5 lb/ft^3)
before digestion to 770 kg/m^3 (9.9 lb/ft^3) after 97 days of
digestion.

The performance of the methane-phase anaerobic filter,
which was packed with burl saddles, improved markedly within
three days after start up. Methane production rate increased
from 0.21 vol/day-vol at start-up to about 4.0 vol/day-vol
on the sixth day of operation. As indicated by the VA profiles in Figure 4, operation of the filter was so adjusted
that filter-effluent VA's closely paralleled the bed-effluent
VA's. The methane-phase filter was able to accept influent
VA and COD conversion efficiencies up to 81% and 80%, respectively. Depending on the stage of operation, methane
content of the gases ranged between 50 mol % and 79 mol %,

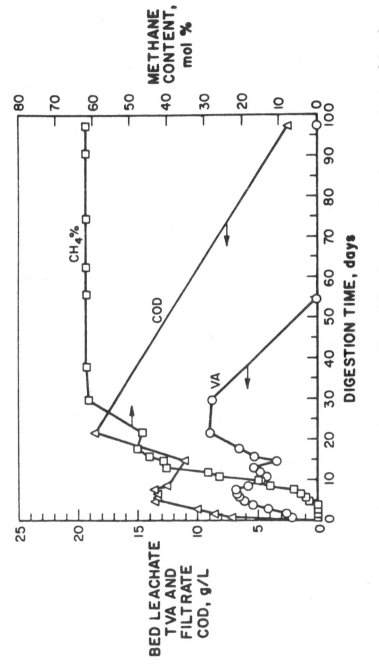

Figure 3.   Time profiles of leachate volatile acid and COD, and methane content of head gases for MSW solid bed.

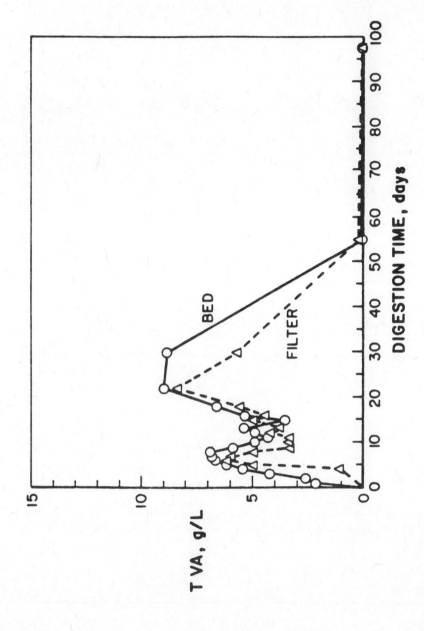

Figure 4.   Time profiles of volatile acids in MSW bed leachate and methane-phase filter effluent.

and methane production rate after start-up varied from 4 vol/day-vol to 7 vol/day-vol. These performances obtained with the refuse bioleachate were superior to filter performances observed with other feeds.

The overall two-phase system exhibited a biogas yield of 0.38 std m^3/kg VS added (6 SCF/lb VS added) and a methane yield of 0.21 std m^3/kg VS added (3.3 SCF/lb VS added) after about 97 days of two-phase solid-bed digestion. Comparison of this methane yield with the value of 0.26 m^3/kg VS added observed during the long-term bioassay test indicated that about 81% of the biodegradables were gasified after 97 days of digestion. The total mass of gases produced during this period accounted for about 45% of the raw refuse VS indicating a VS reduction of higher than 45%.

CONCLUSIONS

Data reported in this paper showed that conventional single-stage complete-mix anaerobic digesters experience unbalanced acidogenic and methanogenic fermentations and exhibit low stabilization efficiencies when they are operated with high solids-content feeds and at low HRT's. These limitations of the conventional process were overcome by the two-phase digestion process.

Upflow two-phase anaerobic digestion of primary sludge at 5.9-day HRT exhibited a mesophilic methane yield of 0.48 SCM/kg VS added representing a 74% VS reduction and complete conversion of the biodegradable sludge VS; this yield was higher than those reported in the literature for sludge. The methane production rate of 3 volumes per day per unit culture volume observed with a 5.8 wt% TS-content sludge was 4-5 times that achieved with conventional single-stage digestion. Under the HRT and feed solids-concentration conditions mentioned above, a conventional single-stage process does not produce any surplus methane; by comparison, 83% of the methane from the two-phase process could be available as surplus.

For low-moisture (or "dry") organic feeds, a two-phase process consisting of a leach-bed acid digester, a slurry-phase acid digester, and an upflow methane digester was more appropriate. A leach-bed digester containing MSW was utilized to conduct solid-phase acidic fermentation to produce a high VA-content aqueous effluent. The slurry-phase acid digester was most needed during start-up of the leach bed. The methane digester received the low SS-content acidic

effluents from the leach bed, and exhibited methane production rates up to 7 vol/day-vol, and VA and COD conversion
rates up to 81% and 80%, respectively, at an HRT of 8 hours
under mesophilic (35°C) conditions. Considerable gasification of the MSW occurred in the leach bed under conditions
of batch operation. A biogas yield of 0.27 SCM/kg VS added
was observed under ambient-temperature fermentation conditions compared with an overall system biogas yield of 0.38
SCM/kg VS added. The leach-bed two-phase digestion process
afforded a methane yield of 0.26 SCM/kg VS added. About
81% of the biodegradable VS was converted by this process
after 97 days of batch digestion. This performance was
superior to those of other landfill gasification processes.

ACKNOWLEDGMENTS

This research was supported in part by Bay State Gas
Company, Boston Gas Company, Brooklyn Union Gas Company,
Cogenic Systems, Inc., Essex County Gas Company, Valley
Resources, Inc., the Executive Office of Energy Resources,
Commonwealth of Massachusetts, and the Institute of Gas
Technology. The efforts of Messrs. A Sajjad and M.P. Henry
in helping with the experimental work are gratefully ack-
knowledged.

LITERATURE CITED

1.   Ghosh, S., "Gas Production by Accelerated In-Situ
     Bioleaching of Landfills," U.S. Patent 4,323,367,
     April 6 (1982).
2.   Ghosh, S., J.R. Conrad, and D.L. Klass, "Anaerobic
     Acidogenesis of Sewage Sludge," *J. Water Pollut. Con-
     trol Fed.*, *47*(1), 30-45 (1975).
3.   Ghosh, S. and D.L. Klass, "Two-Phase Anaerobic Digestion,"
     U.S. Patent 4,022,665 (1977).
4.   Ghosh, S. and D.L. Klass, "Two-Phase Anaerobic Diges-
     tion," *Process Biochemistry, 13,* 15-24 (1978).
5.   Ghosh, S. and F.G. Pohland, "Kinetics of Substrate
     Assimilation and Product Formation in Anaerobic
     Digestion," *J. Water Pollut. Control Fed.*, *46*(4),
     748-759 (1974).

22
ENGINEERING STUDIES ON MSW AS SUBSTRATE FOR METHANOGENESIS

L.F. Diaz, G.M. Savage, and C.G. Golueke

Cal Recovery Systems, Inc.
Richmond, California

ABSTRACT

This paper is a summary of three studies conducted and reported by the authors on the anaerobic digestion of raw MSW. The first study (1968-72) was on the digestibility of individual organic components of MSW and of organic refuse as a whole. In the second study (1973-76) was explored the feasibility of lowering the cost of digestion and simultaneously increasing its efficiency by having it constitute a unit process in a waste treatment-resource recovery facility. Loading rates, and potential advantages from "co-digesting" sewage sludge and MSW were investigated. The third study (1981-82) was a determination of the digestibility of outputs of unit processes designed to separate MSW into feedstocks for the recovery of inorganic resources and of organic resources destined for energy recovery. It was a part of a study on the integration of biogasification into an operation designed to thermally recover energy bound in the combustible fraction of the processed wastes. A 70% energy recovery is possible with such an integration.

INTRODUCTION

The present paper is a summary of three studies conducted at various times during the past 15 years on the anaerobic digestion ("digestion", "biogasification") of municipal refuse (MSW), and in which the authors had a part. The first study took place during the period 1968-1972 (Golueke and McGauhey, 1969; McFarland et al., 1972); the second study, from 1973-1976 (Diaz, 1976; Diaz, et al., 1981); and the third study, from 1981-1982 (Golueke et al., 1982).

Aside from the fact that to do so seemed appropriate for the
Symposium, the rationale for incorporating the three studies
into a single paper was threefold: 1) The sequence and nature
of the studies exemplify the evolution of the existing knowledge on the anaerobic digestion of MSW.   2) They reflect the
effect on the course of the research on anaerobic digestion of
refuse had by the shift in national priority from mere disposal to energy recovery.   3) They typify the logical success-
ion expected for research on MSW digestion begun at a time
when information was confined to that on the disposal of the
garbage fraction into the sewerage system by way of a kitchen
garbage grinder.

The first study was conducted at the Sanitary Engineering
Research Laboratory (University of California, Berkeley); the
second, at the Mechanical Engineering Laboratory (University
of California, Berkeley); and the third, at the Cal Recovery
Systems research facilities (Richmond, California).

MATERIALS AND METHODS

Digester Start-Up:

In the three studies, digesters were "started" by filling
them to full working volume with mesophillically digested
sewage sludge. The digester cultures were then adapted to the
test feedstock by gradually decreasing the sludge content in
the refuse-sludge loading. (Reactors and feedstocks are described in the sections on the individual studies.)

Digester Operation:

The digesters were fed on a semicontinuous basis, i.e.,
once each day to three times per week, depending upon the
demands of the experiment. Loading rates were a function of
the experiment. The other principal operational variable was
hydraulic retention time. In all studies the temperature was
maintained within the range of 30° to 35°C. No attempt was
made to adjust or control pH level.

Parameters:

Stability of gas production and composition, organic
acid concentration, pH level, and buffer capacity served
as the four operational parameters. Digester performance was
judged on the basis of gas produced per gram volatile solids

(VS) introduced, composition of the gas, rate and extent
of volatile solids desctruction, chemical and physical
characteristics of the sludge solids and liquor, and energy
conversion efficiency.

RESULTS

   First Study:

   Objective. The principal objective of the first study
was to determine the digestibility of each of the individual
organic components of MSW and of organic refuse as a whole.
At the time (1960's), little research had been devoted to the
subject, and hence the available knowledge was minimal. Interest in the digestion of MSW was almost entirely in its use
as a means of disposal--energy recovery had not as yet acquired an urgency in the national consciousness. The principal concern of practically all research conducted previously
was about conjectured adverse effects on the digestion process itself (Babbitt et al., 1936; Schlenz, 1944) and on the
capacity of existing digesters to accommodate the increased
loadings (Watson and Clark, 1962).

   Results. Initial experiments were concerned with permissible ratios of garbage and other refuse components to
sewage sludge. Reactors used in these experiments were 3.78
liter bottles.

   In all experiments it was found that acclimation of the
digester culture to the component or components being tested
was a prerequisite to successful performance. In experiments
with paper pulp added to raw sewage solids, as much as 90%
of the cellulose content was degraded. Digestion proved to
be feasible at loadings of paper as much as 60% of the total
solids in the digester. Digester failure occurred when the
C/N ratio reached 52:1. In experiments with a mixture consisting of 10% newsprint and 90% sewage sludge, 83% of the
total cellulose and only about 55% of the newsprint cellulose
was destroyed at a retention time of 30 days. Increasing the
newsprint content of the mixture of 30% resulted in a destruction of 63% of the total cellulose and 42% of the newsprint cellulose. Grass clippings proved to be 73% digestible.
Under the conditions of the experiment, wood was impervious
to digestion, but it did aggravate the scum formation.

The experiments on organic refuse as the sole feedstock were conducted with the use of a 1.5 m^3 reactor. The feedstock was size reduced refuse from which inert components had been removed manually. From 86 to 96% of the loading was cellulosic in nature. Allowing for the introduced volatile solids (about 18%) and cellulose (about 22%) present in the scum layer, volatile solids destruction amounted to about 64.5%; and cellulose destruction, about 77%. Gas production averaged 0.62 liters/g VS destroyed, and 0.43 liters/g VS introduced. The ratio of CH_4 to CO_2 was on the order of 61 to 39. The major problem was scum formation.

Second Study:

Objective. In the second study was explored the feasibility of lowering the cost of digesting MSW and simultaneously increasing its efficiency and reliability by having anaerobic digestion constitute a unit process in a waste treatment-resource recovery facility. The rationale for the approach was based in large part upon the premise that it is possible to integrate and arrange the various unit processes such that residue from one process becomes the input to a succeeding one.

Procedure. The feedstock was a highly organic material segregated from MSW by way of a dry separation system developed by the authors (Diaz et al., 1982). The system is diagrammed in Figure 1. Refuse, direct from the collection vehicle, was size reduced and air classified. The "lights" were passed through a trommel screen (13 mm openings). In the study, the material of interest was the fraction that passed through the openings of the trommel screen, i.e., the "rejects". (Materials retained on the screen can serve as a refuse derived fuel or can be processed through a fiber recovery system.) The rejects are termed "light rejects" so as to distinguish them from the heavies output of the air classifier. The light rejects were subjected to a series of screenings. The composition of the light rejects at various ranges of screen opening sizes is indicated by the bar graph in Figure 2. The light fraction that passed the 12.7 mm screen and was retained on the 7-mesh screen was used as the feedstock in the second and third studies.

Results. The result showed that digester feedstock produced by the system used in the study served as a satisfactory substrate for anaerobic digestion, and that supplementation with other materials was not necessary.

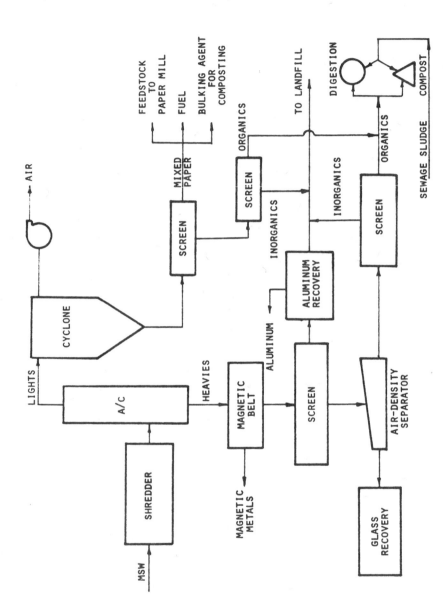

Figure 1. Schematic diagram of resource recovery process.

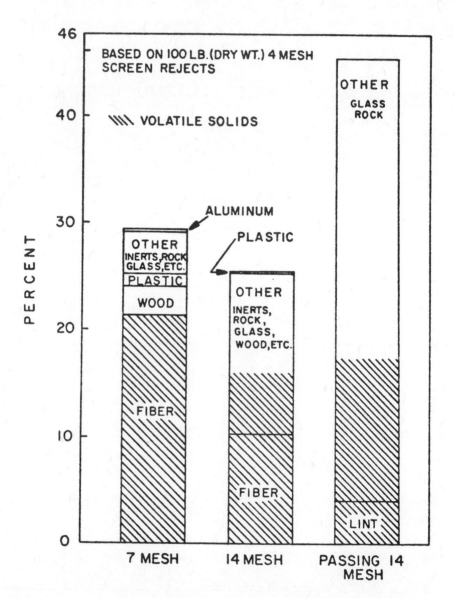

Figure 2.   Composition of light rejects at various screen opening sizes.

At a loading rate of 1.12 g VS/liter/day and a retention time of 30 days, gas production was independent of the refuse to sludge ratio. Average gas production (57% CH_4, 32% CO_2, 11% N_2) was 0.42 liter of gas/g VS added. At a 4:1 ratio of refuse to sludge loading, VS destruction was 65%. Increasing the loading to 1.6 g VS/liter/day combined with shortening the retention time to 15 days brought about a drop in gas production to 0.31 liter/g VS added. Volatile solids destruction was 62.7% when the refuse content of the feedstock was increased to 100%. The maximum permissible loading rate seemed to be on the order of 6.4 g VS/liter/day.

Third Study:

Objective. The third study was conducted principally to determine the digestibility of outputs of unit processes designed to separate MSW into feedstocks for the recovery of inorganic resources and of organic resources destined for energy recovery.

Procedure. In the laboratory studies, four 289-liter reactors were used; and in the pilot study, an especially constructed 1.6 m^3 digester. A feature of the pilot digester was the positioning of the stirring blades of a mechanical stirrer in the surface layer of the culture.

The feedstock was prepared in the manner followed in the second study. Loadings in the laboratory phase ranged from 1.04 to 3.7 g VS/liter/day; and in the pilot study, from 1.6 to 4.8 g VS/liter/day. Retention times in the laboratory and pilot phases were 15 and 20 days.

Results and Discussion. The experimental results indicated that when the loading was no greater than 2.4 g VS/liter/day, about 0.3 or 0.4 liters of medium heating value digested gas (55-60% CH_4) could be produced from a gram of volatile solids in MSW processed as described herein. A major problem encountered in the study was the buildup of a dense scum layer. Although increasing the intensity of mixing and prolonging its duration and frequency might have lessened scum formation in the pilot scale digester, doing so would have been self-defeating in terms of net energy recovery.

DISCUSSION AND CONCLUSIONS

Results obtained by others and in the three studies described herein demonstrate that, although MSW could serve as a substrate in the use of anaerobic digestion to recover energy from MSW, doing so would by no means be a trouble-free endeavor. With respect to the digestion process itself, some of the problems are related to the suitability of MSW as a feedstock and some to the operation and maintenance of the culture due to reactor inadequacies. With respect to MSW as a feedstock, front-end processing with the system developed by the authors could produce a substrate suitably low in C/N and devoid of plastics and other objectionable low-density components. The particularly troublesome operational problem of scum formation probably could be kept under control by means of an appropriately designed and operated mixing device. Unfortunately, thus far no such device seems to be available; and if one were at hand, energy considerations would place an upper limit on its operation.

A major problem is that of maximizing the energy return (Corder et al., 1984). Researchers have found that because of the refractory nature of certain constituents, only a fraction of the energy contained in the organic fraction of MSW can be recovered through biogasification. With our approach, this problem is compensated by diverting a sizeable fraction of the organic refuse to RDF production. The result is a potential 70% recovery of the energy in the raw refuse instead of the 25 to 30% recovery attainable by digestion alone.

REFERENCES

Babbitt, M.E., B.J. Leland, and F.H. Whitley, Jr., "The Biological Digestion of Garbage with Sewage Sludge," Bull. No. 287, Engineering Experiment Station, University Illinois, Urbana, IL, p. 112 (1936).

Corder, R.E., A.M. Hill, H. Lindsey, M.Z. Lowenstein, and R. P. McIntosh, "SERI Biomass Program FY 1983 Annual Report," SERI/TR-231-2159 DE 84004487, Solar Energy Research Institute, Golden, CO, pp. 47-67 (February 1984).

Diaz, L.F., "Energy Recovery through Biogasification of MSW and Utilization of Thermal Wastes from an Energy-Agro-Waste Complex", Doctoral Dissertation, Department of Mechanical Engineering, University of California, Berkeley, p. 242 (1976).

Diaz, L.F., G.M. Savage, and C.G. Golueke, Resource Recovery from Municipal Solid Wastes, Vol. I, CRC Press, Inc., Boca Raton, FL, p. 166 (1982).

Diaz, L.F., G.M. Savage, G.J. Trezek, and C.G. Golueke, "Bio-gasification of Municipal Solid Waste," *Transactions of the ASME, 103,* 180-185 (June 1981).

Golueke, C. G., L.F. Diaz, G.M. Savage, D.J. Lafrenz, and D. B. Jones, "Laboratory and Pilot Studies of Refuse Digestion," report prepared by Cal Recovery Systems for Southern California Edison Co., Rosemead, CA, p. 114 (March 1982).

McFarlane, J.M., P.H. McGauhey, S.A. Klein, and C.G. Golueke, "Comprehensive Studies of Solid Waste Management," SERL Rep. 72-3, Sanitary Engineering Research Laboratory (SERL), Univeristy California, Berkeley, pp. 139-150 (1972).

Schlenz, H.E., "Controlled Garbage Digestion," *Journ. Western Society of Engineers, 49,* 273 (March 1944).

Watson, K.S. and C.M. Clark, "How Food Waste Disposers Affect Plant Design Criteria," *Public Works, 93,* 105 (June 1962).

23
EVOLUTION OF THE RefCOM SYSTEM

John T. Pfeffer

*Department of Civil Engineering
University of Illinois
Urbana, Illinois*

ABSTRACT

RefCOM is the acronym for the project initiated by the U.S. Department of Energy for demonstration of the biological conversion of refuse to carbon dioxide and methane. In the 1960's and 1970's, a substantial research effort was undertaken to investigate the feasibility of biological conversion of urban refuse to a fuel gas. This research provided basic process parameters needed to develop a conceptual design of such a system. Mass balance and economic evaluations showed that the system had potential. In 1975, a request for proposal was issued by the NSF/RANN program for the design and construction of a proof-of-concept system. Plant startup was initiated in 1978. However, a substantial number of startup problems were encountered. Operation with a modified front-end system was eventually achieved. The biological process was found to function as predicted from the laboratory results. However, numerous operating problems were encountered with the mechanical refuse processing, the feed slurry preparation, the reactor mixing and the slurry dewatering systems. Redesign of these systems was necessary to obtain reliable operation.

INTRODUCTION

The disposal of the millions of tons of urban solid waste that are produced annually has been a costly proposition for municipalities, especially with the advent of more strict environmental quality control regulations. Techniques for recovery of useful material from this mass have been evaluated as a means of reducing these costs. Energy recovery by combustion has been the primary technique for by-product recovery. However, the decreasing reserves of natural gas experienced in the early 1970's made the production of a substitute natural gas attractive.

Early research (in the 1930's) demonstrated that the garbage (food waste) component of solid waste was amenable to anaerobic fermentation. Later work (1950's and 1960's) found that cellulose was a good substrate for this process. It was also recognized at this time that there is a close relationship between the microbiology and biochemistry of an anaerobic digester and that of the rumen in cattle. There is a substantial body of research on the digestion of complex fibers by the rumen. Research in the early 1970's on the production of carbon dioxide and methane from urban refuse developed a technical base for the conceptual design of a process for production of a substitute natural gas from urban refuse.

Prior to the creation of a national energy agency, this research was funded by the NSF Research Applied to National Needs (RANN) program. In 1975 NSF/RANN issued a request for proposal for the design and construction of a proof-of-concept experiment for this system. Before work could be initiated on this project it was transferred to the municipal waste program within the newly created Energy Research and Development Administration (ERDA). ERDA was eventually combined with a number of other agencies and administrations to form the Department of Energy (DOE). This project was funded by DOE at its inception. The system resulting from this R and D effort is known as RefCOM.

BACKGROUND

The typical municipal solid waste (MSW) produced in the United Stated contains a large fraction of organic material (see Table 1). This material falls into three primary categories; paper and paper products, green garbage and other food wastes, and yard and garden debris. Carbohydrates such as cellulose and hemicellulose are the dominant organics present in these waste categories. In the absence of molecular oxygen, i.e., anaerobic conditions, the carbohydrates are fermented to intermediate organic compounds which are finally converted to methane and carbon dioxide. A complete treatment of the biochemical transformations occurring in this process is presented by McInerney and Bryant (1981).

Table 1. Typical MSW Composition

Component	% wet wt.	% dry wt.
Food waste	15	7
Paper and paper products	44	35
Garden trimmings	12	5
Glass	8	8
Metal	9	9
Plastics	3	3
Misc. organics	5	4
Misc. inorganics	4	3
Moisture content	--	26

Source: Resource Recovery and Source Reductions, Second
        Annual Report to Congress, U.S.E.P.A. Pub. No.
        SW-122, Washington, DC (1974).

The application of anaerobic fermentation to the treatment of solid waste in the United States was first investigated by Babbitt et al. (1936). The introduction of home
garbage grinders posed the question of compatibility of
these solids with the anaerobic sewage sludge digestion process. This question was answered with a series of experiments mixing garbage (food wastes) and sewage sludge for
use as substrate in laboratory fermentation systems. The
results demonstrated that garbage was amenable to anaerobic
fermentation.

Ross (1954) reported on the dual disposal of garbage
and sewage sludge at Richmond, Indiana. In this process,
the garbage was shredded and washed for grit removal. The
garbage was then pumped into an enlarged sludge digestion
system. The process operated in the mesophilic temperature
range and worked satisfactorily, with the gas being used
to provide the power for operation of the water pollution
control plant. Ross reported a net profit for the operation
of the system.

In 1965, work was initiated at the University of California at Berkeley on the characteristics of solid waste
and the effect of solid waste on the methane fermentation
process. Golueke and McGauhey (1970) estimated that paper
accounts for about 67% of the refuse that can be converted
biologically to methane. Garbage accounts for only 12% of
the biodegradable refuse. The nitrogen content of MSW was
found to be low. Consequently, supplemental nitrogen was
required to obtain a balanced fermentation.

The importance of cellulose and the rate of cellulose hydrolysis in the anaerobic digestion process was first postulated by Maki (1954) in his study of cellulolytic bacteria isolated from a municipal sewage sludge digester. A study by Chan and Pearson (1970) on cellulose hydrolysis indicated that the hydrolysis of insoluble cellulose to soluble cellobiose appeared to be the rate limiting step in the anaerobic decomposition of cellulose.

In 1969, studies on the digestion of MSW were started at the University of Illinois (Pfeffer, 1974; 1974a). These studies initially concentrated on conditions for achieving maximum conversion of MSW to methane. The major findings of the initial work was the improved conversion efficiencies obtained with digestion at thermophilic temperatures. Subsequent studies developed data on the slurry properties and the dewatering characteristics of the slurry (Pfeffer and Liebman, 1976).

In addition to the above research, Dynatech R/D Co. evaluated the economic and technical feasibility of a system for producing methane from MSW (Kisper et al., 1975). Their analysis showed that the process had good potential for development. Citing the studies that had been conducted on the system to date, the technical staff of NSF recommended that funds be budgeted for a proof-of-concept experiment. Funds were budgeted and a request for proposal was issued in May 1975.

Based on the results of the above studies, the following conceptual design for the methane from refuse plant was developed. The design consisted of the following subsystems: refuse processing, methane fermentation, residue dewatering and disposal, and gas purification.

The objective of the refuse processing subsystem was the production of a feed material that would contain essentially all of the biodegradable organic material and as little of the remaining refuse components as possible. This subsystem contained the following units:

- Primary shredder
- Magnetic separator
- Trommel screen (19 mm openings)
- Secondary shredder
- Air separator

The object of the methane fermentation subsystem was to obtain the maximum feasible conversion of biodegradable organic matter to gas. The following units were part of this subsystem:

- Heated completely stirred tank reactor (CSTR)
- Steam generator
- Feed slurry preparation tank
- Gas handling unit

The object of the residue dewatering and disposal subsystem was the separation of the residual solids from the liquid fraction of the fermenter effluent in a form that will permit incineration of these solids with heat recovery (at least for in-plant use). The following units were part of this subsystem:

- Solids dewatering unit (vacuum filter or centrifuge)
- Fluidized bed furnace
- Waste heat boiler
- Ash handling unit

A gas purification system was included to remove acid gases (carbon dioxide), water vapor, and other impurities in order to satisfy the specifications for pipeline quality gas. An additional objective may be the recovery of carbon dioxide for production of dry ice or liquid CO_2. This subsystem was composed of the following units:

- Absorption tower
- Monoethanolamine (MEA) regenerator
- Steam generator
- Gas compressors
- Gas dehydrator

The above system is one that would have universal application with the product being a pipeline quality gas. Because of budget constraints and the uncertainty of the performance of the methane fermentation system, a complete system was not constructed for RefCOM. Figure 1 shows the schematic of the initial system installed at the Pompano Beach, Florida facility. It was a complete system up through the digester slurry dewatering step. Incineration of the dewatered cake and gas clean up were not included.

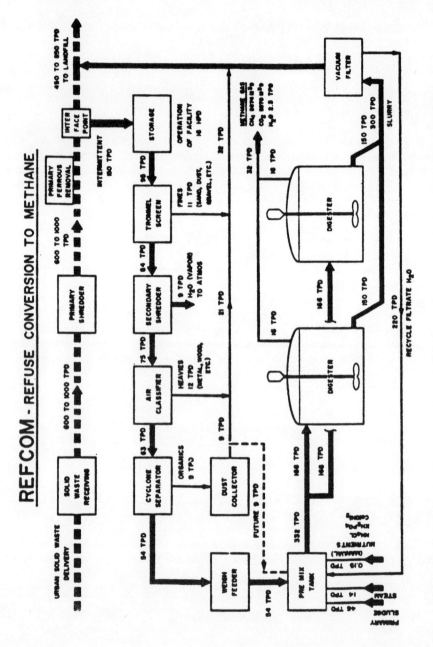

Figure 1. Schematic diagram of the RefCOM process as initially constructed.

OPERATION OF RefCOM

Start-up of the plant was a major problem. Multiple problems existed, ranging from inadequate conveyor systems to a digester that would not retain gas. Many of these problems are documented elsewhere (Anon., 1981; Pfeffer and Isaacson, 1982) and will not be repeated. Essentially, it was impossible to consistently operate the refuse processing subsystem. Consequently, it was very difficult to achieve steady state operation of the fermenter. Additional problems were encountered with the reactor mixing system and with the integrity of the reactor. Resolution of the above problems left two areas that limited the plant operation; 1) feed slurry concentration, and 2) the loading rate on the vacuum filter.

After several years of attempting to make the initial refuse processing subsystem operationally reliable, a decision was made in 1982 to replace the entire refuse processing train between the storage building and the weigh feeder (see Figure 1). The revised system contains two disc screens that separate the refuse into three different streams; greater than 15 cm (6 in.), 1.25 cm to 15 cm (0.5 to 6 in.) and less than 1.25 cm (0.5 in.). A description of the operation of the disc screen is given by Hamilton and Kelyman (1979).

The oversize material is primarily plastic, textiles and some paper. With the proper operation of the primary shredder, this fraction will be about 5% or less of the total refuse stream. The undersize material is primarily dust, dirt, shattered glass, and other inorganic material. This fraction may contain a substantial portion of organic material, especially grass clippings. Both the oversize and undersize streams are rejected. The middle size fraction becomes the feed to the digester. The particle size range of this middle fraction is such that cans and unbroken glass will report here. Magnetic belts and head pulleys can be used to remove the ferrous components. An air knife has been inserted in the process train to separate the aluminum cans, glass and other nonferrous dense material in this size range. The air knife is a horizontal flow of air that aerodynamically separates the light material. As the discharge from the 1.2 cm screen drops through this air flow, the light material is carried horizontally with the air while the dense material continues its vertical fall. The resulting digester feed material is expected to be high quality. Data will be forthcoming on the efficiency of this processing.

The revised design of the refuse processing system has been shown to be operationally reliable. In addition, the horsepower requirements for preparing the feed for the digester have been substantially reduced by replacing the 200 HP on the secondary shredder and the 100 HP on the air classifier with 35 HP used to drive the disc screen and the air knife. The current system is capable of processing refuse at a rate in excess of 10 tons per hour. This equipment was placed in operation in June 1984. Operation has been continuous except when external factors caused shutdown of the plant.

The original facility was operated intermittently until mid-November of 1981. At this time a sustained run was initiated, and steady-state conditions were achieved with an 18-day hydraulic retention time and a fermentation temperature of 60°C (140°F). Gas production during the month of January 1982 was 1706 m^3/d (60,280 ft^3/d) (Mooij, 1982). The gas was 53.3% methane. The variation in gas production was significant, characterized by a standard deviation of 283 m^3/d (10,000 ft^3/d). This variation was the result of many factors, from the tonnage fed to the digester, to the composition of the feed material. A more relevant measure of gas production would be the gas volume per unit of dry volatile solids fed to the digester. The data for January 1982 show a gas yield of 0.51 m^3/kg (8.17 ft^3/lb), with a standard deviation of 0.08 m^3/kg (1.3 ft^3/lb). The gas volume was measured in terms of a saturated gas at a temperature ranging from about 30 to 40°C (90 to 105°F).

One of the most satisfying observations has been the stability of the methane fermentation process. The biological fermentation has not been upset during the four years of intermittent operation. However, the loading rates have been low, so one would not expect any instability. It has yet to be demonstrated how the system will perform at the high loading rates necessary for economic viability. Also, during these 4 years, the 10^3 Mg (1100 tons) of refuse processed by this system did not contain any materials that were toxic to the microorganisms in the digester.

Although the microbial system operated as expected, numerous problems were encountered with the physical components of the digestion system. These problems ranged from mixer shaft failures to gas leaks in the cover to excessive plugging of slurry withdrawal pipes. These problems are

discussed in detail elsewhere (Pfeffer and Isaacson, 1982).
The solutions to these various problems have been developed
and it is possible to design a digester system that will
be reliable.

For reasons of economy, only a partial residue disposal
system was installed. This consisted of a rotary-drum vacuum
filter for dewatering the digester effluent. The resulting
cake was landfilled at the site. The operation of the vacuum
filter has been satisfactory; the only problem encountered
was frequent fabric failure. The synthetic fiber fabric
employed on this filter is susceptible to cuts resulting
from metal and glass fragments present in the digester slurry.
These fragments accumulated in the filter tank and eventually came in contact with the moving filter cloth.

The dewaterability of this slurry was exceptionally
good. During the month of January 1982, the cake solids
averaged 26.9% with a standard deviation of 1.97%. The
total solids were 61.1% volatile (Mooij, 1982). The dewatering process did not use any chemicals to condition
the slurry prior to dewatering. Consequently, the solids
capture was not as high as that experienced with normal
sewage sludge dewatering. The average value of the filtrate
total solids was 1.75% for January 1982. The ash content
of these solids was 35.9%. The low solids capture is not a
problem because the filtrate is recycled as slurry water
for the refuse feed.

The only problem with the vacuum filter is its limited
capacity. It is not possible to process the effluent as
fast as it is produced when the digesters are being fed.
Consequently the slurry dewatering rate limits the capacity
of the system. Alternate dewatering systems are being investigated. A belt filter press has been obtained for installation at RefCOM. Preliminary laboratory results suggest a much drier cake can be produced.

SUMMARY AND CONCLUSION

The results to date have demonstrated that it is possible
to assemble a process train that can produce a significant
quantity of fuel gas from MSW. Critical deficiencies in the
process train have been identified and eliminated and weak
areas have been identified so that they can be corrected in
the next design. The simplified process train reduces both
capital and operating costs, making the economics of the
process much more favorable.

There are two questions that yet need to be addressed in the experimental program. The first is the gas yield from a full-scale plant as a function of retention time. If good gas yields can be obtained with short retention times in the digester, a reduction in capital and operating costs will occur. The trade-off at the shorter retention times is in the value of the gas not produced vs. the reduced costs. If gas is priced at a high value, the economics will support the longer retention items that yield more gas. The second question relates to the long-term stability of the methane fermentation process. Can the biological process function for extended periods without upset? These two questions will be answered during the next year of operation.

LITERATURE CITED

Anonymous, *Advanced System Experimental Facility, Solid Waste to Methane Gas (RefCOM), Plant Modifications April 1978 to January 1981,* Waste Management, Inc., Oak Brook, IL (1981).

Babbitt, H.E., B.J. Leland, and F.H. Whitley, Jr., *The Biological Digestion of Garbage with Sewage Sludge,* Bulletin No. 287, Engineering Experiment Station, University of Illinois, Urbana (1936).

Chan, D.B. and E.A. Pearson, *Comprehensive Studies of Solid Waste Management - Hydrolysis Rate of Cellulose in Anaerobic Fermentation,* SERL Report No. 70-3, University of California, Berkeley, CA (1970).

Golueke, C.G. and P.H. McGauhey, *Comprehensive Studies of Solid Waste Management, Second Annual Report,* U.S. Public Health Service Report, SW 3rd USPHS, Publ. No. 2039, Washington, DC (1970).

Hamilton, F. and J. Kelyman, *Solid Waste Management/RRJ, 22*(5), 40-44 (1979).

Kispert, R.G., S.E. Sadek, L.C. Anderson, and D.L. Wise, *Fuel Gas Production from Solid Waste,* Report No. 1258, Dynatech R/D Co., Cambridge, MA (1975).

Maki, L.R., *Antonie van Leeuwenhoek, 20,* 185-200 (1954).

McInerney, M.J. and M.P. Bryant, *Biomass Conversion Process for Energy and Fuels,* pp. 277-295, (S.S. Sofer and O. R. Zaborsky, eds.), Plenum Publishing Co., New York (1981).

Mooij,H.P., *Advanced System Experimental Facility, Solid Waste to Methane Gas: Technical Status Reports (RefCOM)* (January 1982), Waste Management, Inc., Oak Brook, IL (1982).

Pfeffer, J.T., *Reclamation of Energy from Organic Refuse,* EPA-670/2-74-016, U.S. EPA, National Environmental Research Center, Cincinnati, OH (1974).

Pfeffer, J.T., *Biotech. Bioeng.*, *16*, 771 (1974a).

Pfeffer, J.T. and J.C. Liebman, *Resource Recovery and Conservation, 1,* 295-313 (1976).

Pfeffer, J.T. and H.R. Isaacson, *Biochemical Conversion of Municipal Solid Waste: A Technology Status Report, ANL/CNSV-TM-122,* Argonne National Laboratory, Argonne, IL (1982).

Ross, W.D., *Sewage and Industrial Wastes, 26,* 140 (1954).

24

PRELIMINARY EVALUATION OF THE EFFECTS OF A d-RDF BINDING AGENT, RDF PARTICLE SIZE AND HEAT PRETREATMENT ON EFFICIENCY OF ANAEROBIC DIGESTION OF RDF

V. Kenneth Wright

Cyclic Energies, Inc.
Denton, Texas

Barney J. Venables

TRAC Laboratories, Inc.
Denton, Texas

Kenneth E. Daugherty

Department of Chemistry
North Texas State University
Denton, Texas

ABSTRACT

A version of an upflow anaerobic sludge blanket (UASB) digester for conversion of biodegradable organics to a high Btu biogas is being evaluated at the pilot plant scale at the City of Denton municipal sewage treatment plant. The digester design under study was originally developed, tested and commercially used for stabilization of food-processing wastes. Hydraulic retention time of 24 hours resulted in a 55% reduction in COD and a 95% reduction in suspended solids. In concert with the development of binder systems for densified-refuse derived fuel (d-RDF), under a Department of Energy/ Argonne National Laboratory grant, binding agents and other factors affecting efficiency of anaerobic digestion of potential d-RDF products are being evaluated. Laboratory scale anaerobic digesters demonstrated no effects of two commercially available particle sizes of RDF or the presence of calcium oxide, a potential d-RDF binder, on digestion efficiency. Efficiency was significantly increased by a heated pretreatment.

INTRODUCTION

Daugherty (1982) began investigating the fuel potential of municipal solid wastes (MSW) in Denton in 1980. Initial studies included examination of local municipal solid wastes and test burns of commercially prepared refuse-derived fuel (RDF) in ovens at a local brick plant in a cooperative study including the participation of North Texas State University (NTSU), the City of Denton and ACME Brick. The results established a favorable economic projection for this application of RDF and the great advantage of the production of a densified RDF (d-RDF) product. In late 1983, under DOE Contract #83-56-26 entitled "Densified fuel preparation from municipal solid waste," studies were initiated to evaluate candidate binders and densification systems. Evaluation protocol includes the general criteria headings of cost, storability, strength, environmental acceptability and the effectiveness of the d-RDF as a fuel source for incineration in stoker fed boilers, pyrolysis and anaerobic digestion. Preliminary results of this evaluation indicate that one of the top binder candidates is calcium oxide or inexpensive sources of calcium oxide such as the waste by-products of lignite burning power plants, kiln dust from cement plants and possibly granulated blast furnace slag from the iron and steel industry. Optimal concentrations appear to be in the range of 1-5% dry weight RDF.

Concurrent with the examination of d-RDF binder by NTSU, the City of Denton, in conjunction with Cyclic Energies, was investigating the application of an upflow anaerobic sludge blanket (UASB) digester for rapid stabilization of municipal sewage. This study was being conducted with a 500 gallon pilot plant scale digester located at the City of Denton's Pecan Creek Wastewater Treatment Plant. The long-term goal of the effort is the possible elimination of primary clarification by greatly reducing anaerobic digester hydraulic retention times. The application of this digester design to the food processing industry wastes has been described by Pette (1979) and Nyns and Naveau (1981).

The various parties involved in the activities described above hoped to combine their interests by examining the efficiency of the UASB digester with sewage and sewage in combination with RDF and appropriate quantities of d-RDF binder. Our hypothesis was that inclusion of calcium oxide as a d-RDF binding agent would not significantly inhibit the anaerobic digestion of a d-RDF product. There was reason to suspect that the addition of an alkaline binder might aid the anaerobic digestion of RDF either through neutralization of acid production or enhanced hydrolysis of cellulosic material. Acidic conditions resulting from formation of acetic and propionic acid and carbonic acid from CO_2 can inhibit anaerobic methane production (Pfeffer and Isaacson, 1982).

The rate limiting step in the anaerobic digestion of MSW may well be cellulose hydrolysis (Chan and Pearson, 1970; Pfeffer and Isaacson, 1982; Klass, 1984). This hydrolysis can be enhanced by heat and alkaline treatments (Langton, 1981; Young and McCarty, 1981; Pfeffer and Isaacson, 1982). However, the addition of calcium oxide and the formation of an insoluble calcium carbonate precipitate may interfere with mechanical and biological functions of the digester (Diaz et al., 1982). Additionally, particle size of RDF may affect ease of anaerobic digestion. We had RDF from the Ames Iowa Solid Waste Recovery System with a nominal size of 5 cm. and Madison Wisconsin Resource Recovery Center with a nominal size of 2.5 cm. Thus the specific goals of the preliminary data presented here were (1) establish a baseline efficiency to be expected from the pilot-plant scale version of the UASB digester when fed only sewage, and (2) examine the effects of addition of RDF, heat and calcium oxide pretreated RDF, and calcium oxide on laboratory scale batch digesters. Future studies will focus on the effects of the addition of RDF and binding agents to sewage in UASB and other anaerobic digester systems.

MATERIALS AND METHODS

The pilot-plant UASB digester is a 500 gallon fiberglass cylindrical tank located at the receiving end of the City of Denton's Pecan Creek Wastewater Treatment Plant. The digester is plumbed to receive influent immediately after preliminary rough screening of raw sewage. Biogas produced is metered and compressed in a storage tank. Samples of digester influent and effluent were collected and analyzed for biochemical oxygen demand (BOD), chemical oxygen demand (COD), total suspended solids (TSS) and pH according to Standard Methods (15th Edition, 1980). Representative samples of biogas produced were collected in gas bags and analyzed for methane content by gas chromatography.

Data were collected from the pilot-plant digester from May to July 1984. The digester was seeded with 50 gallons of anaerobic sludge from the city's anaerobic digester. Initial flow rates were regulated to maintain a 48 hours hydraulic retention time. The digester performed well at this flow and consequently the retention time was reduced to 24 hours. After a 2 week acclimation period, daily estimates of influent and effluent BOD, COD, TSS, and gas production were made for six days.

Effects of RDF particle size, calcium oxide and heated pretreatment were examined in a short-term laboratory experiment. Thirty 4-ℓ plastic jug anaerobic digesters were filled with slurry taken from the City of Denton's anaerobic digester. The digesters were divided into ten experimental groups with three replicates per group. The groups were distinguished by a) absence of RDF, b) RDF added at concentration of 1% dry weight/digester volume, c) calcium oxide absent, d) calcium oxide added at a concentration of 5% weight/dry weight of RDF, e) RDF with large particle size from Ames Plant (nominal size 5 cm.), f) RDF with smaller particle size from Madison Plant (nominal size 2.5 cm.), g) RDF unheated, h) RDF exposed to a heated pretreatment consisting of boiling a 10% aqueous slurry for 10 minutes. Thus the ten experimental groups were designed as described in Table 1.

The thirty digesters were maintained in a dark room at 30-32°C, the lower end of the mesophilic temperature range. They were mixed once daily. COD, TVS (total volatile solids) and pH were determined at 5-7 day intervals over the experimental period of 15 days. Effects of the experimental treatments were statistically examined by analysis of variance and tests.

RESULTS

Table 2 presents the results from the pilot plant UASB digestion of raw sewage. With a 24 hour hydraulic retention time, COD was reduced by approximately 55% while 95% of suspended solids were removed. Reactor temperature varied from 24-30°C. Methane concentration of the biogas produced ranged widely from a low of approximately 20% to as high as 70%.

Table 1. Experimental Design for Testing Effects of RDF Particle Size, Binder and Heated Pretreatment on Efficiency of Anaerobic Digestion. Each Experimental Unit had 3 Replicates.

Experimental Unit	RDF Particle Size			Binder		Heated Pretreatment		Code in Text
	None	Small	Large	Present	Absent	Yes	No	
1	X				X		X	NAU
2	X			X			X	NPU
3		X			X		X	SAU
4		X		X			X	SPU
5			X		X		X	LAU
6			X	X			X	LPU
7		X			X	X		SAH
8		X		X		X		SPH
9			X		X	X		LAH
10			X	X		X		LPH

*Table 2. Preliminary Data on Performance of a Pilot Plant
UASB Digester Receiving Raw Sewage With a Hydraulic
Retention Time of 24 Hours. Values are 6-Day Means
(s.d.).*

	BOD (mg/ℓ)	COD (mg/ℓ)	TSS (mg/ℓ)	Gas Production (ℓ/day)
Influent	119(8)	351(19)	277(56)	
				305
Effluent	19(7)	159(19)	13(10)	

The loading rate was equivalent to 0.32 kg COD per cubic meter
of digester volume per day. These results indicate that a 24
hour retention time is sufficient for the relatively dilute
influent received by the digester.

Table 3 summarizes the TVS and COD results of the labora-
tory-scale batch digester experiments. Analysis of variance
indicated a significant effect of time with the average TVS
in all digesters decreasing 61% from 0.78% at time 0 to 0.30%
at 15 days. Similarly COD dropped 49% from 10,821 to 5,475
mg/ℓ. No experimental effects of either the particle size
of the RDF or the presence of the binder were detected at the
95% confidence level. However, the heated pretreatment resulted in lower COD on the last day (unheated mean COD = 6538,
heat pretreated mean COD = 3882, t = 2.32, df = 28, p = 0.03).

Alkalinity was measured once at the mid-point of the
experiment, day 7. Estimates were made based on end points
of pH 5.75 and 4.5 for consideration of acetate alkalinity
(Bravko and Chen, 1977). There were no significant (p = 0.05)
experimental effects on alkalinity based on pH 4.5 endpoint
and the average for the thirty digesters was 2,182 mg/ℓ.
Alkalinity based on pH 5.75 endpoint was not affected by RDF
particle size or the presence of the binder but was significantly lower in those digesters receiving the heated pretreatment (mean unheated = 1,863 mg/ℓ, mean heat pretreated =
1,513 mg/ℓ, t = 3.21, df = 28, p = 0.003). Digester pH
varied between 6.6 and 7.2 with no significant relation to
experimental treatment.

Table 3. Mean (3 Replicates) Total Volatile Solids (TVS) and Chemical Oxygen Demand (COD) for 10 Experimental Treatments Observed Over a 15 Day Period. N= No Refuse derived fuel (RDF), S= Small (2.5 cm) RDF Added, L= Large (5 cm) RDF Added, A= Binder Absent, P= Binder Present, U= RDF Was Unheated and H= RDF was Exposed to Heated Pretreatment.

Experimental Treatment	Day 0 %TVS	Day 0 COD	Day 4 %TVS	Day 4 COD	Day 11 %TVS	Day 11 COD	Day 15 %TVS	Day 15 COD
NAU	.50	7532	.54	9790	.17	3874	.17	4517
NPU	.56	7099	.50	9646	.24	3208	.19	4084
SAU	.83	12465	.82	17615	1.36	13086	.40	8702
SPU	1.15	11965	.83	20357	1.28	7292	.29	5394
LAU	1.03	12065	.73	9357	.40	4606	.48	5283
LPU	.70	10098	.66	11832	.39	4584	.45	11244
SAH	.89	12932	.56	9091	.67	4573	.23	3352
SPH	1.00	18165	.52	12931	.40	5738	.25	4362
LAH	.44	6699	.38	5228	.39	5761	.27	3707
LPH	.65	8799	.43	12376	.39	7692	.29	4107

DISCUSSION

The anaerobic conversion of organic matter to methane is a complex process involving a wide variety of microorganisms. As organic matter biodegrades under "normal" aerobic conditions, it is converted principally to CO_2, H_2O, NO_3, NO_2 and other minerals. When the oxygen is depleted, anaerobic digestion proceeds in two generalized steps. First, the individual classes of organic chemicals, fats, proteins and carbohydrates, are hydrolyzed by a combination of obligate and facultative anaerobes. This yields triglycerides, fatty acids, amino acids and sugars which are further metabolized via fermentation and β-oxidations to form short-chain volatile acids, alcohols and bacterial cell constituents (Benefield and Randall, 1980). The most important products of this first stage in the process are acetic and propionic acids which are responsible for the great bulk of methane production. McInerney and Bryant (1981) extensively reviewed the biochemistry of methanogenesis.

Methanogenesis is carried out by relatively slow-growing bacteria and is generally considered to be the rate-limiting step in the overall process (Benefield and Randall, 1980). An exception to this generalization is probably represented by MSW. Anaerobic digestion of MSW is probably rate limited by hydrolysis of cellulosic material (Klass, 1984). Still it is estimated that approximately 3,700 cubic feet of biogas can be produced for each ton of MSW (Schulz et al., 1976), or 7.6 cubic feet of methane per pound of volatile solids (Diaz et al., 1974). The actual realized yield will depend on a great many factors affecting the quality of the MSW as well as environmental factors affecting the rate and efficiency of the biological processes.

Typical MSW is composed pricipally of water, carbon and oxygen, with about 24% of the wet weight being non-combustible (Jackson, 1974). Of the combustible material, the largest single contributor is paper, averaging about 42% of total wet weight. Thus, MSW differs considerably from the chemical composition of the more traditional anaerobic feedstock, municipal sewage sludge, which is largely proteinaceous. The composition of MSW varies geographically and seasonally, but expected means and standard deviations for the various components have been published (Baum and Parker, 1974; Daugherty, 1982). The composition of our local Denton MSW was also described on a seasonal basis by Daugherty (1982). The average

Btu content of MSW is increasing and is projected to reach 6,500 Btu per pound in the year 2000. To maximize conversion of this energy content to methane, a number of basic design/environmental requirements for anaerobic digestion should be met.

The UASB digester design has been under development for several years in the Netherlands (Lettinga et al., 1979; Van der Meer et al., 1980; Nyns and Naveau, 1981). It has been used principally for digestion of food processing wastes and is capable of 90% reduction of organic loads of up to 10 kg COD per cubic meter digester volume per day with retention times in the order of hours rather than weeks. The sludge remains in a blanket at the bottom of the digester while a baffle at the top of the digester separates gases from settleable sludge particles, returning the particles to the bottom of the digester.

The pilot-plant UASB digester at Denton's sewage treatment plant under light loading conditions equivalent to 0.32 kg COD/cubic meter reactor volume/day achieved only a 55% reduction in COD. However, TSS was greatly reduced to approximately 5% the original concentration. We anticipate some problems in reducing particle size of RDF sufficiently to permit use of the pilot plant digester. This limitation would presumably not be as critical in a full production scale UASB digester.

The results of the laboratory scale batch digesters indicated that the relatively dilute RDF organic load was digested at a rate similar to that found in previous studies. TVS was reduced by 61% and COD by 49%. Pfeffer (1973) reported 43-51% reduction in 30 days at 35°C. Particle size, within the range examined here representing RDF available from two commercial sources, did not influence efficiency of digestion. The heated pretreatment resulted in significantly lower COD values on day 15, presumably due to enhanced digestibility of cellulosic material. Although pH 4.5 alkalinity was the same for pretreated and unheated groups, pH 5.75 alkalinity was lower in the pretreated group due to greater acetate alkalinity also resulting from enhanced digestibility of cellulosic material. The heat pretreatment used here was relatively mild and its effect could presumably be increased by more severe pretreatment conditions (see Young and McCarty, 1981).

The addition of calcium oxide at low concentrations required as a possible binding agent for RDF did not alter digestion efficiency. This is an important criterion in our overall evaluation of the potential value of a d-RDF binding agent. The results of our work to date indicate that the binding mechanism between calcium oxide and RDF is as follows. Calcium oxide solubilizes and then hydrolyzes with moisture in the RDF to form calcium hydroxide. Ames, Iowa, RDF, for example, varies in moisture content from as low as 6% to as high as 50%. Normally it ranges from 20% to 26% (Personal Communication Mr. Rob Chapman, Utilities Engineer, Ames Resource Recovery Plant). The calcium hydroxide reacts with various acidic components of the RDF forming calcium complexes. Some of the calcium hydroxide gradually carbonates upon exposure to carbon dioxide in the atmosphere. The complexation and carbonation processes result in strong binding to RDF and lead to hard and durable pellets of d-RDF. Our investigations have indicated that the pellets become more durable as they age. The weight percentage of calcium oxide necessary to accomplish the above is in the range of 1 to 5%.

CONCLUSIONS

Preliminary data from a pilot plant UASB digester indicate that the design is effective with dilute raw sewage influent and hydraulic retention times as low as 24 hours, though efficiencies observed to date are not as high as those previously reported for food wastes. This design may be usable with increased organic loading from addition of RDF. Laboratory experiments designed to test for effects of two commercially available RDF particle sizes, a potential d-RDF binder and heated pretreatment showed only the latter significantly affected digestion efficiency. Addition of calcium oxide as a binding agent does not appear to pose any special problems for the use of a d-RDF product in an anaerobic digestion system.

REFERENCES

Baum, B. and C.H. Parker, Solid Waste and Disposal, Vol. 1 Incineeration and Landfill, Ann Arbor Science Publishers, Inc., Ann Arbor, Michigan (1974).

Benefield, L.D. and C.W. Randall, Biological Process Design for Waste-Water Treatment, Prentice Hall, Inc., Englewood Cliffs, New Jersey (1980).

Chan, D.B. and E.A. Pearson, Comprehensive Studies of Solid
    Waste Management - Hydrolysis Rate of Cellulose in
    Anaerobic Fermentation, SERI Report No. 70-3, University
    of California, Berkley, California (1970).

Daugherty, K.E., Urban Waste as a Potential Energy Source for
    Brick Plants, DOE/CS/24311-1 (1982).

Diaz, L.F., G.M. Savage, and C.G. Golueke, "Biogas Production,"
    in Resource Recovery From Municipal Solid Wastes, (L.F.
    Diaz, G.M. Savage, and C.G. Golueke, eds.), Vol. II,
    pp. 57-76, CRC Press, Inc., Boca Raton, Florida (1982).

Diaz, L.F., F. Kurz, and G.J. Trezek, Methane Gas Production
    as a Part of a Refuse Recycling System," *Compost Science*,
    7, 13 (1974).

Jackson, F.R., Energy from Solid Waste, Noyes Data Corporation,
    Park Ridge, New Jersey (1974).

Klass, D.L., "Methane from Anaerobic Fermentation," *Science*,
    223(4640), 1021-1028 (1984).

Langton, E.W., "Digestion Design Concepts," *in* Fuel Gas Pro-
    duction from Biomass, (D.L. Wise, ed.), Vol. II, pp. 63-
    109, CRC Press, Inc., Boca Raton, Florida (1981).

Lettinga, C., A.F.M. van Velson, and S.W. Hobma, "Feasibility
    of the Upflow Anaerobic Sludge Blanket (UASB) Process,"
    Proc. 1979 Nat'l. Conf. on Environmental Engineering,
    ASCE, San Francisco, California, July 9-11, 1979.

McInerney, M.J. and M.P. Bryant, "Review of Methane Fermenta-
    tion Fundamentals," *in* Fuel Gas Production From Biomass,
    (D.L. Wise, ed.), Vol. I, pp. 19-46, CRC Press, Inc.,
    Boca Raton, Florida (1981).

Nyns, E.J. and H.P. Naveau, "Status of European Work in Methane
    Fermentation from Biomass," *in* Fuel Gas Production From
    Biomass, (D.L. Wise, ed.), Vol. I, ppl 135-145, CRC Press,
    Inc., Boca Raton, Florida (1981).

Pette, K.C., Anaerobic Wastewater Treatment at CSM Sugar
    Factories, Proc. 16th General Assembly Commission Inter-
    nationale Technique de Sucrerie, Amsterdam, May (1979).

Pfeffer, J.T., Reclamation of Energy from Organic Refuse,
    Grant No. EPA-R-80076, April (1973).

Pfeffer, J. and R. Isaacson, Biochemical Conversion of Munici-
    pal Solid Waste: A Technology Status Report, ANL/CNSV-
    TM-122 (1982).

Schulz, H., J. Benziger, B. Bortz, M. Neomatalla, R. Szostak,
    G. Tong, and R. Westerhoff, Resource Recovery Technology
    for Urban Decision Makers, Technical Report, Technology
    Center, Columbia University (1976).

Standard Methods for the Examination of Water and Wastewater,
    15th Edition, American Public Health Association, Wash-
    ington, DC (1980).

Van der Meer, R.R., K.C. Pette, P.M. Hoertjes, and R. de-
    Vletter, The Upflow Reactor for Anaerobic Treatment of
    Wastewater Containing Fatty Acids, Proc. 3rd International
    Congress on Industrial Waste Water and Wastes, Stock-
    holm, February (1980).
Young, L.Y. and P.L. McCarty, "Heat Treatment of Organic
    Materials for Increasing Anaerobic Biodegradability," *in*
    Fuel Gas Production From Biomass, (D.L. Wise, ed.), Vol.
    II, pp. 133-176, CRC Press, Inc., Boca Raton, Florida
    (1981).

25
NEW CONCEPT FOR THE PRODUCTION OF HIGH-BTU GAS FROM ANAEROBIC DIGESTION

Thomas D. Hayes and H. Ronald Isaacson

Gas Research Institute
Chicago, Illinois

ABSTRACT

A major economic consideration in the use of anaerobic digestion to convert municipal wastes to pipeline quality methane and commercial grade CO_2 is the cost of separating CO_2 from the product gas. Physical/chemical processing for upgrading digester gas (55-65% CH_4) to a methane gas stream meeting Gas Industry specifications ($\leq$ 95% CH_4) is costly with estimates typically ranging from $1.50-3.00/mmBtu. Certain anaerobic digestion process designs have the potential of achieving higher methane concentrations in the product gas thereby reducing gas cleanup costs. There are at least three inherent features that make anaerobic digestion uniquely suited to gas separations: 1) the presence of a gas/liquid interface, 2) the phased production and utilization of volatile acids, and 3) the capability of anaerobic bacteria to produce methane against elevated gas pressures. Because the solubility of CO_2 in water is many times greater than that of methane, it should be possible to develop reactor schemes capable of achieving efficient separations of these two gases. An equilibrium model has been developed that predicts the methane and CO_2 content of the product gas of a steady state digester as a function of commonly-measured influent and effluent parameters. This model, when tested with existing reactor data, showed good correlation between predicted and observed values for methane content. On the basis of the model and the literature, a two-stage, pH-swing/pressure-swing digestion scheme has been conceptualized that has the potential of directly producing pipeline-quality methane and high-grade CO_2 from municipal wastes and biomass.

INTRODUCTION

Although anaerobic digestion is a biological process long employed by industry and municipalities for pollution control and sludge stabilization (24), it is also applicable to the conversion of various kinds of biomass (plant tissue) to methane (1,10,12,14,17). Preliminary assessments of advanced biomass-to-methane systems, including farming, harvesting, transportation and conversion, indicate that a total methane cost of $5 to $6/GJ is achievable without major breakthroughs in the technology (4,23).

The product gas from anaerobic digesters in use today typically contains 55-65% methane (CH_4), 35-45% carbon dioxide (CO_2) and less than 1% hydrogen (H_2) and hydrogen sulfide (H_2S). If anaerobic digestion is to be used to convert biomass and waste to pipeline quality gas ($\leq$ 95% CH_4), gas cleanup is required to remove CO_2, moisture and H_2S. The cost of CO_2 separation alone amounts to about $1.50 to $2.50 per GJ, depending on the technique used and the scale of application. Any improvements that can be made in reactor design to increase the methane content of the product gas can thus result in substantial gas cleanup savings. Recent studies reported in the literature indicate that two phase anaerobic digesters are generally capable of producing a gas of a higher methane content (70-80% CH_4). Although various reasons for this improved performance are hypothesized, the most likely mechanism involves the absorption and desorption of CO_2 from liquid streams (9). The objective of this study was to use simple CO_2 distribution relationships to identify reactor design strategies that have the potential of improving the quality of gas generated by anaerobic digestion.

BACKGROUND

Methane enrichment in digester gas has not been of major concern in the past since applications of anaerobic digestion have emphasized waste treatment rather than energy production. Nevertheless, there are a number of observations that can be drawn from theory and the literature that shed light on possible strategies for increasing the methane content of biologically-produced gas.

The gas/liquid interface in anaerobic digestion has a pronounced effect on the CO_2 concentration in the product gas. This effect is evident from anaerobic digestion data in the literature which report methane-to-carbon dioxide ratios in the product gas that are significantly higher than ratios predicted from stoichiometry. This phenomenon may be explained in terms of the difference in solubilities of methane (CH_4) and carbon dioxide (CO_2) in water. At a pH of 7.0 and 35°C, for example, CO_2 is 40 times more soluble than methane. An aqueous stream entering a reactor that is not saturated in CO_2 (such as is the case with many feed-streams) will have a capacity to absorb a significant portion of the CO_2 produced in the anaerobic conversion of organic matter, but will only absorb a small fraction of the in-soluble methane product. This solubility differential results in a partial separation of the CO_2 from the methane gas stream thereby raising the Btu content of the product gas.

The amount of CO_2 absorbed in the aqueous phase of a digester is influenced by a number of factors such as tem-perature, partial pressure, pH, influent stream CO_2 alkalin-ity, and ionic strength, (2,20). The equilibrium solubility of methane and CO_2 may be described by simple Henry's Law and aqueous CO_2 dissociation equations.

Henry's Law

$$C_x = K_H P_x \tag{1}$$

where: K_H = Henry's Law constant, moles/l-atm*, C_x = con-centration of compound X in water, moles/l, P_x = partial pressure of compound X in the gas, atm.

Carbon Dioxide Solubility Reaction

$$CO_2(gas) \xrightarrow{K_H} H_2CO_3 \underset{K_1}{\overset{H^+}{\rightleftarrows}} HCO_3^- \underset{K_2}{\overset{H^+}{\rightleftarrows}} CO_3^{-2} \tag{2}$$

Methane Solubility Reaction

$$CH_4(gas) \xrightarrow{K_H} CH_4(dissolved) \tag{3}$$

These relationships have been used in steady state and dynamic models to describe the performance of various types of reactors (5,15,19). Such models can be used to conceptualize and evaluate reactor designs that have potential in achieving higher methane concentrations in the product gas. As yet, there is a dearth of literature on this kind of analysis.

A number of investigators have noted that phased diges-tion is capable of producing a gas of substantially higher methane content (70-85%) than is possible with a single stage reactor (6,7,9,13,21,22). An example of this concept is a scheme where two continuous stirred tank reactors are arranged in series: the first reactor converts a substrate suspended in a slurry to volatile fatty acids (VFA's), the second converts VFA to methane and CO_2. Gas from the second stage of this scheme applied to MSW conversion has been shown to contain up to 75% methane (14). Another example of methane enrichment using two phase digestion can be found in the experimental operation of the leaching bed/packed bed system. In this scheme, a high solids (20-40% total solids) biomass feedstock is introduced into an acid-phase where it is converted to VFA's. Water is continuously per-colated through the biomass solids to wash VFA's out of the acid phase reactor; this acidic stream is continuously fed into a high-rate methanogenic reactor, such as the packed bed, which converts VFA's to methane CO_2. The effluent water stream can be recycled back to the leaching bed for use in percolation. This reactor scheme has been successfully demonstrated in the laboratory with MSW (1,13,14,22), grass clippings (7), and straw (7). Conversion efficiencies of up to 80% have been achieved within 20 days of solids re-tention using MSW and grass as feedstocks (7,14). Equally important, the methane content of the gas generated by the methanogenic reactor typically ranges from 70 to 85%, again, a significant improvement in gas quality over single-stage digestion.

This improvement in gas quality achieved by two-phase digestion is due in large part to the pH difference that naturally occurs between phases. As a liquid stream is passed through the acid phase, organics are converted to volatile acids, the pH decreases to pH 5.0 to 6.5, and dissolved CO_2 is shifted toward the formation of the H_2CO_3 species and CO_2 gas in accordance with Eq. (2). As this stream passes through the methane reactor, volatile acids are converted to less acidic products (methane and CO_2) causing the pH to be increased to 7.2 to 8.0. This increase in pH causes dissolved CO_2 to predominate in the bicarbonate

(HCO_3^-) form and increases the absorption capacity of the liquid feed stream for biologically produced CO_2. The result is methane enrichment of the methane-phase product gas without significant inputs of chemicals required for pH control.

There is, therefore, ample evidence from the literature that the anaerobic digestion process can be modified in design to improve the methane content of the product gas. The goal of this study was to identify what further improvements could be made in two-phase digestion to boost the methane concentration in the product gas from the currently-achieved 70-80% range to near-pipeline quality (90-95% methane).

APPROACH

The general approach used in this study was to develop a computer model describing CO_2 distributions in anaerobic digestion and to use the model to identify design improvements in two-phase processes that have potential for substantially increasing methane in the product gas.

A simplistic steady-state model that estimates the distribution of CO_2 in anaerobic digestion was developed from existing models and data reported in the literature. Equations comprising the model are summarized in Figure 1 and variables and constants used in the model are defined in Table 1. One of the most important assumptions used in the model is that CO_2 separation in anaerobic digestion approximates equilibrium conditions. This would appear to be a reasonable assumption in view of the long retention times (≤ 5 days) typically applied to anaerobic digestion. The model incorporated CO_2 and VFA equilibria and Henry's Law for partitioning of CO_2. The central relationship of the CO_2 management model is the mass balance of CO_2 around the liquid fraction of the anaerobic digestion at steady state, as shown below.

$$R_G = R_O - R_1 + R_B \tag{4}$$

where R_G = Rate of CO_2 entering the gas phase, moles/l/day

R_O = Rate of dissolved CO_2 entering the reactor, moles/l/day

R_1 = Rate of dissolved CO_2 leaving the reactor, moles/l/day

R_B = Rate of CO_2 biologically produced from the organic feedstock, moles/l/day

Table 1.
Glossary of Symbols Used in the
CO_2 Management Model

Symbol	Definition
C_O	Influent total CO_2 moles/l
C_1	Effluent total CO_2 moles/l
F	Flow of feed, l/day
Ha	Henry's law constant for CO_2, moles/l-atm
Hm	Henry's law constant for CH_4, moles/l-atm
K_1	First dissociation constant for Co_2
K_2	Second dissociation constant for CO_2
O	Hydraulic retention time, days
R_G	Rate of CO_2 evolution into gas phase, moles/day
R_O	Rate of soluble CO_2 entering reactor with influent stream, moles/day
R_1	Rate of soluble CO_2 leaving reactor with effluent stream, moles/day
R_B	Rate of CO_2 produced in the biological conversion of organics, moles/day
R_{MB}	Rate of CH_4 produced in the biological conversion of organics, moles/l
P_T	Total gas pressure, atm
P_C	Partial pressure of CO_2, atm
P_W	Vapor pressure of water, atm
S_O	Influent substrate concentration, g/l BVS
S_1	Effluent substrate concentration, g/l BVS
K_S	First order kinetic coefficient, days^{-1}
$Y_{C/S}$	Yield of CO_2 in substrate conversion, mole/mole
$Y_{M/S}$	Yield of CH_4 in substrate conversion, mole/mole
V	Volume of reactor, l

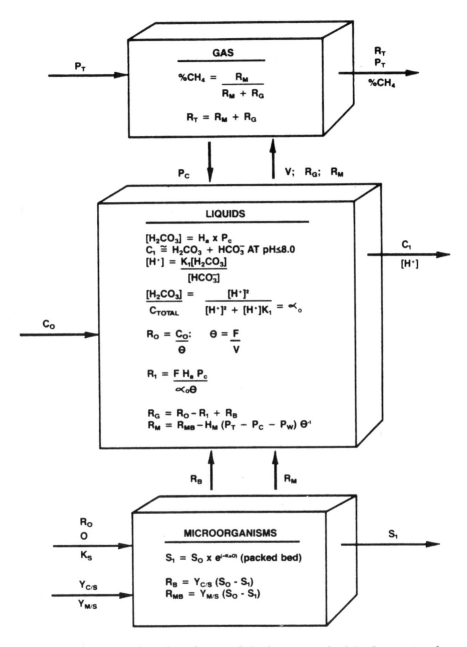

Figure 1. CO_2 distribution model for a packed bed at steady state.

The R_O term is calculated from the CO_2 alkalinity and the pH of the influent stream and R_B is calculated from the reaction stoichiometry and kinetics of the bioconversion of organic matter to methane and CO_2. The R_1 term is calculated from the dissolved CO_2 concentration in the digester liquid contents in equilibrium with the CO_2 partial pressure of the gas phase in accordance with Henry's Law and aqueous CO_2 speciation at equilibrium. The rate of methane release into digester gas is calculated from the following equation:

$$R_M = R_{MB} - R_{M_1} \tag{5}$$

where R_M = Rate of methane entering the gas phase, moles/l/day

R_{MB} = Rate of methane biologically produced from the organic feedstock, moles/l/day

R_{M_1} = Rate of dissolved methane leaving the reactor in the effluent stream, moles/l/day

Again, the term R_{MB} is calculated from reaction stoichiometry and microbial kinetics, while R_{M_1} is calculated from the concentration of dissolved methane in the reactor liquid contents in equilibrium with the partial pressure of methane in digester gas. The percent methane in the digester gas on a dry basis is calculated as follows:

$$\%CH_4 = R_M \times 100/(R_M + R_G) \tag{6}$$

The overall stoichimetric relationship used to describe the conversion of organic matter to CO_2 and CH_4 is given in the literature (10).

$$C_n H_a O_b + [n - a/4 - b/2]H_2O = \tag{7}$$

$$[n/2 - a/8 + b/4] \, CO_2$$

$$+ [n/2 + a/8 - b/4]CH_4$$

This equation is applied to the biological occurring in methane-producing reactors. Theoretical yields of methane (Ym/s) and CO_2 (Yc/s) from organic substrates were calculated in the model from this equation. In actuality, multiple reactions are involved in the conversion of complex organics to methane. Major reaction steps include the initial formation of volatile fatty acids (VFAs) and conversion of VFAs to methane and CO_2. In two-phase digestion these two reaction steps can be performed in separate stages as shown in Figure 2.

To test the validity of the model, eleven experiments from seven researchers were selected from the literature. The basis for selection was to choose the first eleven reactor runs with complete sets of input data required by the model; these data sets are shown in Table 2. Model predictions of methane content in digester gas correlated well to actual methane content results reported in the literature as shown in Figure 3. The high correlation coefficient (r^2 = 0.98) suggested that CO_2 distributions in available literature on digestion approximate equilibrium conditions, even in reactors with intermittent mixing and high throughputs, and that the equilibrium model may be considered a useful tool for the estimation of the methane content of digester gas for a variety of reactor designs and operating conditions. Once validated, the model was used to evaluate the effects of certain operating parameters on gas quality in single stage and two-phase reactors. Following these simulations, it was possible to conceptualize reactor designs and operating modes that have potential in maximizing the methane content in digester gas.

RESULTS

The effects of pressure, effluent pH, influent biodegradable volatile solids concentration (BVS), hydraulic retention time (HRT), and CO_2 alkalinity were examined for a packed bed reactor operated on a soluble feedstock at steady state. Results from those simulations are presented in Figures 4-5. Of all the factors considered, digester gas pressure and pH had the greatest impact on gas quality. As shown in Figure 5, elevated pressures above 2.0×10^5 gauge Pa (30 psig) at a pH near 8.0 result in a reactor product gas of greater than 90% methane. These operating conditions seem compatible with biological methanogenesis in view of successful animal waste fermentations reported at pHs above 7.8 and in light of the fact that hydrostatic pressures exceeding 1.4×10^5 gauge Pa (20 psig) are exerted in the bottom zones of many working commercial digesters without apparent detrimental effect.

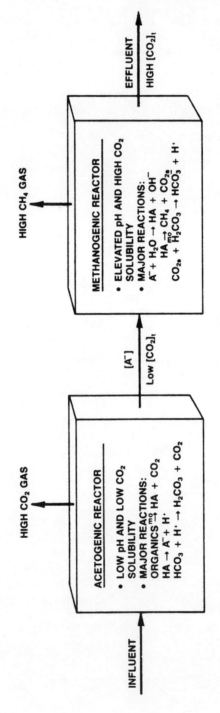

Figure 2. Major reactions affecting pH and CO₂ solubility in two-phase anaerobic digestion.

Table 2. CO_2 Management Model Verification

Reference	Feed	Ft (STM)	T2 (°C)	Influent Substrate So(g/l)	Effluent Substrate S (g/l)	HRT O(days)	C_A (mg/l) as $C_A CO_3$	Infl. pH	Effl. pH	Methane Content of The Product Gas % Reported	Methane Content of The Product Gas % Predicted by Model
Boone, 1982 (3)	Cattle Manure	1.0	35	57.5	37.5	10	1,550	6.88	7.15	55	59.9
Jewell, 1978 (17)	---	1.0	35	74.6	41.4	20	8,660	7.77	7.3	54	56.2
Jewell, 1980 (18)	---	1.0	35	103.6	70.8	30	13,430	6.6	7.7	58	60.4
Jewell, 1980 (18)	---	1.0	35	112.1	80.9	15	12,020	6.6	7.7	55	61.4
Jewell, 1978 (17)	---	1.0	61	35.2	28.4	5.0	2,554	7.1	6.8	51	53.3
Jewell, 1978 (17)	---	1.0	60	35.2	27.7	10	2,796	7.7	7.3	61	61.9
Fong, 1973 (11)	---	1.0	32	24	16	10	1,580	7.0	6.9	62	62.2
Gramms, 1971 (16)	---	1.0	32.5	19	6.4	10	3,760	7.2	7.2	58	62.4
Dalrymple & Proctor, 1967 (8)	---	1.0	35	19.2	10.5	12	650	7.5	7.43	77	76.9
Colleran, 1982	Grass	1.0	30	10.15	2.6	3.0	400	6.5	7.85	90	89.9
Ghosh and Henry, 1982 (14)	Hyacinth + sludge + MSW	1.0	35	3.22	1.22	2.3	470	5.71	6.88	73.0	72.2

* 1) Reactors used by Jewell, Boone, Fong, Gramms and Dalrymple were continuous stirred tank reactors (CSTRs)
2) The Celanese process is a packed bed system
3) Colleran's process is a leaching bed followed by conversion of the leachate to methane by a packed bed process
4) Ghosh & Henry process is a leaching bed followed by conversion of the leachate to methane with an anaerobic filter

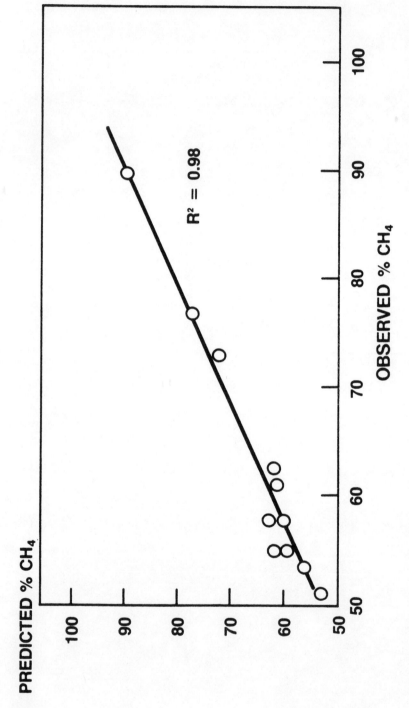

Figure 3.   Correlation between observed literature values and predicted methane content in the product gas from anaerobic digestion.

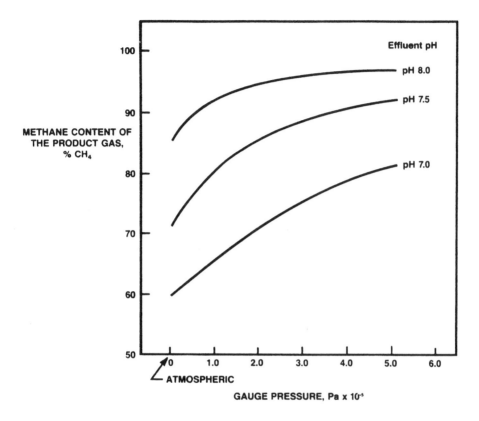

GAUGE PRESSURE, Pa x 10⁻⁵

BASELINE CONDITIONS

INFLUENT SUBSTRATE: 20 g/l TVS
INFLUENT BiCARBONATE
ALKALINITY: 500 mg/l
INFLUENT pH: 6
EFFLUENT pH: VARIABLE
DIGESTER TEMP.: 35°
HRT: 3 days
FIRST ORDER KINETIC
RATE COEFFICIENT, K: 0.95 days⁻¹
DIGESTER PRESSURE: VARIABLE

Figure 4.   Effect of digester total gas pressure on the methane
            content of digester gas.

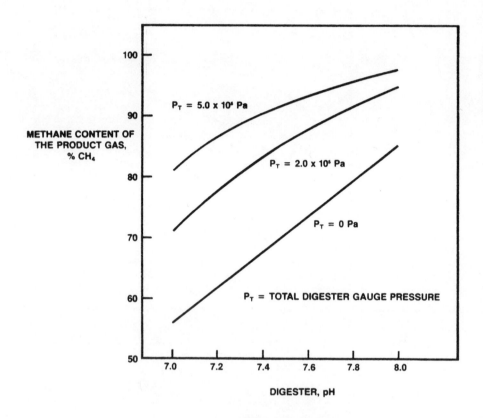

BASELINE CONDITIONS

INFLUENT SUBSTRATE: 20 g/l TVS
INFLUENT BICARBONATE
ALKALINITY: 500 mg/l
INFLUENT pH: 6
EFFLUENT pH: VARIABLE
DIGESTER TEMP.: 35°
HRT: 3 days
FIRST ORDER KINETIC
RATE COEFFICIENT, K: 0.95 days⁻¹
DIGESTER PRESSURE: VARIABLE

Figure 5.   Effect of digester effluent pH on the methane content of the product gas.

In the initial analysis, it was assumed that pH would be controlled to a desired value. At pHs between 7.0 and 8.0, soluble CO_2 exists predominately in the bicarbonate ion form (HCO_3^-). As pressure is increased, more CO_2 is solubilized from the digester gas phase into the liquid causing the pH to decrease. Maintaining high pH under elevated pressure, would require caustic addition which can be costly. For example, if lime is used at \$83/tonne (\$75/ton) to achieve 90% CH_4 at 2.0 x 10 gauge Pa (30 psig), chemical costs alone could amount to more than \$2.50/GJ.

In two phase digestion, however, much of this caustic demand for pH control can be supplied through the biological conversion of VFAs to methane as shown in Figure 2. Since C_1 to C_4 VFAs exist predominately in the ionic form at pHs above 5.5, and since these acids must be in the protonated form before they can be biologically metabolized to methane and CO_2, one mole of hydroxide is released per mole of monoprotic acid converted to methane and CO_2.

These considerations led to the conceptualization of a two-phase, pressure-swing anaerobic digestion scheme shown in Figure 6. This process consists of a leaching bed hydrolysis reactor placed in tandem with a pressurized packed bed methane reactor. A liquid stream transports VFAs formed in the leaching bed to the packed bed where the volatile acids are converted to methane and where CO_2 is absorbed at an elevated pH and pressure. The liquid stream, saturated with CO_2, is then recycled back to the leaching bed where CO_2 is desorbed at a reduced pH and pressure. In order to achieve still higher methane concentrations in the product gas, the option of air sparging the leachate prior to feeding the packed bed was considered.

The CO_2 model was applied to this process to evaluate its performance under various hypothetical operating conditions. The pHs of both phases were calculated from acetic acid and CO_2 equilibria, mass balances and charge balance relationships. Under the operating assumptions of Figure 6, the CO_2 management model predicted that a product gas containing 92-95% methane could be generated without the addition of chemicals for pH control. Figure 7 compares this level of methane enrichment with typical methane contents achieved with non-pressurized single and two-phase digestion. Maximum methane enrichment would require air sparging of the acidic leachate stream to remove at least 75% of the residual CO_2. Without the sparging feature, the

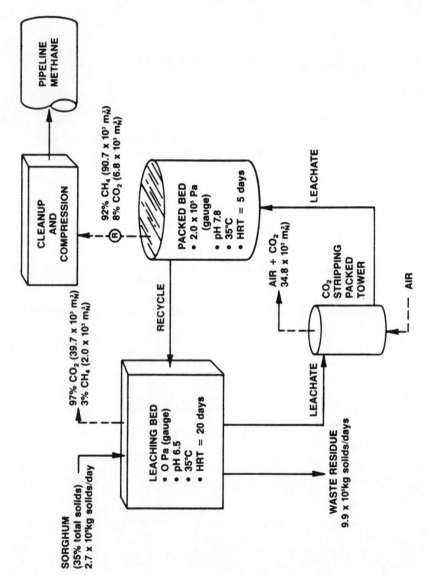

Figure 6.   Schematic of the pressure-swing, two-phase digestion process.

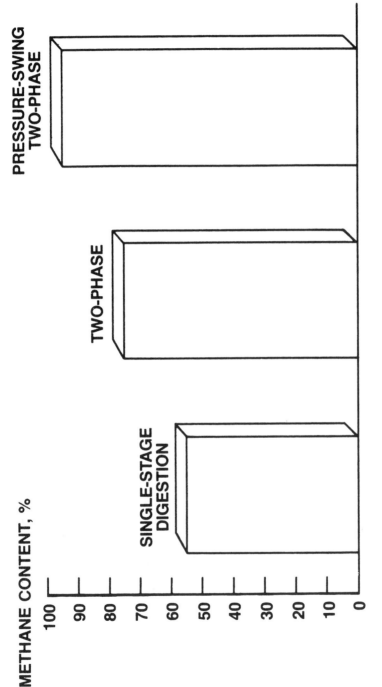

Figure 7. Results from CO₂ distribution model simulations.

methane content reaches 87%. Interestingly, a gas stream
containing more than 95% CO_2 can be generated from the acetogenic stage, raising the possibility of recovery of this gas
for a by-product credit. The model also predicts that H_2S
can be reduced by more than 92% in the methane product gas
using this process scheme.

ENGINEERING CONSIDERATIONS

On the basis of theoretical considerations, the concept
of pressure-swing, two-phase digestion appears to have potential value in reducing the gas cleanup costs associated with
producing pipeline quality methane from biomass and waste.
This process design may also be useful as a complement to gas
cleanup systems currently being developed for low-cost, small-
scale applications. Obvious drawbacks to the concept include the added cost of reinforcing methane phase tanks to
withstand elevated gas pressures of $1.4-2.0 \times 10^5$ gauge Pa
(20 to 30 psig) and the lack of a significant data base on
the actual operation of such a system. Preliminary cost data
from tank manufacturers indicate that the added cost of tank
construction could amount to \$0.25-\$0.35/GJ, depending on
the retention times and tank diameters used in the design.

CONCLUSION

Two phase digestion that incorporates features of a
pressurized methane phase (2.0×10^5 gauge Pa or 30 psig)
and CO_2 stripping has the potential of achieving enrichment
of methane in the product gas to 90-95% without chemical
additions for pH control. The ability to generate high-Btu
gas directly from anaerobic digestion is expected to reduce
gas cleanup and compression costs by \$1.50 to \$2.50/GJ.

A patent application on the pressure-swing, two-phase
process has been prepared and submitted by GRI staff. The
effects of various operating parameters affecting CO_2 distribution in single and two-stage laboratory digesters are
under investigation.

LITERATURE CITED

1.   Buswell, A.M. and W.O. Hatfield, "Anaerobic Fermenta-
     tions," Bulletin No. 30 of the Illinois State Water
     Survey, Urbana, IL (1936).
2.   Stumm, W. and J. Morgan, "Aquatic Chemistry," Wiley-
     Interscience, John Wiley & Sons, Inc., New York (1970).

3. Boone, D.R., "Terminal Reactions in the Anaerobic Digestion of Animal Waste," *Appl. and Environ. Microbiol., 43*(1), 57 (1982).

4. Brehany, J.J., "An Economic and Systems Assessment of the Concept of Nearshore Kelp Farming for Methane Production," *R.M. Parsons Co. Final Report to the Gas Research Institute,* GRI Report No. GRI 82/0067, Chicago (1983).

5. Christensen, D.R. and P.L. McCarty, "Biotreat: A Multi-Process Biological Treatment Model," paper presented at the Annual Conference of the Water Pollution Control Federation, Denver, CO, October 8 (1974).

6. Cohen, A., "Two-Phase Digestion of Liquid and Solid Wastes," Proceedings of the Third International Symposium on Anaerobic Digestion, p. 123, Boston, MA, August 14-19 (1983).

7. Colleran, E., M. Barry, and A. Wilke, "The Application of the Anaerobic Filter Design to Biogas Production from Solid and Liquid Agricultural Wastes," *Symposium Papers, Energy from Biomass and Wastes VI,* Institute of Gas Technology, pp. 443-478, Chicago (1982).

8. Dalrymple, W. and D.E. Proctor, "Feasibility of Dairy Manure Stabilization by Anaerobic Digestion," *Water and Sewage Works, 114*(9), 361 (1967).

9. Fannin, K.F. et al., "Anaerobic Processes," *J. Water Pollution Control Federation, 5516,* 623 (1983).

10. Fannin, K.F., et al., "Effects of the Interaction Between Biomass Composition and Reactor Design on Anaerobic Digestion Process Performance," paper presented at Energy from Biomass and Wastes VII, Lake Buena Vista, Florida, January 30-February 3 (1984).

11. Fong, W., "Methane Production from Animal Wastes by Anaerobic Digestion," M.Sc. Thesis, University of Manitoba (1972).

12. Wise, D. (ed.), *Fuel Gas Production from Biomass,* CRC Press, Inc., Boca Raton, Florida (1981).

13. Ghosh, S., "Microbial Production of Energy," paper presented at the Seventh International Biotechnology Symposium, New Delhi, India, February 19-25 (1984).

14. Ghosh, S., M.P. Henry, and D.L. Klass, "Bioconversion of Water Hyacinth-Coastal Bermuda Grass-MSW-Sludge Blends to Methane," *Second Symposium on Biotechnology in Energy Production and Conservation,* (C.D. Scott, ed.), pp. 163-187, John Wiley & Sons, New York (1980).

15. Graef, S.P. and J.F. Andrews, "Mathematical Modeling and Control of Anaerobic Digestion," *Water, American Institute of Chemical Engineers, Vol. 70,* p. 101, New York (1974).

16. Gramms, L.L., L.B. Polkowski, and S.A. Witzel, "Anaerobic Digestion of Farm Animal Wastes," *Trans. Am. Soc. Agr. Eng., 14*(1), 7 (1971).

17. Jewell, W.J., et al., "Anaerobic Fermentation of Animal Residue: Potential for Improvement and Implementation," U.S. Department of Energy Report No. EY76S0229817, NTIS, Springfield, VA (1978).

18. Jewell, W.J., et al., "Anaerobic Fermentation of Agricultural Residue," U.S. Department of Energy Report No. DOE/ET/20051-T2, NTIS, Springfield, VA (1980).

19. Kleinstreuer, C. and T. Poweigha, "Dynamic Simulator for Anaerobic Digestion Processes," *Biotech. Bioeng., 24,* 1941 (1982).

20. McCarty, P.L., "Anaerobic Waste Treatment Fundamentals," *Public Works, 95,* pp. 107, 123, and 191 (1964).

21. Messing, R.A., "Immobilized Microbes and a High-Rate, Continuous, Anaerobic Waste Processor," *Symposium Papers, Energy from Biomass and Wastes, VI,* pp. 425-442, Institute of Gas Technology, Chicago (1982).

22. Rijkens, B.A., "A Novel Two-Step Process for the Anaerobic Digestion of Solid Wastes," *Symposium Papers, Energy from Biomass and Wastes, V,* pp. 463-475, Institute of Gas Technology, Chicago (1981).

23. Warren, C.S. et al., "Evaluation of the Lake Apopka Natural Gas District," Document No. PB84-184969, NTIS, Springfield, VA (1984).

24. Metcalf and Eddy, Inc., *Wastewater Engineering,* McGraw-Hill Book Co., New York (1972).

26
ETHANOL AND CHEMICALS FROM CELLULOSICS
IN SOLID WASTES

George T. Tsao

Laboratory of Renewable Resources Engineering
Purdue University
West Lafayette, Indiana

ABSTRACT

Cellulosic materials contain in average 75% by weight carbohydrates with cellulose and hemicellulose being the major components. Cellulose can be hydrolyzed with either an enzymatic or an acidic catalyst to yield glucose. Hemicellulose is relatively easy to hydrolyze yielding xylose, galactose, mannose, arabinose, and so on. Fermentation can convert hexoses and xylose into ethanol. Meanwhile, many other fermentation processes can also be utilized to produce other useful chemicals.

Cellulosic materials can also be converted into valuable products by reactions of organic chemistry. Heterogeneous catalytic reactions can also yield additional useful chemicals. This paper will present discussion on various possibilities of production of fuels and chemicals from cellulosic materials.

ACID HYDROLYSIS

Acid hydrolysis of cellulosics can be divided into two groups: high temperature and low temperature. Approximately, 140°C can be considered the demarcation of the two types of processes. The low temperature acid process ranges from 80 to 140°C; most high temperature processes operate from 160 to 240°C. The main difference in the reaction results caused by the difference in the reaction temperature is the formation of furfural and its derivatives at a high

temperature in the acidic solution. Furfural and related
by-products represent a loss of sugar yield and create
problems in subsequent fermentations due to the inhibitory
effects of these by-products on living cells. The acid is
also much more corrosive at a high temperature and the required special metal for equipment fabrication is costly.

Concentrated mineral acids including phosphoric acid
(70% strength or more), hydrochloric acid (40%), sulfuric
acid (62% or more) and aqueous zinc chloride (a Lewis acid,
72% or more) are known to be able to swell and dissolve
cellulose (1). After pretreatment by a concentrated mineral acid, the decrystallized cellulose can be readily
hydrolyzed. Generally, a concentrated mineral acid acts
more as a "solvent" for cellulose dissolution rather than
an acid catalyst for hydrolysis. After addition of water,
the diluted acid will then become an effective catalyst
for cellulose hydrolysis. Purdue's low temperature acid
process involves the use of a mineral acid in the concentrated form to first decrystallize and then after dilution
with water to hydrolyze cellulose.

Another version of a dilute acid process of cellulose
hydrolysis involves heating the cellulosics at a high temperature in the presence of water. Under these conditions,
organic acids, mostly acetic acid, will be released from
the plant materials. The released acid can then promote
additional hydrolysis. This is a mild acid hydrolysis with
little or no external addition of acid and thus it has been
known as the autohydrolysis.

SOME DETAILS OF A LOW TEMPERATURE ACID HYDROLYSIS PROCESS

Hemicellulose Hydrolysis

Acid Pre-conditioning. Cellulosic materials are generally low in bulk density. If one adds a sufficient amount
of water to submerge a pile of wood chips or cornstalks in
a container, a minimum of liquid to solid ratio will be 5
to 1. After hydrolysis, even at 100% sugar yield, the obtainable sugar concentration in the hydrolysate will be low.
The well known Madison acid hydrolysis, for instance, gives
a solution of less than 5% by weight of total reducing sugars.
To overcome this dilution problem, a "roasting-leaching"
technique can be used to achieve a high concentration of
xylose and other soluble sugars in the hemicellulose hydrolysate.

For most farm residues such as cornstalks, the moisture content can be anywhere from 15% to 50% when they are collected and delivered to the plant site. Green wood chips usually contain about 50% moisture. The first step of the "roasting-leaching" processing is to add a dilute acid to the biomass and uniformly distribute the acid in biomass. For agricultural residues, spraying of a predetermined amount of sulfuric acid on the cellulosic solids can be easily done. A gentle rotating action will help to distribute the acid.

Roasting. The acidified plant biomass containing 50% moisture will appear "dry" and feel "dry" unpon touching; it does not contain enough moisture to show any "freely flowing" liquid. This material is then heated with either live steam or some type of waste flue gas to 80-100°C. The sulfuric acid is not very corrosive at this temperature range. With no freely flowing liquid, the contact between the acidic wet solids and the container inner wall is also at minimum.

After the 80-100°C treatment for several hours, 90% of the hemicellulose is solubilized. However, the monomeric and oligomeric products are still inside the cavities and pores in the bulky biomass due to the presence of only a limited amount of moisture allowed in the system at this point. The "leaching" step that follows will extract the solubles out with warm water.

The 80-100°C low temperature is desirable because at this temperature, xylose, arabinose, and uronic acids in the reaction mixture are stable for at least 24 hours without detectable furfural formation. According to published literature, in order to reduce furfural formation from acidified pentose solutions, one should use a high temperature but a short reaction time to carry out the hydrolysis. This is apparently true for the high temperature range of 130°C or so, judging from the published values of the activation energy and kinetics constants of the involved chemical reactions. However, that preference apparently cannot be extrapolated to the low temperature range of 80-100°C.

Leaching. The solubilized hemicellulose carbonydrate in the internal cavities and pores of the plant biomass can be extracted with added warm water at 60 to 100°C. Depending upon the level of oligomers in the mixture, a slightly higher temperature will finish the acid-promoted

hydrolysis to monomers once an extra amount of hot water
is added. The roasting-leaching technique will allow the
production of a hemicellulose hydrolysate of a relatively
high concentration of solubles. This can be seen as follows.
A typical plant biomass is assumed to contain about 35%
by weight of hemicellulose. In a reaction mixture of 50%
moisture, the concentration of solubles after hydrolysis will
be about:

$$100 \ \frac{35 \text{ lbs solubles of 100 lbs original biomass}}{100 \text{ lbs moisture} + 35 \text{ lbs solubles}} = 26\%$$

When we add hot water to extract the solubles, a dilution
is inevitable. However, by operating the leaching in a
fashion like a plug flow, trickling bed, the first portion
of the exiting effluent will be high in sugar concentration.
As additional volumes of effluent are collected, the sugar
concentration will be progressively lowered. We can thus
take two cuts: one above a preselected cumulative concentration of, say, 15% by weight of dissolved solids and one
below this concentration. The weak juice can be recycled
to extract sugars in the next column. Thus, by the roasting-
leaching techniques, one can obtain a fairly strong hemicellulose hydrolysate for subsequent processing.

Cellulose Hydrolysis

Pretreatment. Cellulose in the lignocellulosic solid
residue, after roasting and leaching to remove hemicellulose,
will be hydrolyzed by a dilute acid to yield glucose with
or without a prior ethanol extraction to isolate oligomeric
lignin. As we have experienced, the presence or absence of
lignin does not greatly affect the hydrolysis catalyzed by
an acid as much as in the case where enzyme catalysts are
used.

After hemicellulose removal, the physical integrity of the
solid residue is greatly reduced. The lignocellulose solids
can be easily crushed to very fine powders with little power
input. At this point, however, the crystalline structures
of the cellulose remains, essentially, intact. The cellulose
crystallinity could even be increased beyond that of the
native biomass due to the possible, partial selective removal of the "amorphous" cellulose by the roasting-leaching
treatment. In order to achieve a high level of cellulose
conversion, a pretreatment of some sort is necessary.

The hemicellulose-removed lignocellulose residue can be crushed to fine powder easily. This wet powder can be dewatered to 40% to 45% moisture by pressing. Before filtering and pressing, we may purposely add fresh acid to adjust the acid strength to a predetermined level and then dewater it. The wet cake by then has a substantial amount of acid uniformly distributed throughout the particles. The wet cake is then gently heated to remove water by evaporation, and thus gradually increase the acid strength in the solid. The gradually concentrating acid will decrystallize cellulose in a uniform manner. A suitable combination of the level of drying, the particle size, the amount of acid, the drying temperature and the length of drying time will allow an optimal level of cellulose decrystallization by this pretreatment.

Cellulose hydrolysis after the pretreatment. The wet mixture described above will be added with a sufficient amount of water to reduce acid strength in the liquid. At the diluted level, the acid becomes more a catalyst for hydrolysis. The slurry is heated at 80 to 140°C to produce glucose. The effectiveness of the pretreatment, the temperature, and the time of reaction will determine the glucose yield. On the other hand, one may want to purposely leave a sufficient amount of unreacted cellulose to be used together with lignin as the boiler fuel for the whole plant. In this case, the targeted level of desired cellulose conversion will dictate the temperature and the time of the hydrolysis step and also the extent of the pretreatment. Generally speaking, a very high level of cellulose conversion into glucose will require a disproportionately high level of pretreatment and thus also the addition of a large amount of concentrated sulfuric acid. This situation of diminishing return by the applied solvent can be quantified in an economical study and the optimal levels of pretreatment and hydrolysis can, hopefully, be determined. In some cases, perhaps one would want to convert all the cellulose into glucose, and prefer to use a different fuel for steam generation.

Glucose is much more stable than xylose when heated in the presence of an acid. Furthermore, because of the decrystallization pretreatment, the subsequent cellulose hydrolysis can be done at a relatively low temperature of 80 to 140°C. Therefore, the usual problem of decomposition of glucose to form hydroxymethyl furfural and other undesirable by-products can be avoided in this process.

The hydrolysate can be easily separated from residual cellulose and lignin by filtration and cake-washing. The cake rich in lignin no longer holds much water. The acidic glucose solution can be neutralized and then fermented to chemical products. In case we do not object to combining glucose from cellulose and xylose and other sugars from hemicellulose, and then fermenting or otherwise utilize them together, we can use the process option shown in Figure 1. In this case, the acidic glucose solution from cellulose hydrolysis is used to impregnate, roast and leach the original cellulosic raw material to hydrolyze hemicellulose. Even though this would mean a prolonged exposure of glucose to an acidic condition at a moderate temperature of 80 to 100°C (roasting temperature) for several additional hours, there is no serious problem of glucose decomposition because glucose is stable under those conditions. In this process option, the same acid is utilized three times; once in the concentrated form to decrystallize cellulose, once after dilution in cellulose hydrolysis as a catalyst and, once again as a catalyst for hemicellulose hydrolysis.

UTILIZATION OF HEMICELLULOSE HYDROLYSATE

The single factor that will have a decisive effect on the overall process economics in conversion of cellulosics is the utilization of hemicellulose hydrolysate. A typical chemical composition of North American hard wood is shown in Table 1, where the total hemicellulose consisted of pentosans, hexosans, polyuronics, and acetyls is about 35% of the total biomass weight. Among the various components, xylan, the polymer of anhydro-xylose is the most prominent. Only until recently, xylose was considered non-fermentable, meaning that the common yeast, *Saccharomyces cerevisiae*, could not ferment it to ethanol. About three years ago, it was discovered that xylulose, an isomer of xylose, could be readily fermented to yield ethanol, and xylose could be converted into xylulose by a treatment with an enzyme called xylose isomerase which is also known as glucose isomerase (Gong et al., 1981). In this article, we will examine four options given below for hemicellulose utilization.

Production of Furfural

Pentoses upon heating in the presence of an acid will lose water and form furfural (Eq. I):

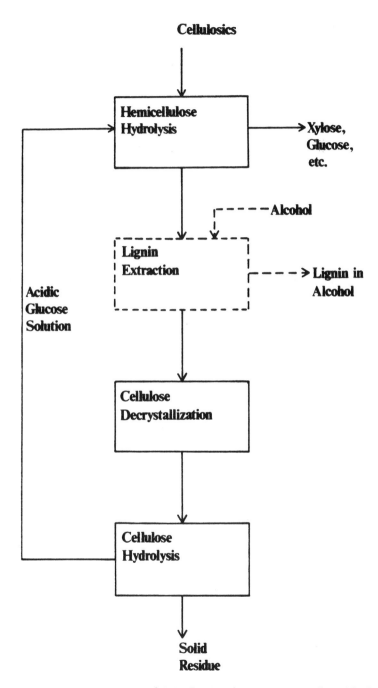

Figure 1. The overall acid hydrolysis process of cellulosics
with recycle of acidic glucose solution.

Table 1.   Chemical Composition of Wood

Components		Weight Percent
Cellulose		45.0
Hemicellulose		34.5
Pentosans		19.4
Xylan	18.9	
Araban	0.5	
Hexosans		7.1
Glucan	3.9	
Galactan	0.8	
Mannan	2.4	
Uronics		4.1
Acetyl		3.9
Lignin		20.2
Ash		0.3

$$C_5H_{10}O_5 \xrightarrow{-3H_2} > C_5H_4O_2 \longrightarrow \text{Decomposition products} \quad (I)$$

Pentose   Furfural

$$\underset{\text{Uronic Acids}}{\begin{matrix} \text{CHO} \\ | \\ (\text{HCOH})_4 \\ | \\ \text{COOH} \end{matrix}} \xrightarrow[\text{CO}_2]{\text{Heat}} \underset{\text{Pentoses}}{\begin{matrix} \text{CHO} \\ | \\ (\text{HCOH})_4 \\ | \\ \text{H} \end{matrix}} \longrightarrow \text{Furfural} \quad (II)$$

Uronic acids upon decarboxylation by heat will yield pentoses, and thus, are also potential raw material for furfural formation (Eq. II). The theoretical maximum yield is 64 lbs of furfural per 100 lbs of pentoses consumed.

Furfural is a chemical of considerable uses as a solvent and also as a feedstock. The very first nylon was, in fact, made from furfural. Furfural is currently sold at above 50 cents per pound. At this high price, it cannot compete effectively in the chemical market. By the technique of roasting and leaching a fairly concentrated acidic

solution of pentoses and uronic acids can be produced relatively cheaply. As we can recall, in this processing sequence, there has been relatively few expensive cost items
before arriving at the concentrated hemicellulose hydrolysate. In fact, with the acid already in the hydrolysate we
can convert pentoses and uronic acids into furfural by simply
heating it to about 260°C in a tubular reactor. As shown
in Eq. (I), furfural under the acidic, high temperature condition can form decomposition products of unknown identity.
Simple heating with no special precaution will give a furfural yield of about 65% of the theoretical maximum, i.e.,
about 64 x 0.65 + 42 lbs of furfural per 100 lbs of xylose
equivalent (PE = pentose equivalent). Since xylose, arabinose, and uronic acids can all be converted into furfural as
shown in Eqs. (I) and (II), the term PE is to express the
total available raw material for furfural formation. With
a special process precaution by, for instance, removing
furfural from the aqueous solution as soon as it is formed
to prevent its decomposition, a higher yield of 90% of the
theoretical maximum is possible (Sproull, 1984).

Furfural, besides being useful as a chemical and a
solvent, is also an anti-knocking agent for improving octane
ratings of gasoline products. However, a furfural being an
unsaturated aldehyde has a tendency to form a gummy polymeric
mess under engine combustion conditions. By a selective
hydrogenation, furfural can be converted into furfural
alcohol which is also a good anti-knocking agent and also
useful as a chemical.

Production of Single Cell Protein as a Feed By-Product

The common *Saccharomyces* and *Candida* yeasts cannot use
xylose readily under anaerobic conditions. However, with
oxygen adequately supplied, both *Saccharomyces* and *Candida*
can grow very fast utilizing xylose, arabinose, uronic acids
and acetic acid as the carbon source to yield new yeast
cells. The cells of *Saccharomyces* can be used as the
seed for ethanol fermentation of glucose in cellulose hydrolysate, and the yeast cells can also be harvested, dried and
marketed as a feed by-product. In this case, practically
all the available carbon source in the hemicellulose hydrolysate is utilized and thus it minimizes the need of subsequent
waste water treatment. For a small-scale operation, this approach of hemicellulose utilization could be the simplest in
terms of process design, product marketing and also meet the
requirements of environmental regulatory agencies.

Production of 2,3-Butanediol and Methyl Ethyl Ketone

A bacterial culture, *Klebsiella oxytoca,* can utilize xylose, arabinose and also uronic acids under anaerobic conditions (Jansen, 1982). From xylose and arabinose, the bacterial cells can produce 2,3-butanediol in a high yield (90% of the theoretical maximum) in a fairly high concentration of 9% by weight in the final fermentation broth. This diol has a number of possible uses of its own such as being a good anti-freeze agent. This fermentation product, however, is difficult to purify by distillation due to its high boiling point of 180°C. A method was developed for converting 2,3-butanediol in the fermentation broth directly into methyl ethyl ketone (MEK) by acidic dehydration in nearly a quantitative yield (Emerson, 1981). MEK, which boils at about 78°C can be easily purified by distillation. MEK also has a well established market.

Ethanol from Xylose After Isomerization

After a great deal of research effort, we can now convert xylose into ethanol (Gong et al., 1981). Other components including arabinose, acetic acid, and uronic acids cannot yet be utilized for this purpose. Currently, there are two parallel approaches of research and technology development dealing with the problem of xylose conversion into ethanol. One of them can be incorporated in this article. The second approach based upon gene splicing work has yet to be investigated further. Briefly, the conversion is shown in Figure 2. Xylulose can be fermented to ethanol via the biochemical metabolic reaction sequence known as the Pentose Phosphate Cycle. Xylose cannot be converted to ethanol unless it is first converted into xylulose. The common *Saccharomyces* yeast lacks the enzyme, xylose isomerase, which is responsible for the conversion of xylose to xylulose and thus cannot ferment xylose to ethanol. The enzyme, xylose isomerase, is also known as glucose isomerase, which happens to be the commercially available enzyme extensively used in the industrial production of high fructose corn syrup (HFCS). Therefore, the first approach for ethanol production from xylose is to conduct a simultaneous enzyme reaction and yeast fermentation. *Saccharomyces* yeast with the help of externally added glucose isomerase will convert xylose into ethanol. The yield of this process option is about 80% of the theoretical maximum, i.e., about 0.4 pound of ethanol can be produced from each pound of xylose. However, the external addition of an industrial enzyme designed for a food product, HFCS,

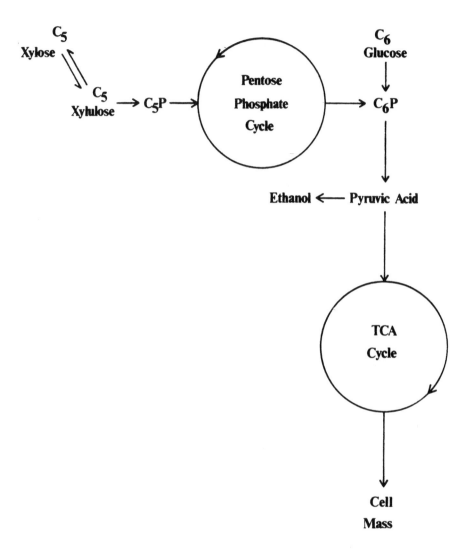

Figure 2. Metabolic pathway from xylose to ethanol.

adds expense to ethanol production. There are several large corn processing companies in the United States of America, which apply this enzyme for producing huge amounts of HFCS. The usual industrial practice is to run an immobilized enzyme complex for only 3 half-lives. As the definition of half-life suggests, at the end of the 3 half-lives, the enzyme activity becomes $(1/2)^3 = 1/8$ of that of the fresh enzyme complex. However, since xylose is the real natural substrate of this enzyme which should really be named xylose isomerase rather than the common commercial name, glucose isomerase, the isomerase is much more active (about 300% more) towards xylose than glucose. Therefore, even the spent enzyme after 3 half-lives in HFCS production can still be very active for xylose conversion into xylulose. For companies with large HFCS operations, the spent glucose isomerase which otherwise means a cost of solid waste disposal can be used for ethanol production via this first process option. This approach is likely to be too costly for companies who have to purchase fresh enzyme specifically for ethanol production from xylose.

REFERENCES

1. Tsao, G.T., M. Ladisch, C. Ladisch, T.A. Hsu, B. Dale, and T. Chou, "Fermentation Substrates from Cellulosic Materials: Production of Fermentable Sugars from Cellulosic Materials," *Ann. Reports of Fermentation Processes, 2,* 1-22 (1978).
2. Gong, C.S., L.F. Chen, M.C. Flickinger, L.C. Chiang, and G.T. Tsao, "Production of Ethanol from Xylose by Using Xylose Isomerase and Yeasts," *Appl. Environ. Microbial., 41,* 430-436 (1981).
3. Sproull, R., "Furfural: Reaction Kinetics, Mechanism and Process Design," Ph.D. Thesis, 1984, Purdue University, West Lafayette, IN.
4. Janse, N.B., "Application of Bioenergetic Principles to Modelling the Batch Fermentation of Xylose to 2,3-Butanediol by *Klebsiella oxytoca,*" Ph.D. Thesis, 1982, Purdue University, West Lafayette, IN.
5. Emerson, R.R., "A Study of the Kinetics of the Dehydration of Aqueous 2,3-Butanediol to Methyl Ethyl Ketone," M.S. Thesis, 1981, Purdue University, West Lafayette, IN.

27

THE PRODUCTION·OF CHEMICALS AND FUELS BY ACID HYDROLYSIS AND FERMENTATION OF MSW

E.C. Clausen and J.L. Gaddy

Department of Chemical Engineering
University of Arkansas
Fayetteville, Arkansas

ABSTRACT

Concentrated mineral acids may be used to hydrolyze ligno-cellulosics, such as MSW, to sugars. These sugars may then be fermented to ethanol, organic acids, or other industrial chemicals and fuels. This paper presents data and results for a two-step acid hydrolysis of MSW, followed by the fermentation of the resulting sugars to ethanol. Yields, kinetics and by-products are presented and discussed. Preliminary design and economic projections for a commercial facility to produce 20 mm gallons of ethanol per year using the combined hydrolysis-fermentation process are developed.

INTRODUCTION

As the cost of crude oil and natural gas increases and the proven reserves decrease, the search for alternative sources of raw materials for chemical and energy production must be expedited. Biomass, especially in the form of industrial, municipal and agricultural residues, is one source of raw material that can be used as a substitute for petroleum. It is estimated that approximately one billion tons of these various residues are generated annually in the United States (Stephens and Heichel, 1975). If converted into chemical intermediates, these residues could supply the raw material for all of the chemical and petrochemical production in this country.

Biomass materials are comprised of three major constituents: cellulose, hemicellulose, and lignin. The composition of various biomass materials is shown in Table 1. As
noted, the composition of biomass materials varies widely.
A typical biomass material, however, has 20-30% hemicellulose,
35-45% cellulose, and 10-25% lignin. The balance of the
material is ash and cell constituents.

MSW is shown to be an atypical biomass material, having
a composition that is relatively high in cellulose and low
in hemicellulose and lignin. Prior to analysis, the MSW
was processed to remove most of the metals and glass. Approximately 16% of the processed MSW consisted of these
refractory materials. Over 50% of the MSW is carbohydrate
which is available for conversion to sugars, followed by
subsequent conversion to fuels and chemicals.

Table 1.
The Composition of Selected Biomass Materials

Material	Percent Dry Weight of Material		
	Hemicellulose	Cellulose	Lignin
Tanbark Oak	19.6	44.8	24.8
Corn Stover	28.1	36.5	10.4
Red Clover Hay	20.6	36.7	15.1
Bagasse	20.4	41.3	14.9
Oat Hulls	20.5	33.7	13.5
Newspaper	16.0	61.0	21.0
Processed MSW	5.0	46.6	12.0

The hemicellulose of biomass contains varying quantities
of glucan and xylan. Paper hemicellulose, for example, contains about 50% glucan and xylan. Corn stover hemicellulose
contains 75-85% xylan. The cellulose fraction is almost entirely glucan. The constituents exist as polymers, which must
be reduced to monomeric sugars before they can be converted biologically to useful chemicals. Therefore, the biological conversion of biomass is actually a two-stage process: hydrolysis followed by fermentation.

The carbohydrate hydrolysis can be carried out by contact with cellulase or xylanase enzymes, or by treatment with mineral acids. Enzymatic hydrolysis has the advantage of operating at mild conditions and producing a high-quality sugar product. However, the enzymatic reactions are quite slow, and the biomass must be pretreated with caustic or acid to improve the yields and kinetics. Acid hydrolysis is a much more rapid reaction, but requires higher temperatures or high acid concentrations to achieve good yields. Furthermore, under these conditions, xylose degrades to furfural and glucose degrades to 5-hydroxymethyl furfural (HMF), both of which are toxic to microorganisms. A two-stage acid hydrolysis process enables the hemicellulose and cellulose to be degraded separately under conditions appropriate for each reaction. High yields are obtained and mild conditions are required, with very little resultant degradation. Large quantities of acid are necessary, however, and acid recovery is an important part of this process.

Various combinations of temperature and acid concentration may be used in hydrolyzing lignocellulosics to sugars. The two major classes of hydrolysis processes employ either dilute acid at high temperatures or concentrated acid at low temperatures. For example, complete conversion of the hemicellulose in corn stover into monomeric sugars may be obtained by treating with 2N acid at 100°C, or with 10N acid at 30°C (Clausen and Gaddy, 1982). Cellulose requires more drastic hydrolysis conditions. To obtain complete conversion an acid concentration of 10N at 100°C or an acid concentration of 14N at room temperature is required. Lignin, the other major constituent in biomass, is non-carbohydrate, and cannot be converted to sugars by hydrolysis. Work is presently underway to determine economical ways of converting lignin to phenols and cresols, as well as other chemicals.

Both single- and two-stage acid hydrolysis processes have been developed in the University of Arkansas laboratories. These processes have been found suitable for the production of fermentable sugars from agricultural and municipal residues. The purpose of this paper is to describe a process for producing fuels and chemicals from MSW. Data for the acid hydrolysis of MSW are presented along with the fermentation results of the hydrolyzate to alcohol. The economics of this process are determined, and areas for potential improvements in the process are examined.

ACID HYDROLYSIS

The two-stage acid hydrolysis process involves a pre-hydrolysis to remove the hemicellulose fraction, followed by a hydrolysis reaction to break down the cellulose.

Pre-hydrolysis

The two major factors which control the pre-hydrolysis reaction are temperature and acid concentration. Studies in the University of Arkansas laboratories have been made to determine the effect of these variables on the pre-hydrolysis of biomass. In deciding upon which conditions to use for acid hydrolysis, several factors must be considered. The acid cost would be prohibitively high, even at low concentrations (2N), unless the acid was recovered for reuse. The degradation of xylose to furfural is more readily promoted by high temperature than by high acid concentration (in this range). Therefore, the preferred operating range is a moderate acid concentration and mild temperature.

Hydrolysis

The hydrolysis of cellulose to glucose is also highly dependent upon acid concentration and temperature. The effect of these variables on the hydrolysis of biomass residue from the pre-hydrolysis has also been studied. Total conversion can be achieved at room temperature with an acid concentration of 14N. Higher temperatures are required to give suitable conversions at lower acid concentrations. High temperatures also promote glucose degradation and re-polymerization.

The sugar concentrations and yields from a typical two-step and one-step hydrolysis of MSW are given in Table 2. Very dilute (2-7%) sugar concentrations result from these reactions. The pre-hydrolysis step yields 8% of the initial MSW as xylose. The combined yield of glucose is 60%. A single-step hydrolysis yields 9% of the MSW as xylose and 50% as glucose. These yields represent nearly complete conversion of hemicellulose and cellulose to sugars.

As shown, a two-step hydrolysis process has the advantage of slightly increased yields and the separation of xylose and glucose in the sugar product. Sugar separation is particularly important if the hydrolyzate sugars are to be fermented to fuels and chemicals. Most organisms are able to use glucose

Table 2. MSW Acid Hydrolyzates

	Percent
Two-Step Hydrolysis:	
Prehydrolyzate	
Xylose, g/ℓ	9.5
g/100g	8.0
Glucose, g/ℓ	18.5
g/100g	16.0
Hydrolyzate	
Xylose, g/ℓ	0
g/100g	0
Glucose, g/ℓ	67.8
g/100g	44.0
Combined	
Xylose, g/100g	8.0
Glucose, g/100g	60.0
One-Step Hydrolysis:	
Xylose, g/ℓ	11.1
g/100g	9.0
Glucose, g/ℓ	59.1
g/100g	50.0

as a carbon source, but few are able to utilize both glucose and xylose simultaneously. Also, since xylose is particularly susceptible to degradation under hydrolysis conditions, separation of the xylose prior to hydrolysis could prevent undesirable furfural formation.

The proposed process for the stagewise acid hydrolysis of MSW to sugars is shown schematically in Figure 1. MSW is fed continuously to the pre-hydrolysis reactor. Residual solids are separated by filtration, washed with fresh acid and fed to the hydrolysis reactor. Solids from the hydrolysis

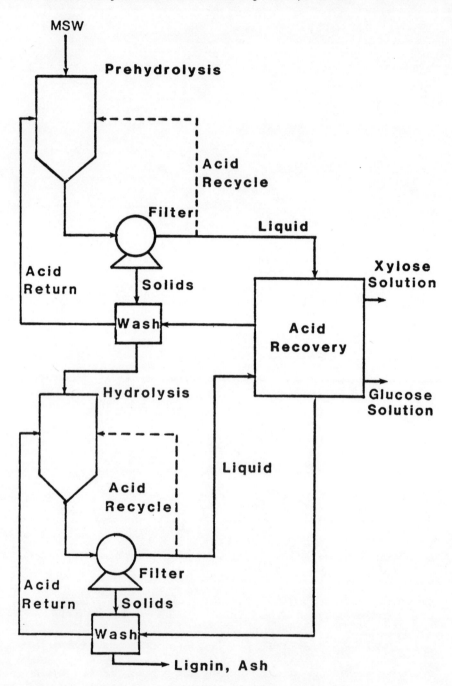

Figure 1.   Schematic of acid hydrolysis process.

reactor are filtered, washed and discarded. Acid and sugars
are separated and the acid is returned to the reactors. Two
dilute sugar streams result: the pre-hydrolyzate, containing
primarily xylose, and the hydrolyzate, containing only glucose.
These sugars can then be fermented to ethanol or other chemicals.

Sugar Decomposition

The fermentability of the sugars is dependent upon the
sugar decomposition that occurs during hydrolysis. Xylose
decomposes to furfural and hexoses decompose to HMF, which
are both toxic to yeast. Tolerance can often be developed,
and toxicity is difficult to define. However, the toxic limit
of furfural on alcohol yeast is reported to be .03-.046%
(Banerjee and Vishwanathan, 1972). HMF is reported to inhibit yeast growth at .5%, and alcohol production is inhibited
at .2% (Banerjee and Vishwanathan, 1974).

The rate of decomposition of xylose to furfural and
hexoses to HMF were studied at varying sugar concentrations.
Using the method of initial rates, these reactions were found
to be first-order. The ratio of rate constants for decomposition to formation are given in Table 3. These ratios
appear to be small, and subsequent calculations and experiments show that the rate of HMF appearance is insignificant.
However, the rate of furfural appearance can reach toxic
limits, especially if acid recycle is utilized.

*Table 3. Ratio of First-Order Rate Constants for Sugar
Decomposition to Formation Under Pre-Hydrolysis Conditions*

Sugar	Acid Concentration	Rate of Decomposition/ Formation
Glucose	2N	0.0053
	3N	0.0090
	4N	0.0074
Xylose	2N	0.0257
	3N	0.0402
	4N	0.0374

ETHANOL FERMENTATION

The acid hydrolysis of cellulosic residues, if performed at high temperature, can produce a sugar/degradation product mixture that is difficult, if not impossible, to ferment. However, if the hydrolysis conditions are mild, only small quantities of toxic substances are observed, and normal fermentation is expected.

Batch fermentation experiments were carried out at 30°C to compare the production rates of ethanol from hydrolyzates and synthetic glucose. *Saccharomyces cerevisiae* ATCC 24860 was used in the study. Inocula containing approximately 5×10^7 cells/ml from a 24-30 hour old seed culture were used in all experiments. The analysis of glucose and ethanol were made on a YSI 27 Industrial Analyzer.

Figure 2 shows the comparison of a batch fermentation with *S. cerevisiae* of both a synthetic glucose and hydrolyzate at a 4% sugar concentration. Identical results are found when fermenting synthetic glucose and hydrolyzate in the presence of yeast extract. Ethanol yields were also nearly identical. The concentrations of furfural and HMF in the hydrolyzates were found to be negligible. These very low levels of by-products are believed to be the major reason for the successful fermentation.

When the sugar concentration in the hydrolyzate exceeded 10%, slight deviations in the fermentability of hydrolyzate from that of synthetic glucose were seen. Studies are presently underway to investigate substrate toxicity and other reasons for these differences.

ECONOMIC PROJECTIONS

To illustrate the economics of this process, a design has been performed for a facility to convert MSW into 20 million gallons per year of ethanol, utilizing the acid hydrolysis procedures previously described. The capital and operating costs are summarized in Table 4.

MSW would be collected and delivered to the plant site as needed. Feedstock preparation consists of metal and glass removal, shredding, grinding and conveying to the reactors. The cost of the removal of glass and metals is not included in the feed processing cost, as reports indicate that resale of these materials will offset the capital and operating costs of separation. The hydrolysis section, as shown

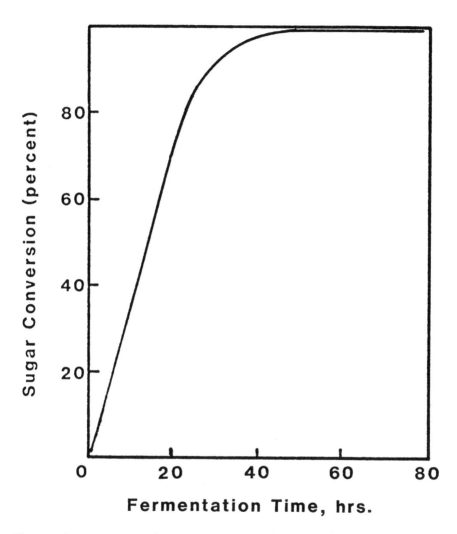

Figure 2. Fermentation of hydrolyzate and synthetic glucose.

Table 4.
Economics of 20 Million Gallon Per Year Ethanol Facility

A. Capital Cost

	Million $
Feedstock Preparation	2.1
Hydrolysis	6.5
Acid Recovery	10.0
Fermentation & Purification	6.6
Utilities/Offsites	11.6
	$36.8

B. Operating Cost

	Million $/yr	$/gal
MSW	-	-
Utilities	1.5	0.08
Chemicals	2.0	0.10
Labor	1.1	0.06
Fixed Charges		
Maintenance (5%)	1.8	0.09
Depreciation (10%)	3.7	0.18
Taxes & Insurance (2%)	0.7	0.03
Profit (57%)	21.2	1.06
	$32.0	$1.60/gal

in Figure 1, consists of continuous reactors. Acid resistant
materials of construction are necessary for this equipment.
Acid recovery is accomplished by electrodialysis; although,
other methods, such as evaporation, are equally applicable
and cost about the same. The typical batch fermentation and
ethanol distillation units are included. The total capital
cost for this plant is $36.8 million, including all utilities,
storage and offsites.

The annual operating costs are also shown in Table 4.
These costs are also given on the basis of unit production
of alcohol. As mentioned previously, no cost is included
for MSW. The energy costs are substantial because of the
dilute sugar and alcohol solutions resulting from the hydrolysis. A lignin boiler is used to reduce the energy requirements, and energy costs are $0.08 per gallon. Fixed charges

are computed as a percentage of the capital investment and
total 75% or $1.36 per gallon. The total cost, including
profit, to produce alcohol by this process is $1.60 per
gallon.

It should be noted that this process does not include
utilization of the pentose stream. Acid recovery is included,
but fermentation of the pentoses is not provided. Xylose
could be fermented to alcohol, acids or other valuable chemicals, which would improve the economics. However, since
this technology is not perfected, such products have not
been included.

In examining these economic figures to determine where to
improve the process, it is evident that yields cannot be substantially improved, nor raw material costs reduced. Other
significant cost areas are utilities, chemicals (including
acid and denaturants) and fixed charges. Therefore, if economic improvements in this process are possible, reductions in
the energy and capital costs must be sought.

PROCESS IMPROVEMENTS

Perhaps the single most detrimental factor to the economics
of this process is the very dilute solutions that result from
acid hydrolysis. Dilute concentrations increase both the
equipment size and the energy required for purification.
Another pernicious economic factor in this, and other fermentation processes, is the use of large conventional batch
fermentors. Methods for overcoming these problems will be
considered next.

Solids Concentration

The ultimate sugar and alcohol concentrations are direct
functions of the initial solids concentration in the hydrolysis. Since fluidity in a stirred reactor is a requirement,
a 10% mixture has been considered maximum. Therefore, the
resultant sugar concentrations have been only 2-7% and alcohol
concentrations only half as much.

If the limiting factor is considered to be fluidity in
the *reactor*, instead of the feed mixture, the feed concentration could be increased by roughly the reciprocal of one
minus the solids conversion in the reactor. Of course, solids
and liquid would have to be fed separately, which would also
save equipment cost. For MSW containing 24% hemicellulose

and 44% cellulose, the reactor sizes could be reduced by
24% and 58% for pre-hydrolysis and hydrolysis, respectively.
Attendant reductions would also result in the filtration
and washing units.

Equally important are the resultant increases in sugar
concentrations. The xylose concentration from the pre-hydrolysis reactor would be increased only slightly to about 3.5%.
However, the glucose concentration in the hydrolyzate would
be more than doubled to about 15%. Energy and equipment
costs in the fermentation area would be reduced proportion-
ately.

This simple alteration in the process has a profound
impact on the economics. It is estimated that the capital
cost would be reduced by 40% in the hydrolysis and acid recovery sections and 60% in the fermentation and utilities
areas. Furthermore, the energy requirements for distillation
are reduced by 60%.

Acid Recycle

Another method to increase the sugar concentration is
to recycle a portion of the filtrate (acid and sugar solution) in each hydrolysis step. The acid would catalyze
further polysaccharide hydrolysis to increase the sugar concentration. Of course, recycle of the sugars increases the
degradation to furfural and HMF.

Experiments have been conducted in our laboratories to
determine the enhancement possible with acid recycle. Various
amounts of the acid and sugar solution from the filtration
were recycled to determine the resulting sugar and by-product
concentrations. Acid and solids concentrations and temperatures were kept constant.

Figure 3 gives the sugar concentration as a function
of the quantity of filtrate recycled for the pre-hydrolysis
reactor. As noted, the xylose concentration is increased
nearly four-fold and the glucose concentration is increased
nearly six-fold at total recycle. Final concentrations of
8.7 and 2.8% are obtained for xylose and glucose, respectively.
It should be noted that not all the filtrate can be recycled,
since a portion adheres to the solids in filtration. Therefore, the yield of xylose begins to decrease due to losses
with the filter cake and due to decomposition.

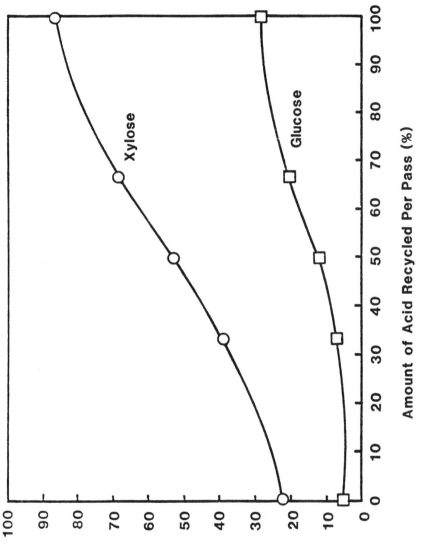

Figure 3.   Sugar concentration with various amounts of recycle.

As expected, the concentrations of by-product furfural
and HMF increase as a function of the quantity of filtrate
recycled. With total recycle, the composition of these by-
products reaches concentrations of .2 and .17% for furfural
and HMF, respectively. These values are above the toxic
limit of furural on most microorganisms; therefore, high
recycle rates would not be possible. The xylose yield is
maximized with a 50% recycle and the furfural level is not
excessive in this range.

With 50% recycle, a steady-state concentration of 6%
xylose is achieved, or about 2.5 times higher than with no
recycle. Similar results were obtained for 50% acid recycle
in the hydrolysis step. No furfural is present in this
stream, since xylose is not present in the hydrolyzate.

The effect of acid recycle on the economics is significant. A recycle rate of 50% in both pre-hydrolysis and
hydrolysis reactors, coupled with high solids concentrations,
would result in a xylose concentration of nearly 9% in the
pre-hydrolyzate, and a glucose concentration of almost 38%
could be achieved in the hydrolyzate. Practically, this
concentration should not exceed 25%, so a smaller recycle
fraction is required. It should be noted that these concentrations have been exceeded in the laboratory, while maintaining furfural and HMF less than .05%. These high concentrations reduce the equipment size in the acid recovery
section by 50% and in the fermentation section by another
60%. Energy consumption is also reduced another 60%.

Continuous Fermentation

One serious problem with the large scale production of
industrial chemicals by fermentation is the high cost of
providing for batch fermentors and insuring sterile conditions. Most industrial fermentations, including ethanol
for fuel usage, are conducted in batch equipment so that contamination and mutation can be controlled. Sterilization
between batches and the use of a fresh inoculum insure efficient fermentation. However, most batch alcohol fermentations are designed for thirty hours (or more) reaction time,
which results in very large and expensive reactors.

The reactor size can be reduced substantially by using
continuous flow fermentors. When fermenting acid hydrolyzates, the problems with maintaining sterile conditions are

substantially reduced. Therefore, the use of continuous fermentation is a natural application for producing alcohol from MSW hydrolyzate.

A number of continuous fermentation schemes are being studied, including the CSTR (Cysewski and Wilke, 1978), cell recycle reactor (Elias, 1979), flash fermentation (Cysewski and Wilke, 1977) and immobilized cell reactors (Sitton and Gaddy, 1980). Typical laboratory data for the fermentation of a synthetic MSW hydrolyzate in a CSTR are shown in Figure 4. A sugar concentration of 3% is utilized, and a maximum alcohol concentration of 1.4% is achieved. The maximum productivity of 1.7 gm/ℓ-hr occurs at a dilution rate of .2 hr^{-1}, or a 5 hour retention time.

Figure 5 shows the results of fermentation in the University of Arkansas laboratories of the same hydrolyzate in a 36 inch column with immobilized *S. cerevisiae* (ATCC 25858). The column was packed with 1/4 inch Raschig rings, coated with gelatin and cross-linked with glutaraldehyde. Dilution rates as high as 1.75 hr^{-1} were attained in the immobilized cell reactor (ICR), or 6 times the washout rate in the CSTR. Alcohol productivities of 15.9 gm/ℓ-hr were achieved, or about 10 times the CSTR. Furthermore, alcohol inhibition and furfural toxicity were greatly reduced in the ICR.

Compared to a batch fermentation time of 30 hours, the ICR could perform the same fermentation in 2 hours with 93% less reactor volume. An estimated savings of 60% of the capital cost for batch fermentors should result when using an ICR.

ECONOMIC IMPACT OF PROCESS IMPROVEMENTS

It is instructive to determine the consequences on the economics of incorporating the above process modifications. The process, using concentrated feed, acid recycle and continuous fermentation results in capital costs as given in Table 5. The plant investment is reduced by over 50%. Adjusted operating costs are also shown. On the basis of an ethanol price of $1.60 per gallon, a pre-tax profit of 144% results. Without interest or profit, the cost of producing alcohol is $0.33 per gallon. Sufficient margin should remain for commercialization.

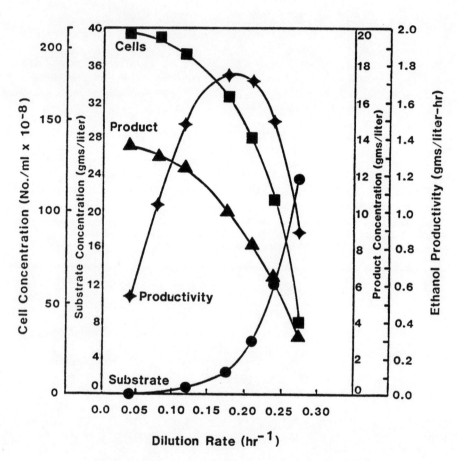

Figure 4. Hydrolyzate fermentation in a CSTR.

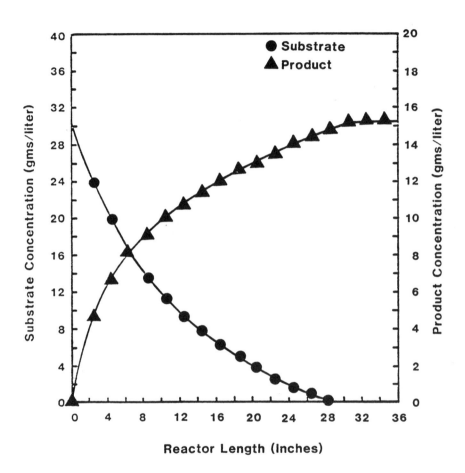

Figure 5. Hydrolyzate fermentation in an ICR.

Table 5.
Improved Economics of 20 Million Gallon Per Year Ethanol Facility

A.   Capital Cost

	Million $
Feedstock Preparation	2.1
Hydrolysis	3.9
Acid Recovery	5.0
Fermentation & Purification	1.6
Utilities/Offsites	5.0
	17.6

B.   Operating Cost

	Million $/yr	$/gal
MSW	-	-
Utilities	0.7	0.03
Chemicals	1.8	0.09
Labor	1.1	0.06
Fixed charges		
Maintenance (5%)	0.9	0.04
Depreciation (10%)	1.8	0.09
Taxes & Insurance (2)	0.4	0.02
Profit (144%)	25.3	1.27
	$32.0	$1.60/gal

CONCLUSIONS

The two-stage acid hydrolysis of residues, such as MSW, require mild temperatures, but large quantities and high concentrations of acid.   The resulting hydrolyzates, containing primarily xylose and glucose, can be fermented to ethanol or other chemicals.   The economics of this process are marginal in the present economy.

The process economics can be improved by increasing the sugar concentration through high solids reactions and acid recycle, and by using continuous fermentation.   These modifications result in reducing the capital cost by 50% and improving the profitability to over 70%, after taxes. It should be noted that these process modifications could be employed to improve the economics of similar processes to produce other products, as well as ethanol.

REFERENCES

Banerjee, N. and L. Vishwanathan, *Proc. Annual Sugar Conv. Tech. Conf.*, (CA), *61*, 1228f (1972).
Banerjee, N. and L. Vishwanathan, *Proc. Annual Sugar Conv. Tech. Conf.*, (CA), *82*, 123277W (1974).
Clausen, E.C. and J.L. Gaddy, "Production of Ethanol from Biomass," *Biochem. Engr.* (1982).
Cyseweki, G.R. and C.R. Wilke, *Biotech. Bioeng.*, *20*, 1421 (1978).
Elias, S., *Food Engineering*, 61 (October 1979).
Cysewski, G.R. and C.R. Wilke, Lawrence Berkeley Laboratory Report 4480, March (1976); *Biotech. Bioeng.*, *19*, 1125 (1977).
Sitton, O.C. and J.L. Gaddy, *Biotech. Bioeng.*, *22*, 1735 (1980).

28
THE BIOCONVERSION OF MUNICIPAL SOLID WASTE AND SEWAGE SLUDGE TO ETHANOL

Daniel R. Coleman, Thomas J. Laughlin, Melvin V. Kilgore, Jr.,
Chris L. Lishawa, and William E. Meyers

Southern Research Institute
Birmingham, Alabama

Michael H. Eley

University of Alabama in Huntsville
Huntsville, Alabama

ABSTRACT

We have investigated a system for the bioconversion of
municipal solid waste (MSW) and sewage sludge to ethanol.
The system has three primary components: MSW processing/fract-
ionation, cellulose hydrolysis to fermentable products, and
fermentation/distillation.

The novel pretreatment process developed in this study
effectively separated the organic and inorganic fractions of
unsorted MSW. The cellulose in the organic fraction was
readily hydrolyzed to glucose by enzymatic and acid hydrolysis
processes. This material was demonstrated to be fermentable.
The overall yield was about 40 gallons of ethanol per ton
MSW (as received).

Our economic analyses indicate that a plant processing 1500 tons of MSW per day would require a $90,000,000 capital outlay and that the breakeven price of ethanol would be $1.50 per gallon (based on ethanol production alone).

INTRODUCTION

With the realization that mankind's resources are limited, it has become apparent that societies must learn to recycle their waste. Two of the largest waste streams are municipal solid waste and sewage sludge, and one of the limited resources is liquid fuel. In this report, we review our development of a process for converting MSW and sewage sludge to fuel-grade ethanol (1-4). And we present the conceptual design of a 1500-ton/day MSW plant for which we have demonstrated technical and economic feasibility.

BASIC COMPONENTS

The system we investigated consists of three basic components: MSW/sewage-sludge processing, cellulose hydrolysis, and fermentation/distillation. We placed the greatest emphasis on the two less developed technologies (MSW/sewage-sludge processing and cellulose hydrolysis) and on interfacing the main components.

The MSW/sewage-sludge processor used in this study was a modified version of a processor patented by Urban Waste Resources of Birmingham, Alabama (5). The 2- to 4-ton/day MSW unit operated in this study, termed the Holloway Processor, is shown in Figure 1. A schematic of the unit is shown in Figure 2. The unit operated much like a pressurized clothes dryer. MSW and sewage sludge were mixed with steam under pressure and tumbled for about an hour. The processed material was dumped through the 16" x 9" exit port onto a double-deck vibrating screen. The material was fractionated into three categories: overs (greater than 2" x 2" minimum area), middles (greater than 1/2" x 1/2" minimum area), and fines (material with less than 1/2" x 1/2" area).

The fines were sent to several outside laboratories for hydrolysis and fermentation. These results were confirmed by out in-house experimentation.

Figure 1.   The MSW pilot plant operated by Southern Research Institute.

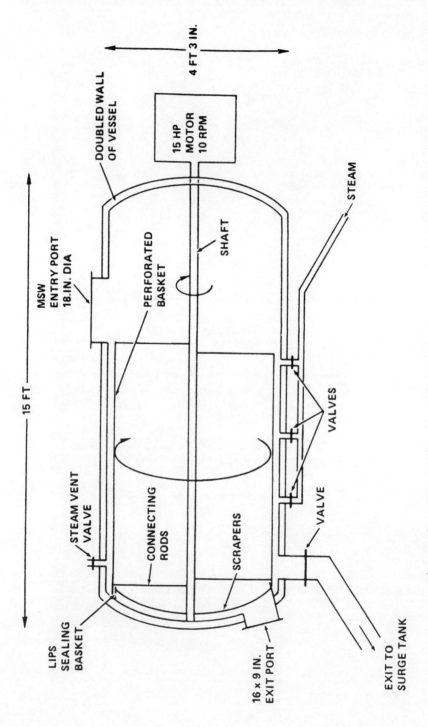

Figure 2.   Schematic diagram of the 2- to 4-ton/day MSW processor.

MSW/SEWAGE-SLUDGE PROCESSING

We processed 30 tons of MSW and 7.5 tons of sewage sludge over a six-month period. The MSW was from a middle-class suburb of Birmingham, Alabama. The only presortment performed was the removal of items which could not go through the 18"-diameter entrance port. The primary sewage sludge, obtained from the Shades Valley Treatment Plant (Birmingham, AL), consisted of 3.0% organics and 1.1% inorganics.

The MSW unit could handle up to 1000 lb of MSW plus 500 lb of sewage sludge or water per run. It typically took 1 h to load, 1 h to run, and 1 h to screen. The unit was usually operated under 60 lb pressure at a temperature of about 300°F.

The processed material was fractionated into overs, middles, and fines (Figure 3). A mass balance for a typical run was:

600 lb MSW	800 lb fines
360 lb sewage sludge	250 lb middles
250 lb steam	60 lb overs
	120 lb steam

The overs were primarily bottles and cans. Most of this fraction could be sold as recyclables. The middles were mostly organic. When recycled through the MSW unit, about 70% of the total mass came out in the fines fraction. The fines were a soggy mass of organic material. As indicated in Table 1, they were about 30% solids with about half of the solids being cellulose. All three fractions were sterile.

HYDROLYSIS

The cellulose in the fines was found to be readily hydrolyzable by several acid and enzymatic processes (Table 2). Two processes, one acid (6) and one enzymatic (7,8), gave cellulose-to-ethanol yields of over 80%. In the acid process, concentrated HCl is used to hydrolyze the cellulose to glucose which is then fermented to ethanol. In the enzymatic process, the cellulose is simultaneously hydrolyzed (enzymatically) and fermented.

Figure 3.   The three fractions from the MSW processor.

Table 1.
COMPOSITION OF THE FINES

MOISTURE	68.6	4.6
FEED ANALYSIS		
(wt %, dry basis)		
Cellulose	54.5	2.7
Lignin	16.0	1.7
Hemicellulose	1.3	1.6
Protein	5.0	0.5
Fiber	70.0	4.6
Fat	4.6	1.1
Ash	12.8	6.4
Total Carbohydrates	10.0	0.7
TOC	36.5	6.3
Organic N	0.7	0.1
Inorganic N	0.2	0.1
Kjeldahl N	0.8	0.1
Total Solids	31.4	2.7

ECONOMIC ANALYSIS

We performed an extensive economic analysis on an acid-based process (6). Although we felt that an enzymatic-based process (7) would be equally feasible, at the time of the study we lacked sufficient data to evaluate it. The study was based on the 500-ton MSW/day plant shown in Figure 4. Then, using standard scale-up procedures, we projected the economics for both a 1000- and 1500-ton/day plant. The results are listed in Table 3. The breakeven price for ethanol in a 1500-ton/day MSW plant (as required for Birmingham, AL) is $1.50 per gallon. Of course, credits for tipping fee, recyclable metals and glass, and animal feed could further reduce this figure.

CONCLUSION

We have developed a system for the bioconversion of MSW and sewage sludge to ethanol. In spite of the capital-intensive nature of this process, it appears to be both technically and economically viable.

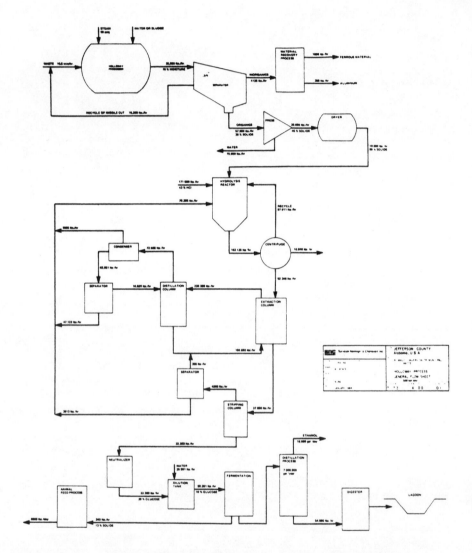

Figure 4.   Process flow diagram for a 500 ton/day plant.

TABLE 2. Comparison of Hydrolysis Conditions and Yields from SCO

Process	Glucose concentration, g/L	Glucose/SCO, g/g	Actual conversion, % cellulose to glucose	Theoretical EtOH/ton MSW,[a] gal	Actual EtOH MSW,[a] gal	Overall conversion efficiency, %
12% H_2SO_4[b] 10°C; 3.5 min	49	0.25	43	20	20	43
45% HCl[c] 50°C; 30 min	ND[d]	0.47	71	33	ND	ND
4% SO_2 gas 150–190°C; 50 min	ND	0.34	55	26	ND	ND
2-step HCl process[f] (30°C)	83.8	0.58	100	46	43	93
2-step HCl process[f] (30°C)	86.3	0.60	104	48	45	97
1-step HCl process[f] (30°C)	59.1	0.50	86	40	38	80
0.93% H_2SO_4[g] 240°C; 8.41 sec	8.45	0.22	38	18	ND	ND
1.03% H_2SO_4 (pretreatment)[g] 220°C; NOVO C-30	18.3	0.45	78	36	ND	ND
Enzymatic hydrolysis PSR[h] (pretreatment)[i] SSF 48 h	43.7	0.48	81	38	37.5	81
10% H_2SO_4[j] 100°C; 2 h	4.5	0.10	17	8	ND	ND
37% HCl[j] 28°C; 75 min	17.5	0.12	20	9	ND	ND
37% HCl[j] 27°C; 2 h	31.3	0.27	46	21	ND	ND
44% HCl[j] 28°C; 90 min	44.2	0.36	61	28	ND	ND
37% HCl, 28°C; 75 min[j] +44% HCl, 25°C; 75 min	28.1	0.49	84	39	ND	ND
PSR (pretreatment)[j] SSF 74 h	24	0.35	61	28	28	61

[a]As received.
[b]Dr. Y.Y. Lee, Dept. of Chem. Eng., Auburn University.
[c]Dr. I.S. Goldstein, Dept. Wood & Paper Sci., North Carolina State University.
[d]No data.
[e]Dr. M. Wayman, Dept. of Chem. Eng., University of Toronto.
[f]As described by Dow (2).
[g]Dr. A.O. Converse, Thayer School of Eng., Dartmouth College.
[h]Particle-size reduction.
[i]Dr. G.H. Emert, Biomass Research Center, Univ. of Arkansas.
[j]Southern Research Institute, Biotechnology Section.

Table 3.
Annual Cost (in thousands)

PLANT SIZE	500 Tons/Day	1000 Tons/Day	1500 Tons/Day
Materials	$ 1,200	$ 2,438	$ 3,657
Labor	1,321	2,107	2,577
Electrical Energy	215	431	646
Fuel	2,224	4,448	6,672
Water	22	44	66
TOTAL DIRECT COSTS	5,002	9,468	13,618
Capital Cost	42,081	66,846	90,429
Finance Charges (18y, 10%)	5,130	8,150	11,026
Total Annual Costs	10,133	17,618	24,645
Annual Ethanol Production	5,460	10,920	16,380
Break-Even Ethanol Cost ($/Gal)	$1.86	$1.61	$1.50
Cost Including $5.00/Ton Tipping Fee Credit	$1.73	$1.49	$1.38

REFERENCES

1.   Coleman, D.R., W.E. Meyers, T.J. Laughlin, C.L. Lishawa, M.H. Eley, and M.V. Kilgore, Jr., "The Production of Fuel-grade Ethanol from the Cellulose in Municipal Solid Waste," 4th Ann. Solar Biomass Energy Workshop, U.S.D.A., Southern Agricultural Energy Center, Tifton, GA, pp. 45-48, Atlanta, GA, April 17-19 (1984).

2.   Coleman, D.R., M.V. Kilgore, Jr., T.J. Laughlin, C.L. Lishawa, and W.E. Meyers, "Biomass Conversion of Municipal Solid Waste", Proc. of Miami International Symposium on the Biosphere, Clean Energy Research Institute, Coral Gables, FL, pp. 132-137, Miami, FL, April 23-24 (1984).

3.   Laughlin, T.J., D.R. Coleman, M.H. Eley, M.V. Kilgore, Jr., C.L. Lishawa, and W.E. Meyers, "Development of a Process for the Conversion of Municipal Solid Waste to Ethanol," *Biotechnol. Bioeng.* (submitted).

4.   Lishawa, C.L., R.E. Lacey, D.R. Coleman, M.H. Eley, M.V. Kilgore, Jr., and T.J. Laughlin, "The Economics of Biomass Conversion of Municipal Solid Waste to Ethanol," *Biotechnol. Bioeng.* (submitted).

5.   Holloway, C.C., L.B. Holloway, and W.B. Holloway, "Process for Separating and Recovering Organics and Inorganics from Waste Material," U.S. Patent #4,342,830 (1982).

6.   Foster, A.V., L.E. Marty, and D.E. Ling, (Dow Chemical Company), "Process for Separating and Recovering Concentrated Hydrochloric Acid from the Crude Product Obtained from the Acid Hydrolysis of Cellulose," U.S. Patent #4,237,220 (1980).

7.   Blotkamp, P.J., M. Takagi, M.S. Pemberton, and G.H. Emert, "Enzymatic Hydrolysis of Cellulose and Simultaneous Fermentation to Alcohol," *AIChE Symposium Series No. 181*(74), 85-90 (1978).

8.   Becker, D.K., P.J. Blotkamp, and G.H. Emert, "Pilot-scale Conversion of Cellulose to Ethanol," Fuels from Biomass and Wastes, (D.L. Klass and G.H. Emert, eds.), Ann Arbor Science, Ann Arbor, MI, pp. 375-391 (1981).

29

THE USE OF IN VITRO ANAEROBIC LANDFILL SAMPLES FOR ESTIMATING LANDFILL GAS GENERATION RATES

Richard L. Jenkins and John A. Pettus

Getty Synthetic Fuels, Inc.
Long Beach, California

ABSTRACT

The estimation of landfill gas generation rates is required for environmental management and energy recovery.
This paper describes a technique for predicting gas generation rates using refuse samples which have been taken at
depth under anaerobic conditions in the landfill and transferred to the laboratory for incubation.

Sampling operations are conducted in a pilot bore adjacent to the main well bore. The refuse is retrieved from
auger loads which have been filled at 10-100 ft below the
landfill surface. In the laboratory, anaerobic cylinders
are incubated at $100 \pm 1°F$ (mesophilic).

During the initial stages of incubation, the cylinders
produce abnormally high concentrations of H_2 and CO_2. The %
CH_4 in the gas increases gradually with a commensurate decrease in % CO_2. During this time the % H_2 decreases and
eventually becomes non-detectable ($< 0.1\%$). Following these
gas composition changes, a steady state gas composition is
achieved which is typical of landfill gas (50-55% CH_4, 40-45%
CO_2, $< 0.1\%$ O_2, $< 0.1\%$ H_2, and 0.5-2% N_2). Biochemically,
these observations indicate a shock or upset of methanogenesis
as a result of the sampling process. Metabolism of glucose
and other precursors continues to produce intermediates, CO_2
and H_2 as the methanogens acclimate gradually to a new
environment.

For those samples reaching the stady state, average production rates of 0.1-0.7 ft^3/lb dry refuse/yr have been observed. To date, moisture content of the refuse appears to be the predominant limiting factor in gas production.

INTRODUCTION

The recovery of landfill gas is unique in that all aspects of the process are beneficial. It provides a clean and economical energy source commensurate with positive environmental effects. These environmental improvements include: 1) reduced landfill emissions of odorous, toxic and photochemically reactive compounds, 2) enhanced vegetation growth which accelerates the reclamation of the site, and 3) reduced gas migration and explosion hazards.

The biochemical pathways leading to the production of landfill gas are part of the carbon cycle as shown in Figure 1, where refuse would represent complex organic compounds. Table 1 provides a breakdown of the major constituents of U.S.A. refuse. As shown by the equation below, it is generally accepted that cellulose is the primary precursor for methane formation in landfills.

$$-C_6H_{10}O_5 - + \eta H_2O \longrightarrow 3\eta CH_4 + 3\eta CO_2$$

Traditionally, landfill gas has been recovered with the use of vertical wells of a design similar to that shown in Figure 2. The nature of each landfill dictates the depth, number and type of the recovery wells.

The capital investment to develop a landfill is substantial. For instance, Getty Synthetic Fuels, Inc. (GSF) currently operates the world's largest landfill gas recovery facility at the Fresh Kills Landfill, Staten Island, New York. General data on this project are shown in Table 2.

Table 3 presents some of the major steps involved in developing a landfill for gas recovery. The testing phase is very crucial since it determines 1) the size of the processing plant, 2) the gas delivery rate, 3) the project lifetime, and 4) the economics.

Landfill testing usually requires numerous vertical borings to install test wells and pressure probes. These operations provide the opportunity to retrieve refuse samples from a variety of locations within the landfill.

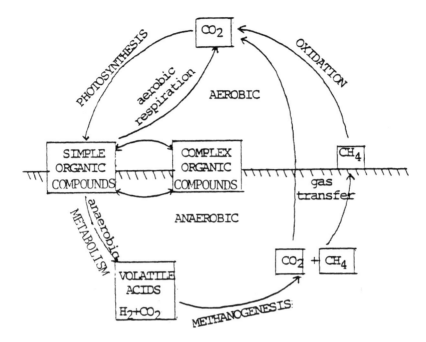

Figure 1. The carbon cycle.

Table 1. Typical Refuse Composition

	% By Weight
Food Wastes	15
Paper	40
Cardboard	4
Plastics	3
Textiles	2
Rubber	0.5
Leather	0.5
Garden Trimmings	12
Wood	2
Glass	8
Tin cans	6
Non-Ferrous Metals	1
Ferrous Metals	2
Dirt, Ashes, Brick, etc.	4
Total	100%

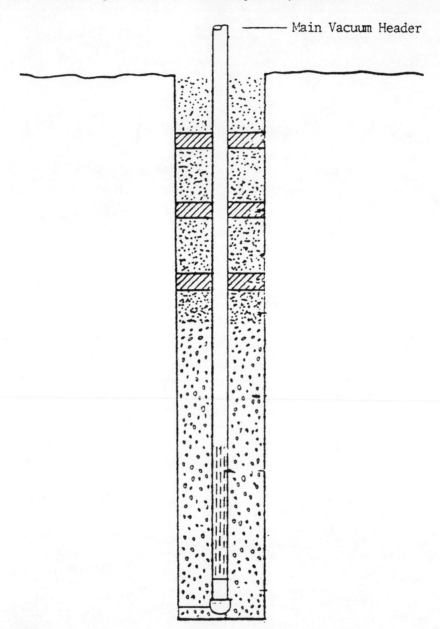

Figure 2.   Typical well design.

Table 2.
GSF's Fresh Kills Methane Recovery Plant

1. $20 MM

2. 10 MMSCFD

3. Operates 24 Hours/Day

4. 150-200 Wells

5. Gas to Heat 10,000-15,000 Homes

6. 1500 Acres

7. 12 Million Tons

Table 3.
Project Development

1. Landfill Size (Present and Potential)

2. Landfill Owner/Gas Rights

3. Landfill Customer

4. Testing of Landfill

5. Economic Analysis

This paper describes our efforts to develop a testing program for anaerobic refuse. The objective of the program was to sample the refuse anaerobically and to assess the gas production potential in the laboratory.

EXPERIMENTAL

A 36" boring is made to the desired depth. An adjacent boring is then made approximately 6 feet deep and 36" in diameter. The shallow bore serves as a ledge where sampling occurs (see Figure 3). The investigator, with air supply and protective clothing, is lowered into the adjacent bore, and the area is covered with plastic. The auger is raised to the sampling elevation. The sample cylinder (6" diameter. 30" long, PVC) is filled with refuse from the

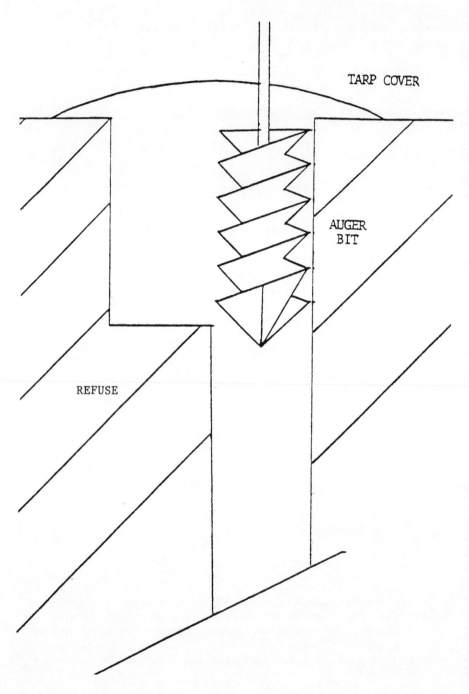

Figure 3.   Anaerobic sampling.

auger, sealed, and shipped to the laboratory. Ten one-pound
aerobic samples are taken from the same auger load for
analytical work-up.

In the laboratory all samples are incubated at a nominal
temperature of 100°F. Unnecessary movement of the cylinders
is avoided. Gas generation volumes are measured in liters
using a precision scientific wet test meter. Gas quality
($\%H_2$, $\%CO_2$, $\%O_2$, $\%N_2$, and $\%CH_4$) is determined with a carle
gas chromatograph. Following this period the samples may be
autopsied which involves opening the cylinder, making visual
observations of the refuse condition, and subjecting the
contents to additional analyses.

RESULTS AND DISCUSSION

From Figure 4, during the initial stages of incubation,
a typical cylinder produces abnormally high concentrations
of H_2 and CO_2. The $\%CH_4$ in the gas increases gradually with
a commensurate decrease in $\%CO_2$. During this phase the $\%H_2$
decreases and eventually becomes non-detectable ($<0.1\%$).
Following these gas composition changes, a steady state gas
composition is achieved which is typical of landfill gas
(50-55% CH_4, 40-45% CO_2, < 0.1% O_2, < 0.1% H_2, and 0.5-2% N_2).
Biochemically these observations indicate a probable upset of
methanogenesis as a result of the sampling process. Metabolism of glucose and other precursors continues and/or
accelerates to produce intermediates, CO_2 and H_2 as the methanogens acclimate gradually to a new environment.

Examination of the gas production curves for many cylinders and landfills did not show the kinetics to follow any
fixed rate law, for example, first or second order. Hence,
an average production rate was computed after the gas composition reached $\%CH_4$ > $\%CO_2$ and $\%H_2$ < 0.1. Using the average
for all cylinders the gas generation rate was computed for
the landfill proportional to the in-place tonnage and average
% moisture.

For a given landfill the range of gas production rates
was generally large (see Figure 5). This wide range of rates
is due primarily to the heterogeneous nature of landfill
refuse. Table 4 presents additional data for the landfill
in Figure 5 which shows a factor of 3 and 10 for the ranges
of % moisture and % cellulose, respectively.

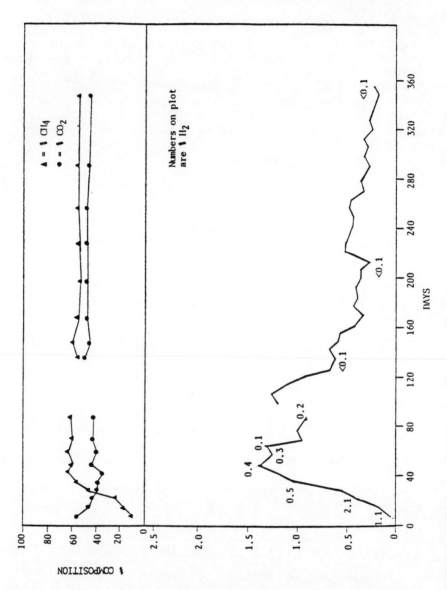

Figure 4.   Gas composition and production versus time.

Table 4. Analysis of Coordinated Aerobics

Well	Depth	% Water	% Cellulose	Generation Rate (Ft^3/dry #/yr)	Production Rate (MMSCFD)
1	30	29.3	14.4	0.34	0.08
2	40	57.9	64.6	0.54	0.07
3	30	21.0	6.6	0.0	0.10
4	30	30.4	11.3	0.57	0.12
5	30	39.5	17.9	0.45	0.12
6	20	25.6	11.3	0.27	0.12
7	20	26.6	8.2	0.69	0.13
8	30	34.1	11.1	0.60	0.12
9	20	37.8	14.2	0.15	0.11
10	30	21.0	6.6	0.22	0.11

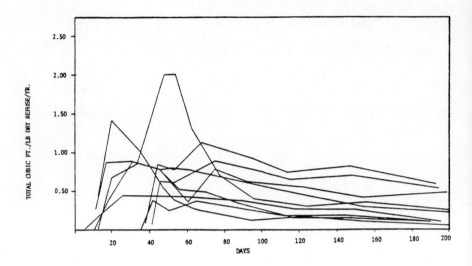

Figure 5.   Variability of sample production rate.

Comparison of the individual cylinder generation rates with the respective single well production rates in Table 4 shows little trend correlation. However, the landfill gas generation of 3.6 mmscfd computed from the overall cylinder average compared favorably with the gas recovery rate of 2 mmscfd from the well tests in the field.

From testing a variety of landfills across the continental U.S.A., the annual rainfall and the degree of infiltration are a strong influence on landfill gas generation rates. This influence is demonstrated in Figure 6 which shows composite gas generation curves for an east coast, a midwest and two southern California landfills. In general the southern California landfills have relatively low gas generation rates which would appear to result from lack of moisture.

To verify this moisture effect, one of the anaerobic samples from Southern California #2 was dosed with sterile deaerated water to a 40% moisture level. In Figure 7 the unmodified growth phase is shown which is characterized by

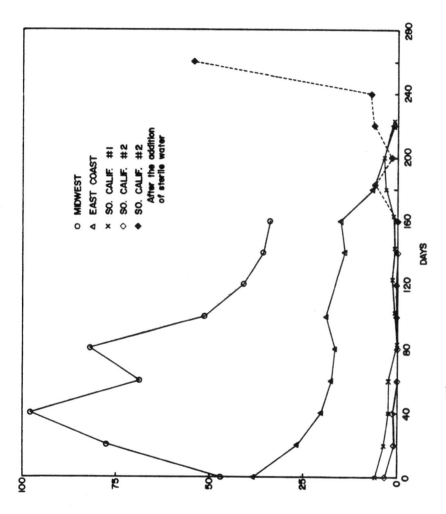

Figure 6.   Gas production curves by region.

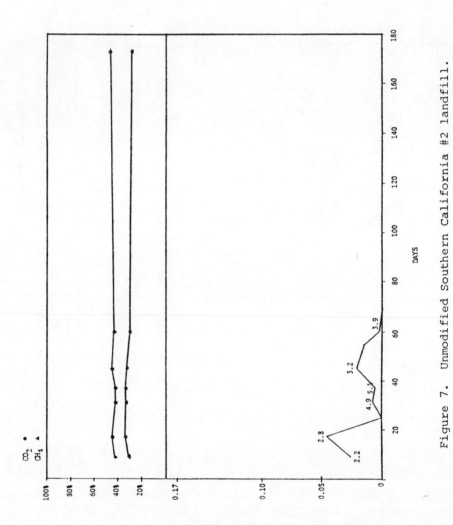

Figure 7.   Unmodified Southern California #2 landfill.

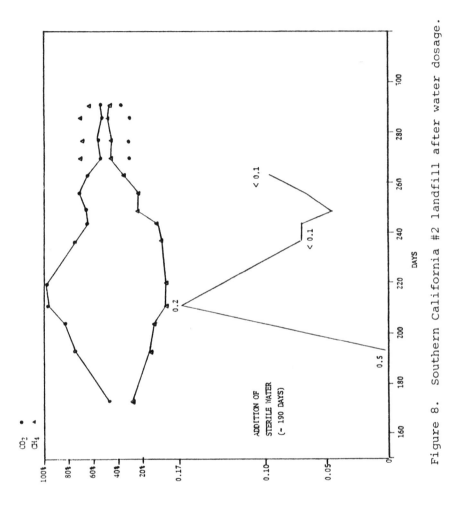

Figure 8. Southern California #2 landfill after water dosage.

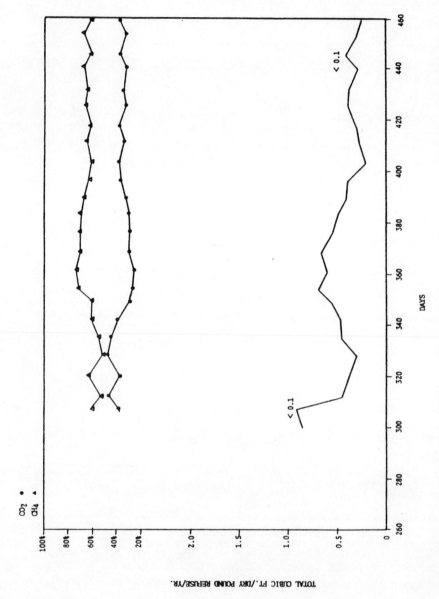

Figure 9.   Longterm gas production of Southern California #2 after water dosage.

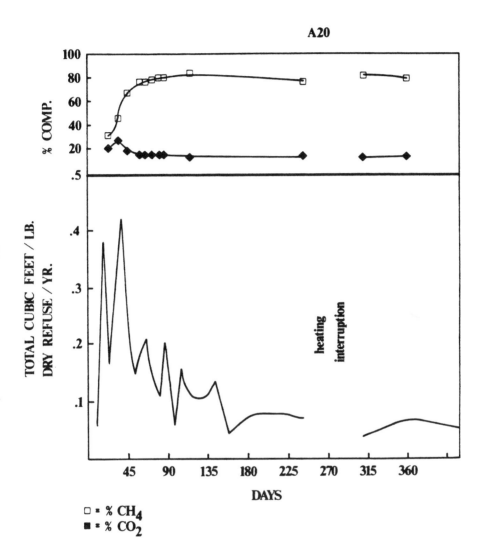

Figure 10.   Anaerobic refuse sample producing unusually high
% CH$_4$ gas.

low gas production and an atypical landfill gas composition ($\%CO_2$ > $\%CH_4$ and $\%H_2$ = 2 - 5). Shortly after the addition of water (see Figure 8) gas production accelerated dramatically, but most of this enhanced production consisted of CO_2. Following this period methane production increased and at about 270 days the partial pressure of CO_2 was high enough to cause a reversal in the CH_4/CO_2 ratio due to significant solubilization of CO_2. At 120 days after dosing, the sample was yielding typical landfill gas at a rate similar to many midwestern landfills (Figure 9).

During the testing of certain midwestern and eastern landfills, portions of these landfills have shown unusually high concentrations of methane in the gas. Anaerobic samples taken in these areas and returned to the laboratory have also shown high methane concentrations (see Figure 10). This phenomenon is currently undergoing further study in our laboratory.

CONCLUSIONS

In-vitro laboratory testing of anaerobic landfill refuse samples is an inexpensive method for obtain additional site-specific information. In many instances the gas generation rates as determined by anaerobic testing have agreed with the results of other testing. However, the accuracy of the method is affected substantially by the difficulty in 1) obtaining representative samples and 2) simulating temperature and moisture movement in the laboratory. The method could be improved significantly by 1) utilizing larger and/or more samples, 2) obtaining more downhole temperature data and 3) developing a coring technique for sampling.

The anaerobic testing technique is particularly useful for special studies which require laboratory investigation of an actual gas producing portion of the landfill.

REFERENCES

1.   Jenkins, R.L., "Methane Recovery from Landfills," presented at California State University, Long Beach Chemistry Department, October 28, 1981, Long Beach, CA.
2.   Rees, J.F., "The Fate of Carbon Compounds in the Landfill Disposal of Organic Matter," presented at the Reclamation of Solid Wastes Commission during 30th IUPAC General Assembly on 2-10 September, 1979, Davos, Switzerland.

3.   Tchobanoglous, G., H. Theisen, and R. Eliassen, Solid
     Wastes, McGraw-Hill Book Company, New York, NY (1977).

30
METHANE-OXIDIZING BACTERIA IN SANITARY LANDFILLS

Rocco L. Mancinelli*, and Christopher P. McKay****

*Washburn University
Topeka, Kansas*

***NASA-Ames Research Center
Moffett Field, California*

ABSTRACT

The relationship between methane-oxidizing bacteria inhabiting the topsoil of sanitary landfills and the methane
content of the soil gas phase has been studied and shown to
be significantly correlated ($p \leq 0.01$). This relationship
can be described by the equation $y = y_0 e^{-b/x}$, where $y = \%CH_4$
in the soil gas phase and x = number of methane-oxidizers per
g of soil. The constants were determined to be $y_0 = 82.6\%$
CH_4 and $b = 8.9 \times 10^6$ methane-oxidizers per g soil. When the
methane content of the soil gas phase, within a sanitary landfill, was greater than 50% the population of methane-oxidizing
bacteria did not significantly increase. Preliminary data
suggest that approximately 10% of the methane produced in a
landfill is consumed and/or transformed by the methane-oxidizing bacteria inhabiting the landfill topsoil. Laboratory
and field experiments clearly indicate that the population
of methane-oxidizers inhabiting sanitary landfill soil is not
limited by its carbon supply (CH_4), but is limited by some
other factor(s).

INTRODUCTION

Organisms capable of growing on methane as a carbon and energy source were first isolated in 1906 by Söhngen (31). Since that time methane oxidizing organisms have been reported to occur in a variety of soil and aquatic environments (1,4-7,11,15,17,20,25,34). It is well known that these organisms play an important role in the carbon cycle of aquatic ecosystems (3,7,12,28), and although they have been isolated from terrestrial environments their importance in the carbon cycle of terrestrial ecosystems has not yet been established.

Anaerobic environments containing organic matter, such as a sanitary landfill, generate methane through microbial action. The methane diffuses throughout the soil where it can be oxidized by aerobic and anaerobic methane oxidizing organisms. Anaerobic oxidation of methane appears to be linked to the metabolism of sulfate reduction in certain bacteria (7,22). Aerobic oxidation of methane is carried out primarily by methanotrophic bacteria (2,7,14). However, a few strains of yeast have also been reported to be capable of oxidizing methane, but their occurrence and importance in ecosystems is uncertain (38,39). Aerobic methane oxidizing bacteria belong to the general category of organisms termed methylotrophs. Methylotrophs are those organisms that have the ability to oxidize compounds that contain no carbon-carbon bonds, although they may contain more than one carbon atom. Several genera of methylotrophs have been characterized and reported (10,24,32,34). The taxonomy of these organisms is discussed by Colby et al. (4), Anthony (2), Whittenbury et al. (33-35), Romanovskoya (26), Romanovskoya et al. (27), and Whittenbury and Dalton (36). *Methylomonas* and *Methylococcus* are the only two genera of strict methylotrophs listed in Bergey's Manual of Systematic Bacteriology (37). Methane oxidizing bacteria have the ability to oxidize methane (10, 24). In addition, several facultative methylotrophs have been isolated (23,24).

The biochemistry and physiology of methane oxidizing organisms has been discussed and reviewed by Large (14), and Anthony (2), Higgins et al. (8), and Colby et al. (4). In general, methane oxidizing bacteria possess both dissimilatory and assimilatory pathways for methane oxidation (Fig. 1). In dissimilatory pathways methane is oxidized completely to carbon dioxide resulting in the production of cellular energy, and none of the carbon becomes cellular material or biomass. The carbon dioxide is given off to the surrounding environment.

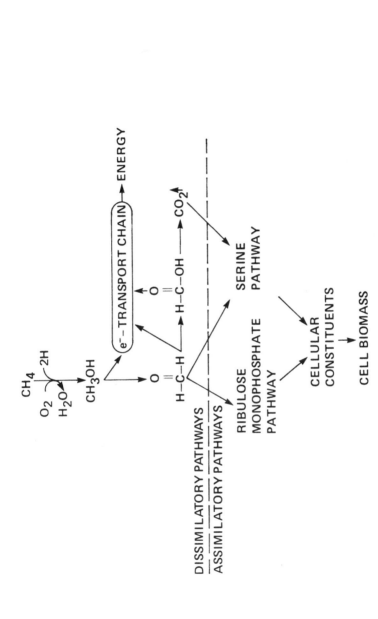

Figure 1.   Schematic diagram of assimilatory and dissimilatory methane-oxidation pathways by methanotrophs.

In assimilatory pathways methane is oxidized and is converted to cellular material. In both pathways methane is first oxidized to methanol, which is in turn oxidized to formaldehyde. The formaldehyde can then either be used as reducing power in the electron transport chain, oxidized to formate, or assimilated by the cell via the ribulose monophosphate pathway and/or the serine pathway. The cell then either oxidizes the formate to carbon dioxide or uses it as reducing power to drive the electron transport chain. The carbon dioxide can either be given off as a gas or be assimilated via the serine pathway. In generaly, most methylotrophic bacteria either possess the ribulose monophosphate pathway or the serine pathway, but not both (2).

Methane generation by methanogenic bacteria within sanitary landfills has been well documented and implicated in several fires and explosions (18). The occurrence of fires and explosions indicates that significant quantities of methane escape from sanitary landfills and are not consumed by methane oxidizing organisms living within the landfill. It has been reported that the number of methane/methanol oxidizing bacteria inhabiting the aerobic soil layer of a sanitary landfill is proportional to the amount of methane in the soil (17). Since it has been shown that these bacteria are sensitive to changes in the oxygen levels in their environment (29,30) the depth of the aerobic zone of a landfill is an important parameter in determining the population density and growth rate of aerobic methane oxidizing bacteria. The depth of oxygen penetration into a landfill has been calculated to be less than one meter (16,19). The purpose of the studies reported here were to determine the relationship between methane oxidizing bacteria and the methane content and flux of sanitary landfill soil.

MATERIALS AND METHODS

Soil samples for gassing were collected as a core 10 cm wide x 15 cm thick beginning at a depth of 10 cm below the surface of the ground. A total of 8 soil samples were collected on 3 separate occasions. The site from which samples were collected was a 3 m x 3 m plot in a closed sanitary landfill that is currently generating methane and is described by Mancinelli et al (17). The soil at this site is a mixture of sand and clay. The sampling plot was one used in a previous study and the soil was known to contain an average of 5% methane (17). A low methane generating plot was chosen in an attempt to determine, in the laboratory, if the population of methane oxidizers exhibited a significant increase when

the methane content of the soil increased. One g of soil was
removed from each core, both before and after gassing, suspended in 100 ml of sterile physiological saline and 10-fold
serial dilutions made. Using the spread plate technique 0.1
ml aliquots of each of the dilutions were plated in triplicate onto two sets of a mineral salts medium, prepared as
described by Mancinelli et al. (17). One set was incubated
under a 30% methane-70% air atmosphere to determine numbers
of methane oxidizers (34). One set was incubated under an
atmosphere to determine numbers of methane oxidizers (34).
One set was incubated under an atmosphere containing only air,
to determine numbers of chemoautotrophs using CO_2 as a carbon
source and numbers of bacteria capable of scavenging carbon
from any impurities that might be present in the medium. In
addition, a set of mineral salts plates containing 0.2% methanol as a carbon and energy source were also inoculated and
grown under an air atmosphere containing 2% CO_2 to determine
numbers of methanol oxidizing bacteria. The petri dishes
were incubated at 22°C for 12 days, and the colonies that
had developed were counted. The number of methane and methanol oxidizing bacteria per gram of soil for each soil core
was determined by calculating the mean of the numbers of
bacterial colonies, colony forming units (CFU), that had
developed on all of the countable plates (those having 0 to
300 colonies per plate). Replica plates of the methane grown
plates were made on mineral salts medium containing 0.2%
methanol to determine if all of the methane oxidizers were
also capable of using methanol as a sole carbon and energy
source. To calculate the dry weight of soil, for every core
collected ten g of soil was collected at a depth of ten cm
below the surface of the ground immediately adjacent to where
the soil core had been collected and dried at 105°C to determine the moisture content of the sample. Ten g of soil was
removed from the soil core and dried at 105°C, after it had
been gassed, to monitor any changes in the moisture content
of the soil.

Each soil core was placed in a closed bottom hollow tube
10 cm x 30 cm which served as a controlled atmosphere gas
flow-through chamber (Fig. 2). The chamber bottom was fitted
with a gas inlet port located in the center and lined along
the inside with a fine mesh nylon screen. A piece of Whatman
no. 1 filter paper was placed immediately above the nylon
screen. The screen and filter paper kept soil from clogging
the inlet port and dispersed the gas so that it did not enter
the bottom of the soil core as a single narrow jet. Once the
soil core was in the chamber a gas tight top fitted with an
inlet port and an outlet port was put in place. Nine scc min^{-1}

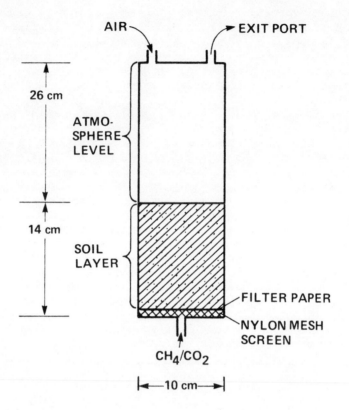

Figure 2.   Flow through soil-gassing chamber.

of a 1:1 mixture of CH_4 and CO_2 was allowed to flow through
the bottom gas inlet port. This corresponds to a flux through
the soil in the chamber of 5.7×10^{-2} scc CH_4 min^{-1} cm^{-2} which
is the approximate rate of methane flow through the soil of
a sanitary landfill (13) and should theoretically result in
aerobic conditions throughout the chamber (19). Air flowed
through one top inlet port at a rate of 32 scc min^{-1}. The
other top port served as an exhaust port. At these flow rates
it took approximately 6 hr to fill a chamber and allow the
gas to come to equilibrium. All times reported are those
measured from the 6 hr gas equilibration time. The soil
cores were gassed for a period of 15 days. The methane concentration of the gas flowing through the exhaust port was

measured every few hours with an MSA model 53 methane analyzer. The methane concentration of the gas flowing into the system was also measured, without disrupting the system, to ensure that it remained constant. As a control a soil core was collected, sterilized by autoclaving, and treated as a test core.

RESULTS

By use of the t-test it was determined that the numbers of methane and methanol oxidizing bacteria colony forming units (CFU) per g dry weight of soil grown on the mineral salts medium under a methane atmosphere or on the mineral salts medium containing 0.2% methanol did not differ significantly (p < 0.01) (Table 1). To accurately determine the numbers of methanol and methane oxidizing bacteria CFU the number of chemautotrophic and carbon scavenger CFU were determined from the mineral salts medium containing no added carbon source and subtracted from the number of CFU on each of the other media. In addition, the replica plates of the methane oxidizing bacteria grown on methanol exhibited the same growth patterns as those seen on the original plates grown under the methane containing atmosphere. These observations indicated that all of the methane oxidizers isolated could also oxidize methanol and that only those methanol oxidizing bacteria capable of oxidizing methane were isolated. When the number of methane oxidizing bacteria CFU of the soil samples before gassing were compared to those obtained after gassing it was found that the number of methane oxidizing bacteria CFU per gram dry weight of soil increased after gassing (Table 1) indicating that the methane was being assimilated by methane oxidizing bacteria.

In the test chambers no change in the methane concentrations exiting the chambers could be detected for an average of about 7 hours. After 7 hours the concentration of methane exiting the chambers began to decrease for a period of 30 hours and then leveled off at an average of 90% of the input, or original concentration. The sterile control soil core exhibited no detectable decrease in the concentration of methane exiting the chamber. This indicates that an average of 10% of the methane within the non-sterile soil cores was consumed in some manner (Fig. 3).

The moisture content of the sanitary landfill soil ranged from 3% to 5%, by weight, for all of the samples. The moisture content of the soil cores after gassing ranged from 1% to 3%, by weight.

Table 1. *Number of Colony Forming Units (CFU) of Methane and Methanol-Oxidizing Bacteria per g Dry Wt. Soil Collected from each Core Before and After Gassing with CH_4/CO_2 and Air.*

	No. CFU x 10^5 g^{-1} dry wt. soil*			
	Methanol-oxidizers		Methane-oxidizers	
Sample #	Before gassing	After gassing	Before gassing	After gassing
1	25 ± 5	120 ± 25	30 ± 8	132 ± 28
2	32 ± 12	148 ± 32	27 ± 10	181 ± 41
3	9 ± 4	175 ± 18	11 ± 9	127 ± 22
4	15 ± 7	256 ± 58	22 ± 8	197 ± 42
5	18 ± 4	118 ± 28	10 ± 9	121 ± 24
6	27 ± 8	105 ± 20	18 ± 11	87 ± 18
7	14 ± 3	178 ± 32	12 ± 5	137 ± 47
8	19 ± 5	95 ± 15	21 ± 4	122 ± 21
Sterile Control	0	0	0	0

*Reported as the mean ±standard deviation of the counts from the triplicate plates, corrected for chemoautotrophs and C-scavengers.

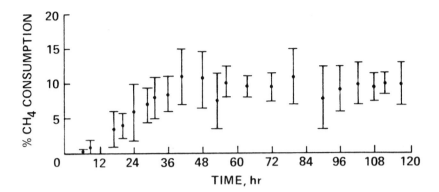

Figure 3.   Percent consumption, or decrease, (mean ± standard deviation) in the methane exiting the gas-flow chamber vs time in hours, measured from the gas equilibration point calculated to be six hours after gas flow was initiated.

DISCUSSION

Field site analyses have determined that changes in the population density of methane oxidizing bacteria inhabiting the aerobic soil layer of sanitary landfills are proportional to the concentration of methane within the soil up to a soil methane concentration of ~50% (17). This relationship can be described by the equation $y = y_o e^{-b/z}$, derived by Mancinelli et al., where y = %CH₄ of the soil gas phase, x = number of methane oxidizers per g of soil, y_o = 82.6% CH_4 and b = 8.9 x 10^6 methane oxidizers per g of soil (16). The curve generated by this equation was shown to be significantly correlated with the data points collected from sanitary landfill field sites (p < 0.01), and is illustrated in Figure 4 (17).

It is clear from the previous study (17) and from the present study that increasing the carbon and energy source, CH₄, leads to an increase in the population of methane oxidizing bacteria in the soil. This is a normal response one would expect to observe when the limiting nutrient supply to a bacterial population is increased. However, these organisms only consumed an average of 10% of the methane that was

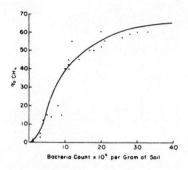

Figure 4.   Relationship between the averaged CH₄ concentration
            (percentage by volume) and the averaged methane/
            methanol-oxidizing bacteria count per gram of soil
            for each site. (Reprinted by permission from
            Mancinelli et al. (17)).

supplied to them under these conditions and at no time consumed all of the available methane. This indicates that
methane is not the true underlying factor limiting the population. The disappearance of 10% of the methane from the soil
core chambers was accompanied by an increase in the methane
oxidizing bacterial population, strongly suggesting that 10%
of the methane was being metabolized by the bacteria. It is
not clear from this study if the microbial consumption rate
of methane is the same for all soil methane concentrations.
In the previous field study the population of methane oxidizers increased with increasing methane concentration, for
methane concentrations less than 50%, suggesting that the
methane consumption rate also increased. For methane concentrations greater than 50% the bacterial population remained
roughly constant. These observations indicate that it is
some additional factor(s) in the soil environment that limits
the methane oxidizing population in sanitary landfills. If
the population were not limited it would consume all of the
available methane, increase proportionally, and allow none
of the methane to escape into the atmosphere. Factors that
may be limiting the methane oxidizing population include,
toxic substances that inhibit growth produced along with
methane from the decomposition of organic matter, competition
with other microbes for some nutrient other than carbon,
predation by other organisms, lack of enough water in the soil
for maximum growth, or displacement of oxygen in the soil by
other landfill gasses leading to a lack of enough oxygen in
the aerobic soil for optimal metabolism of methane. The

phenomenon of methane oxidizers not consuming all of the available methane in an environment is not unique to land-fills and commonly occurs in other habitats such as lake sediments (7), but the reasons why this occurs have not been determined. Total consumption of the methane produced has also been observed (7,9). Hoeks (9) reported complete consumption of methane around point leaks in natural gas pipe-lines. However, the effective flux of methane consumption in the methane oxidation zone of the soil was only about 10^{-6} scc min^{-1} cm^{-2}. As the flux was increased the radial distance of the oxidation zone also increased, which resulted in the rate per unit area of methane consumption being nearly constant.

Clearly, methane oxidizing bacteria play a role in sanitary landfill soil habitats and terrestrial ecosystems in general. Exactly what that role is and its importance to the ecosystem as a whole remains to be elucidated.

REFERENCES

1. Adamse, A.D., J. Hoeks, J.A.M. DeBont, and J.F. Van Kessel, "Microbial Activities in Soil Near Natural Gas Leaks," *Arch. Mikrobiol.*, *83*, 32-51 (1972).
2. Anthony, C., "The Biochemistry of Methylotrophs,"Academic Press, Inc., New York, p. 431 (1982).
3. Cappenberg, T.E., "Ecological Observations on Hetero-trophic, Methane-Oxidizing and Sulfate Reducing Bacteria in a Pond," *Hydrobiol.*, *40*, 471-485 (1972).
4. Colby, J., H. Dalton, and R. Whittenbury, "Biological and Biochemical Aspects of Microbial Growth on Cl Com-pounds," *Ann. Rev. Microbiol.*, *33*, 481-517 (1979).
5. Davis, J.B., "Petroleum Microbiology," Elsevier/North Holland Publishing Co., London (1967).
6. Dworkin, M. and J.W. Foster, "Studies on *Pseudomonas methanicus* (Söhngen) *nov. comb.*," *J. Bacteriol.*, *72*, 646-659 (1956).
7. Hanson, R.S., "Ecology and Diversity of Methylotrophic Organisms," *Adv. Appl. Microbiol.*, *26*, 3-39 (1980).
8. Higgins, I.J., D.J. Best, R.C. Hammond, and D. Scott, "Methane-Oxidizing Microorganisms," *Microbiol. Rev.*, *45*, 556-590 (1981).
9. Hoeks, J., "Effect of Leaking Natural Gas on Soil and Vegetation in Urban Areas," Dutch Agricultural Res. Report 778, Wageningen, p. 120 (1972).

10.   Hou, C.T., R. Patel, A.I. Laskin, N. Barnabe, and I. Marczak, "Microbial Oxidation of Gaseous Hydrocarbons: Production of Methyl Ketones from Their Corresponding Secondary Alcohols by Methane- and Methanol Grown Microbes," *Appl. Environ. Microbiol.*, *38*, 135-142 (1979).

11.   Hutton, W.E. and C.E. ZoBell, "Occurrence and Characteristics of Methane-Oxidizing Bacteria in Marine Sediments," *J. Bacteriol.*, *58*, 463-473 (1949).

12.   Ivanov, M.V., S.S. Belayev, and S.S. Lauinavichus, "Methods of Quantitative Investigation of Microbiological Production and Utilization of Methane", Symposium on Microbial Production and Utilization of Gases (H_2, CH_4, CO), (H.G. Schlegel, G. Göttschalk, and N. Pfennig, eds.), Göttingen, Akademie der Wissenschafter, Göttingen, pp. 63-67 (1976).

13.   Kunz, C.O. and A-H. Lu, "Methane Production Rate Studies and Gas Flow Modeling for the Fresh Kills Landfill," New York State Energy Research and Development Authority Interim Report, New York State Department of Health, Div. of Laboratories and Research, Albany (1980).

14.   Large, P.J., "Methylotrophy and Methanogenesis," American Society for Microbiology, Washington, DC, pp. 88 (1983).

15.   Leadbetter, E.R. and J.W. Foster, "Studies on Some Methane-Utilizing Bacteria," *Arch. Mikrobiol.*, *30*, 91-118 (1958).

16.   Lu, A-H. and C.O. Kunz, "Gas-flow Model to Determine Methane Production at Sanitary Landfills," *Environ. Sci. and Tech.*, *15*, 436-440 (1981).

17.   Mancinelli, R.L., W.A. Shulls, and C.P. McKay, "Methanol-Oxidizing Bacteria Used as an Index of Soil Methane Content," *Appl. Environ. Microbiol.*, *42*, 70-73 (1981).

18.   Martyny, J.W., D. Kennerson, B. Wilson, and C.E. Lott, Jr., "Landfill-Associated Methane Gas a Threat to Public Safety," *J. Environ. Health, 41,* 194-197 (1979).

19.   McKay, C.P. and R.L. Mancinelli, "Gas Flow in an Undisturbed Landfill," (In preparation.)

20.   Naglub, M., "On Methane-Oxidizing Bacteria in Freshwaters. I.   Introduction to the Problem and Investigations on the Presence of Obligate Methane-Oxidizers," *Z. Allg. Mikrobiol.*, *10*, 17-36 (1970).

21.   Nesterov, A.I., Yu N. Mshensky, V.F. Galchenko, B.B. Namsaraev, and V. Yu Ilchenko, "A Comparative Study on Parameters of Growth of Methylotrophic Bacteria," *Mikrobiologiya*, *46*, 10-14 (1977).

22.   Panganiban, A.T., T.E. Patt, W. Hart, and R.S. Hanson, "Oxidation of Methane in the Absence of Oxygen in Lake Water Samples," *Appl. Environ. Microbiol.*, *37*, 303-309 (1979).

23. Patel, R.N., C.T. Hou, and A. Felix, "Microbial Oxidation of Methane and Methanol: Isolation of Methane-Utilizing Bacteria and Characterization of a Facultative Methane-Utilizing Isolate," *J. Bacteriol.*, *136*, 352-358 (1978).

24. Patt, T.E., G.C. Cole, and R.S. Hanson, "*Methylobacterium*, A New Genus of Facultatively Methylotrophic Bacteria," *Int. J. Syst. Bacteriol*, *26*, 226-229 (1976).

25. Quayle, J.R., "The Metabolism of One-Carbon Compounds by Micro-Organisms," *Adv. Microb. Physiol.*, *7*, 119-203 (1975).

26. Romanovskaya, V.A., "Nomenclature of Obligate Methylotrophs," *Microbiology (USSR)*, *47*, 1063-1072 (1978).

27. Romanovskaya, V.A., Y.S. Sadovmikov, and Y.R. Malshenko, "Formation of Taxons of Methane-Oxidizing Bacteria Using Numerical Analysis Methods," *Microbiology (USSR)*, *47*, 120-130 (1978).

28. Rudd, J.W.M. and R.D. Hamilton, "Methane Cycling in Eutrophic Shield Lake and its Effects on Whole Lake Metabolism," *Limnol. Oceanogr.*, *23*, 337-348 (1975).

29. Smirnova, Z.S., "Efficiency of the Utilization of Free Energy by Methane-Oxidizing Bacteria," *Mikrobiolgiya*, *40*, 5-7 (1971).

30. Smirnova, Z.S., "The Material Balance of Methane Oxidation by Microorganisms," *Akad. Nauk. (SSSR) Izv. Ser. Biol.*, *3*, 423-427 (1971).

31. Söhngen, N.L., "Ueber Bakterien Welche Methan als Koklenstoffnakrung und Energiequelle Gebrauchen," *Zentral Bakteriol. Parasitenkd. Infektionskr. Hyg.*, *15*, 513-517 (1906).

32. Wake, L.V., P. Rickard, and B.J. Ralph, "Isolation of Methane Utilizing Micro-organisms: A Review," *J. Bacteriol*, *36*, 92-99 (1973).

33. Whittenbury, R., S.L. Davies, and J.F. Davey, "Exospores and Cysts Formed by Methane-Utilizing Bacteria," *J. Gen. Microbiol.*, *61*, 219-226 (1970).

34. Whittenbury, R., K.C. Phillips, and J.F. Wilkinson, "Enrichment, Isolation and Some Properties of Methane-Utilizing Bacteria," *J. Gen. Microbiol.*, *61*, 205-218 (1970).

35. Whittenbury, R., J. Colby, H. Dalton, and H.C. Reed, "Biology and Ecology of Methane-Oxidizers," Symposium on Microbial Production and Utilization of Gases (H_2, CH_4, CO), (H.G. Schlegel, G. Göttschalk, and N. Pfennig, eds.), Akademie der Wissenschafter, Göttingen, pp. 281-292 (1976).

36. Whittenbury, R. and H. Dalton, "The Methylotrophic Bacteria," The Prokaryotes, (M.P. Starr, H. Stolp, H. Truper, A. Balows, and H.G. Schlegel, eds.), Springer-Verlag, Berlin, pp. 894-902 (1981).

37. Whittenbury, R. and N.R. Krieg, "Family IV. *Methylococcaceae*," Bergey's Manual of Systematic Bacteriology, (N.R. Krieg and J.G. Holt, eds.), The Williams and Wilkins Co., Baltimore, Vol. 1, pp. 256-261 (1983).

38. Wolf, H.J. and R.S. Hanson, "Isolation and Characterization of Methane-Utilizing Yeasts," *J. Gen. Microbiol.*, *114*, 187-194 (1979).

39. Wolf, H.J., M. Christiansen, and R.S. Hanson, "Ultrastructure of Methanotrophic Yeasts," *J. Bacteriol.*, *141*, 1340-1349 (1980).

31

DESIGN OPERATION AND RESULTS OF THE CONTROLLED
LANDFILL PROJECT MOUNTAIN VIEW, CALIFORNIA

Robert E. Van Heuit

EMCON Associates
San Jose, California

ABSTRACT

The Mountain View demonstration project consists of six cells designed to study the effects of water content on gas enhancement, and to verify the synergism between water, buffer, and seed/nutrient content. All cells were constructed in an identical manner and filled with municipal refuse from the City of San Francisco, California. Cells A, B and C received sludge and buffer additions during construction, Cell D received buffer only, Cell E received sludge only, and Cell F (the control cell) received no additions. Despite the fact that no attempt was made to modify the landfilling methods normally employed at the Mountain View Landfill, analysis of cell characteristics after completion of construction indicates that additions were made in a fairly uniform manner. This paper describes the construction and operation of the six demonstration cells, presents the monitoring data to date, provides an analysis of monitoring results and a discussion of problems encountered during the program.

INTRODUCTION

The rising price of natural gas resulting from diminishing world energy reserves has stimulated interest in developing the potential of landfills to generate methane. Interest in commercial use of the methane produced in landfills has, in turn, provided the impetus for research into the enhancement of landfill gas generation and recovery.

A number of studies have been conducted in the United States to define the most important factors influencing the yield and generation rate of landfill-derived methane. Found to be most important were: (1) composition of the refuse; (2) water content of the landfill refuse; (3) the pH of the aqueous phase, and (4) other factors within the landfill, such as temperature conditions in the landfill, the quantity and quality of nutrients, and the density of in-place refuse. In particular, results of experiments by Dynatech R/D Company, as well as other laboratory studies and field observations, indicate that an increase in water content, buffering to a pH of about 7, and the addition of nutrients and bacteria will result in a very short lag time to reach the start of methane generation and a consequent increase in both the yields and rate of methane generation in landfills.

The laboratory and small-scale field studies, however, have provided only qualitative data on the impact of the suggested enhancement parameters. To provide the quantitative data necessary to establish the functional relationships between the methanogenic process and the enhancement parameters, and to assess the enhancement process from a field management point of view, a large-scale field demonstration study was undertaken at the Mountain View Landfill in Mountain View, California.

The nature of this project required the combined efforts of many individuals from a number of organizations. Pacific Gas and Electric Company, Southern California Gas Company and the U.S. Department of Energy/Argonne National Laboratory provided funding and monitoring instrumentation, as well as design reviews and expert personnel. In January 1982, the Gas Research Institute replaced DOE/Argonne as a sponsor of the project. The City of Mountain View, the landfill owner, provided the necessary permission to proceed with the project as well as a crew for the surveying needs of the project. Easley & Brassy Corporation, the landfill operator, constructed the landfill cells and assisted in tasks requiring special equipment and expertise. Finally, the design, supervision of cell construction, installation of all support systems, and monitoring was performed by EMCON Associates, the project's technical manager.

The overall objectives of the Mountain View Controlled Landfill Project were to demonstrate the feasibility of landfill gas enhancement in the field and to collect important

kinetic information which would facilitate the development
of more accurate models of landfill gas production. This
paper presents a brief description of project construction,
operation and monitoring, as well as a preliminary analysis
of results.

DESIGN CONCEPTS AND CELL CONSTRUCTION

The extent of the synergism exhibited between the enhancement parameters is difficult to assess due to the complexity of the processes involved. However, the earlier
studies stressed the importance of this synergism and concluded that simultaneous additions of water, buffer and
seed/nutrients would result in the greatest enhancement benefit.

A six-cell cluster was designed to study the effects of
water content on gas enhancement, as well as to verify the
synergism between water, buffer and seed/nutrient content
(Table 1). Construction of the demonstration cells began in
November 1980. All cells were constructed in an identical
manner. A 5-foot clay base liner was placed to control the
downward movement of liquids and gases, as well as to isolate
the test cells from ground-water intrusion. Cell walls were
also constructed of clay to prevent lateral movement of gas
across cell boundaries. A 30-mil reinforced Hypalon cell
cover was used to isolate the cells from infiltration of precipitation and to contain the gas produced within the cells.

All cells, constructed in successive lifts, were filled
with municipal refuse from the City of San Francisco, California. Neither daily nor intermediate cover were added to
the refuse within the cell limits. Three of the cells (A,
B and C) received sludge (seed/nutrient) and buffer additions.
Cell D received buffer only and Cell E sludge only. Cell F,
maintained as the control cell of the project, received no
additions (Figure 1). Refuse placement and all buffer and
sludge additions were completed by March 1981. The Hypalon
covers were in place on all cells by April 1981.

Analysis of final cell characteristics (Table 2) indicates that parameter additions were made in a fairly uniform
manner during cell construction, despite the fact that no
attempt was made to modify the landfilling methods normally
employed at Mountain View Landfill. Variations which did
occur, largely as a result of precipitation, were anticipated,
since no control was exercised over this parameter. There

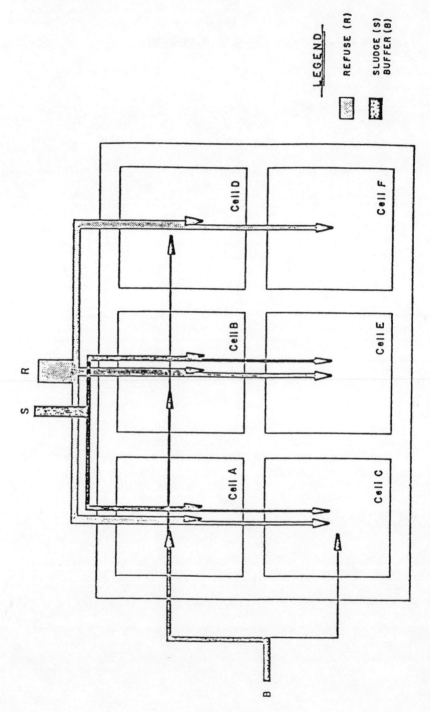

Figure 1.   Schematic of cell ingredients and placement sequence.

Table 1. *Cell Parameter Specifications*

Cell A:	Leachate recirculation	
	Water:	*To exceed field capacity[1] (fast rate)*
	Buffer[2]:	*0.1 - 1.0%*
	Sludge[3]:	*5%*
Cell B:	Buffer:	*0.1 - 1.0%*
	Sludge:	*5%*
Cell C:	Water:	*To achieve field capacity[4] (slow rate)*
	Buffer:	*0.1 - 1.0%*
	Sludge:	*5%*
Cell D:	Water:	*To achieve field capacity (slow rate)*
	Buffer:	*0.1 - 1.0%*
Cell E:	Water:	*To achieve field capacity (slow rate)*
	Sludge:	*5%*
Cell F:	No additions	

Notes:

1. Field capacity is the maximum water content that solid waste can retain in a gravitational field without producing continuous downward percolation.

2. $CaCO_3$ powder is used as buffer in an amount equivalent to 0.1 to 1.0% of the total water weight.

3. Digester sludge is used a seed/nutrient in an amount equivalent to approximately 5% by weight of dry sludge solids.

4. It was originally specified that water would be added to Cell C to achieve field capacity. However, to differentiate the effects of leachate recirculation from mere water addition, it was later decided that Cells A and C would receive equal amounts of water. Consequently, at the end of the water addition period, Cell C had exceeded field capacity.

Table 2. Summary of Cell Mass Characteristics
At Completion of Construction

Parameter	Cell A	Cell B	Cell C	Cell D	Cell E	Cell F
Refuse Solids Mass, tons	5,383	5,980	5,307	6,613	5,473	6,223
Sludge Solids Mass, tons	171	142	73	0	61	0
Buffer Solids Mass, tons	11	11	10	11	0	0
Total Solids Mass, tons	5,565	6,133	5,390	6,624	5,534	6,223
Water Mass, tons	2,911	2,955	2,329	2,364	2,319	2,228
Total Mass, tons	8,476	9,088	7,719	8,988	7,853	8,431
Water Content % (total weight)	34 46*	32 32*	30 44*	26 27*	29 31*	26 26*
Sludge: Refuse Solids Ratio (Dry Weight)	0.032	0.024	0.014	0	0.011	0
Buffer to Water Ratio	0.0036	0.0036	0.0043	0.0047	0	0
Cell Volume (yd^3)	13,151	14,460	12,886	15,390	13,563	14,251
Cell Density (lb/yd^3)	1,289	1,257	1,198	1,168	1,158	1,186

*Cell water content after one year of operation.

is a very narrow spread in many of the characteristic cell
properties. However, at the completion of construction, the
water contents of cells receiving sludge were 3 to 8% above
those receiving no sludge, and buffer-to-water ratios in these
cells also varied.

CELL MANAGEMENT AND MONITORING

Systems for leachate and gas collection and monitoring,
cell temperature and settlement monitoring, and water application were installed in all cells. A ground-water quality
monitoring system and leachate recirculation system were
also installed in Cell A.

Cell management during the first three years of project
operation has consisted of water addition, principally in
Cells A and C, although Cell D and E received small amounts
of added water. Cell A received its first water addition at
a fast rate over a period of 72 hours; water addition in the
other cells was at a slow, incremental rate. Leachate recirculation was also employed in Cell A. To differentiate
the effects of leachate recirculation from those of simple
water addition, a similar cell (Cell C) received a volume of
water equal to that added in cell A. No water was added to
Cell B, so that it could serve as a control in evaluating
the performance of Cells A and C.

To enable the gas production within a given cell to be
compared with that of other cells, and to thereby gain an
understanding of the relationship of the enhancement parameters to the methanogenic process, certain performance parameters were selected for monitoring. These included the
volume and composition of the gas, temperature, settlement,
and leachate levels within the cells. Data on gas production
rate, volume, and composition are used to assess cell methanogenesis, the main objective of the demonstration project.
Cell settlement data are employed to aid in this assessment,
since it has been suggested that settlement is directly related to gas generation. Data on cell temperature and leachate levels can be used in assessing the development of the
aqueous phase, known to influence cell methanogenesis.
Figures 2 and 3 present the parameter data versus time for
all six cells.

Gas turbine meters are used to measure wet gas volume on
a cumulative basis. Gas composition is monitored with an
infrared spectrophotometer and/or gas chromatograph, which

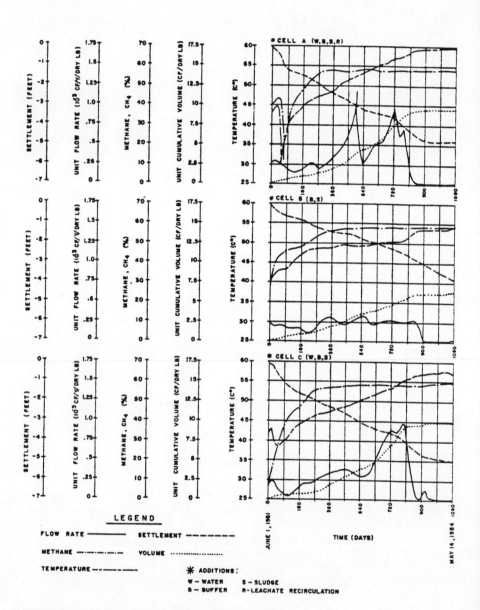

Figure 2.   Summary of data.

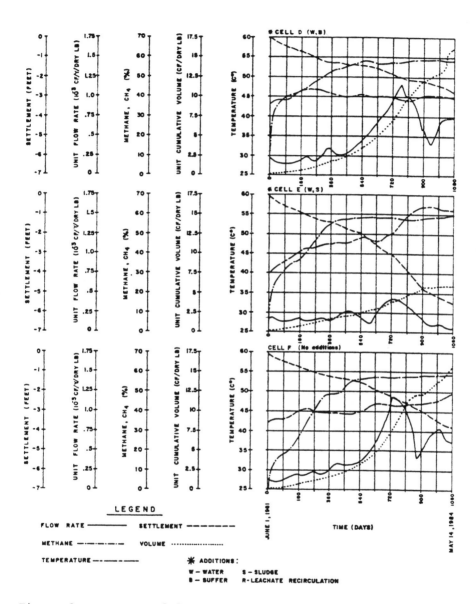

Figure 3. Summary of data.

give results in percent of dry gas. Temperatures in all
cells are monitored at jack panels connected to the thermo-
couples in each cell, and settlement is monitored using
concrete block markers selectively placed over the cell
surfaces.

PRELIMINARY ANALYSIS OF RESULTS

Data indicate that refuse decomposition in the demonstration cells is probably only 15 to 40% complete. Therefore, any assessment of the overall process of methanogenesis
after 36 months of cell management must be viewed as preliminary. Nevertheless, a number of observations can be made
regarding overall system performance, as well as the performance of individual cells.

The mode of water application appears to be significant
over a short time interval. In Cell A, rapid addition of
water over a period of 72 hours resulted in a significant
drop in cell temperatures within hours of water addition.
The drop in cell temperatures was acompanied by a decrease
in methane content and overall gas production, a rise in
leachate level, and considerable settlement--as much as 0.5
foot (15 cm) settlement in individual locations, with an
average settlement of about 0.3 foot (9 cm). The dramatic
reduction in methane content persisted for several months even
after water addition had ceased and cell temperatures had returned to the level considered optimal for methanogenesis.
This suggests a limiting factor other than the initially
observed temperature drop. It is hypothesized that the
cell's increased hydrolysis potential resulted in a pH drop
below the 6.8 to 7.4 range considered optimal for methanogenesis, and the pH was therefore the limiting factor.

This hypothesis seems to be supported by specific conductance and pH data collected during the leachate recirculation period. The specific conductance of the recirculated
leachate ranged from an original low of 5,000 mhos/cm^2 to a
stable high of 24,000 mhos/cm^2 at the end of the recirculation
period (Day 80 to Day 118). During the same period, pH
fluctuated within the narrow range of 5.46 to 5.58, well
below the optimal range. Although the leachate recirculation
period corresponded to the lowest total gas rates observed,
as has been noted, the downward trend was established immediately following water addition and cannot be attributed
solely to leachate recirculation.

In Cell C, the addition of water in small increments over a period of 35 days did not result in the almost immediate cessation of gas production observed in Cell A. On the contrary, landfill gas production increased for approximately 30 days after water addition, followed by a significant drop in the observed rate. Landfill gas production remained low for 60 days thereafter. However, during this same time period, the methane content increased to a stable high value of 40 to 50%.

Leachate production and cell settlement seem to be unaffected by the method of water application. Following water addition, there was significant settlement--1.8 feet (54.8 cm)--in both Cell A and Cell C. Leachate levels in both cells peaked at the end of the water application period, then dropped to stable levels.

The settlement data cannot be directly correlated with gas production, due to the fact that moisture addition frequently results in an associated refuse settlement. A review of cell moisture contents reveals that the greatest unit moisture addition (from all sources) was to Cells A and C. These cells also show the highest rate of settlement. Cells B and E, which have the same moisture content, but significantly less moisture than Cells A and C, have similar settlement rates and, predictably, less settlement to data than Cells A and C. Cell D received only limited added moisture, and Cell F no added moisture. The moisture content and settlement rate of these cells also are similar.

Landfill gas yields through 1,080 days of monitoring range from 0.55 ft^3/lb (0.035 m^3/kg) dry refuse in Cell B to 1.58 ft^3/lb (0.10 m^3/kg) dry refuse in Cell F. Table 3 summarizes landfill gas and methane yields for all cells and normalizes these yields on a unit weight basis for dry refuse. Although values listed in published literature most often are in terms of gas produced per unit weight of wet refuse, it is more useful to normalize landfill gas volume on the basis of a unit weight of dry refuse solids, since this approach accounts for variations in moisture content, known to be an important parameter of landfill methanogenesis.

A review of the data shows that for the first two years Cells A and C produced gas at a higher rate than the other cells. However, during the past year, the temperatures in those cells and in Cells B and E increased to a range of 53°C to 60°C. At approximately day 850, gas production in

Table 3.   Summary of Gas Yield and Production Rates

Cell	Cumulative unit volume $(ft^3/dry\ lb)$ LFG	CH_4	12-Month Unit Volume Production							
			Landfill Gas				Methane			
			6-81 to 5-82	6-82 to 5-83	6-83 to 5-84	Average	6-81 to 5-82	6-82 to 5-83	6-83 to 5-84	Average
A	0.93	0.47	0.21	0.54	0.18	0.31	0.09	0.29	0.09	0.16
B	0.56	0.29	0.20	0.26	0.10	0.19	0.09	0.14	0.06	0.10
C	0.99	0.52	0.19	0.58	0.22	0.33	0.09	0.31	0.12	0.17
D	1.55	0.81	0.19	0.53	0.83	0.52	0.07	0.29	0.45	0.27
E	0.61	0.30	0.13	0.27	0.21	0.20	0.04	0.15	0.11	0.10
F	1.58	0.81	0.17	0.53	0.88	0.53	0.05	0.28	0.48	0.27

Note: LFG volumes are based on meter readings corrected for meter inaccuracies only.

Methane production assumes 5% moisture vapor in LFG by volume.

these four cells essentially stopped. Testing of the leachate
in the cells showed no inhibitive constituents. The preliminary analysis seems to suggest that the high temperature
has resulted in this curtailment of the gas production.

The temperatures for these cells is well within the
thermophilic range, but may have exceeded the temperatures
where stable methane production in a low nitrogen content
substrate occurs.

During the past two winters, significant infiltration of
rain waters has occurred into Cells A and C. This has resulted in an inundation of more than three quarters of the
cell mass in each cell. This vast amount of water may have
not only stimulated the generation of heat but provided a
medium for heat retention.

Cells B and E have comparatively low but similar production curves. However, the average production over the
three-year period is at a rate previously thought by many
experts in the field to be moderate to high. In the future,
attempts to lower temperature by water addition will be considered in order to determine whether resumption of gas production can be accomplished.

The temperatures in Cells D and F remained relatively
constant in the last year while gas production remained high.
This has resulted in landfill gas production of 0.83 to 0.88
ft^3/lb (0.053 to 0.056 m^3/kg) from these cells. The total
gas production from both cells exceeds one and one-half ft^3/lb
(0.095 m^3/kg), and methane production has been recorded at
0.81 ft^3/lb (0.051 m^3/kg). These sustained rates are extremely high. Cell D achieved a stable methane concentration
above 50% in less than 300 days, as did all cells that received buffers. Cell F did not achieve stable methane concentration above 50% until 400 days but has maintained a gas
production rate exceeding 0.4 cu ft/lb-yr (0.025 m^3/kg-yr)
for the last 18 months. During that period, the gas production
of both cells has been equivalent.

The results of the first two years of management and
monitoring were very encouraging with respect to enhancement.
However, the cessation of gas production apparently associated
with high temperatures poses significant challenges during
the fourth year of the project.

All cells which received buffer addition initially showed not only high rates of landfill gas production, but also high methane concentrations, apparently demonstrating the synergistic, if not basic value of pH control. Also demonstrated at that time is the value of limited moisture addition. The value of leachate recirculation, one of the specific management procedures under study, is not clearly indicated. Due to pump malfunctions and maintenance problems, leachate recirculation occurred intermittently for only about 35% of the test period.

It should be stressed that the conclusions presented in this paper are based on preliminary data; only long-term evaluation of the cells can develop sufficient data to support more definitive conclusions.

32
CHEMICAL AND MICROBIOLOGICAL ANALYSIS OF AVELEY LANDFILL, ENGLAND

J.F. Rees

BioTechnica Limited
Cardiff, United Kingdom

ABSTRACT

This paper describes investigations at Aveley Landfill, England which sought to relate very high rates of gas production with the chemical and microbiological conditions encountered at the site. Ideas will be presented which seek to account for the natural establishment of enhanced gas production at this site. Investigative methods will also be discussed. This site investigation forms part of an extensive program of landfill characterization which sought to define the microbiological and chemical conditions present in a range of different landfill types. Ideas for controlling landfill reactions are presented, these include the controlled addition of appropriate microbes at various stages in the landfill's lifetime.

INTRODUCTION

The efficiency of methane production from MSW landfills is uncertain. However, the sheer mass of waste in place and the high organic content of the waste often makes utilization of the gas a commercially attractive proposition.

The current status of gas utilization from US MSW landfills is that about 20 schemes are in operation with 10 large projects. The total gas produced is about 40 million ft^3/day giving an annual revenue of about $30 million.

Landfill technology is extremely important to our society from both an environmental and energy standpoint. This is particularly so, as in most Western countries more than 80% of all MSW is disposed to landfill.

The last 15 years has seen a dramatic increase in our knowledge of the polluting potential of landfills through groundwater contamination from leachate, odor problems, and methane hazards. Also we have recognized that this polluting potential can be controlled and largely eliminated by a combination of microbiological and engineering approaches.

The major thrust of current R & D worldwide is to

1. develop a better understanding of landfill processes,
2. quantify more accurately the rate, duration and yield of gas production, and
3. develop methods to control microbial processes in landfills.

Chemical analysis of leachate and the physical analysis of domestic refuse have been the most common measurements made on landfills. Data on leachate composition in particular have shown that a wide range can be expected from one landfill to another (Pohland and Engelbrecht, 1976) and with the age of the landfill (Stegmann, 1982). However, although important, these measurements alone are inadequate to describe the biochemical behavior of landfill and do not adequately account for the differences between sites. Prediction of biological and chemical reactivities in landfill are complicated by the heterogeneous nature of both the physical and chemical environment of the landfill itself.

Because of the high solids content of landfills a more comprehensive chemical examination of the liquid solid and gaseous phases of landfill was warranted. This would be correlated with the water distribution in the landfill together with rates of microbial activity as measured by rates of methane production and various enzymic activities responsible for the breakdown of organic polymers, i.e., protein, starch and cellulose. This program was initiated in 1980 at the Aveley landfill site, England, and subsequently continued at a variety of landfills which were known to have been engineered in different ways or to have other particular characteristics, i.e., received pulverized/shredded waste/'trade' waste or liquid industrial waste additions, for example.

This paper examines data from the Aveley landfill in England. This landfill is particularly interesting as it can be considered that 'gas enhancement' has occurred naturally. Thus, it is an important model from which to derive information for the design and construction of enhanced gas producers in the future. Detailed studies have been carried out on this landfill and are reported by Rees, 1980; and Rees and Viney, 1982.

METHODS

Boreholes were drilled with a percussion rig. Representative samples of refuse were collected at approximately 1.5 m intervals, placed in airtight plastic containers ~10 ℓ after temperature determination. Gas production was measured from selected samples by incubating the refuse in 1.5 ℓ gas tight jars. Volumes of gas produced were measured by displacement of an acidified solution of sodium chloride, and gas composition by gas chromatography.

Liquid samples were obtained from the refuse by centrifugation and chemical analyses determined as reported by Rees and Viney (1982).

RESULTS

Leachate Composition

The chemical composition of the leachate generally found in the water saturated base of Aveley landfill is shown in Table 1. These data indicate a very bioreactive landfill with particularly low concentrations of carboxylic acids in solutions.

Table 1. *Leachate Composition at Aveley Landfill* *(all concentrations mg.* l^{-1} *except pH)*

pH	8.1
TOC	850
volatile fatty acids	all < 30
SO_4	120
Cl	2600
Na	1850
K	780
NH_4	1400

This reflects the efficient conversion of these intermediates to CH_4 and CO_2. Low concentrations of carboxylic
acids result in pH values being greater than 7.5 and frequently greater than pH 8. The highly bioreactive nature of
the landfill is also reflected in the low sulphate concentrations, generally between 30 and 180 mg l^{-1}. Phosphate and
sulphide ions are rapidly removed from solution at the pH
values found in this landfill and in the presence of high
Fe^{++} concentrations.

Landfill Characterization - Borehole H5

Leachate composition and gas production rates from refuse
samples obtained from borehold H5 are shown in Figs. 1-4.

Carboxylic acid concentrations are clearly influenced by
the water table (Fig. 1) with concentrations of approximately
40,000 mg. l^{-1} in the water-unsaturated refuse, 3-6 m below
the landfill surface, falling dramatically to 10 mg. l^{-1} at
8.5 m below the surface in the water saturated part of the
landfill. Throughout this borehole, the predominant carboxylic acids are acetic, propionic and butyric (Fig. 2).
The chloride profile shows that dilution makes an insignificant contribution to the concentration reduction. Sulphate
concentrations, like the carboxylic acids show a dramatic fall
at the water table. As might be expected, the pH profile is
inversely related to the carboxylic acid profile. All these
data are compatible with microbial utilization of carboxylic
acids and sulphate in energy metabolism and incorporation of
K^+ and NH_4^+ into the microbial biomass.

Figure 3 shows the distribution of certain metal ions
throughout the depth of the landfill in borehole H5. Significant changes occur at the water table, which cannot be
ascribed to dilution effects. Thus the Na^+ concentration of
3000 mg. l^{-1} in the unsaturated refuse, above the water table,
only falls to 1600 mg. l^{-1} below the water table. Peak concentrations of metals in solution occur 4 m below the landfill
surface in the unsaturated refuse. Concentrations of Mn^{2+},
Ca^{2+}, and Fe^{2+} then fall by 2 to 3 orders of magnitude below
the water table. Mg^{2+} and Co^{2+} concentrations fall by 1 order
of magnitude. These changes in concentration are associated
with the pH rising from 5.3 at 4 m to 8.5 at 10 m below the
surface. Figure 2 shows the dramatic increase in rate of gas
production at the water table which coincides with the removal
of carboxylic acids from the leachate. The peak in gas production at the water table is to be expected in a system where

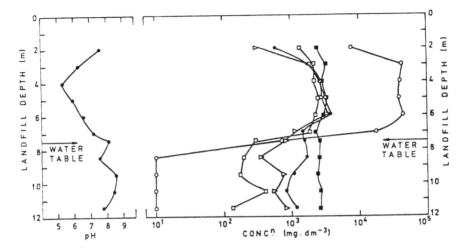

Figure 1.   Aveley landfill site, borehole H5, 1980. Distribution of total carboxylic acids, C2-C6 o; NH_4^+ ● ;
SO_4^{2-} ☐ Cl^- ■ ; K^+ Δ; and pH with depth in the
landfill.

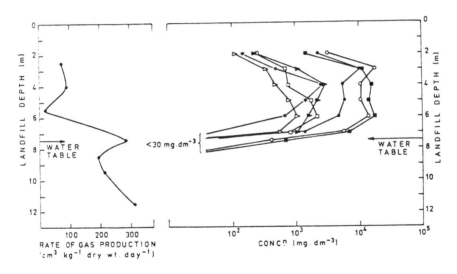

Figure 2.   Aveley Landfill Site, borehole H5, 1980. Rate of
gas production (CH +CO) and distribution of carboxylic acids with depth in the landfill. Acetate
O; Propionate ●; Isobutyrate ☐ ; Butyrate ■; Isovalerate Δ; Valerate ▲; Caproate *.

a gradually rising water table continually encounters new refuse. This peak reflects the greater availability of soluble
fermentable proteins and starch in the previously unwetted
refuse (Table 2).

Total gas yields from various refuse samples range from
0.5 to 13.1 Kg^{-1} refuse (Fig. 4). These yields are significantly lower than expected from fresh refuse. These lower
yields from Aveley are to be expected and in fact become
indicators of the extent of the decomposition process in the
landfill. Thus gas yields generally decrease with depth in
the landfill.

DISCUSSION

Aveley is a very bioreactive landfill as judged by the
criteria of gas production rates, leachate composition and
refuse solids composition. This is to be expected as the
landfill possesses a high water content. However, it is also
clear that at Aveley the removal of volatile fatty acids from
leachate is well established. This can also be correlated
with lower $SO_4^=$ concentrations. In fact, the Aveley leachate
is most unusual in containing virtually zero concentration of
volatile fatty acids, Aveley temperatures ranging from 37-45°C
are also high and suitable for good microbial activity.

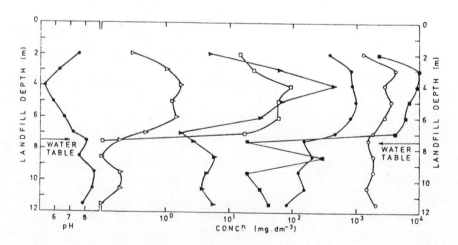

Figure 3. Aveley Landfill Site, borehole H5, 1980. Distribution of Na^+ o; Mg^{2+} ●; Mn^{2+} ◻; Ca^{2+} ■; Co^{2+} △;
Fe^{2+} ▲; and pH with depth in the landfill.

Table 2. *Chemical Composition of Refuse*
Aveley Landfill (% dry wt)

Borehole Depth	Lignin	Cellulose	Cellulose/Lignin	Starch	Starch/Lignin	Protein	Protein/Lignin
Borehole 5							
2–3 m	7.5	28.2	3.75	0.8	0.1	2.8	0.4
4–5 m	8.1	42.2	5.2	1.0	0.1	3.1	0.4
7.5 m	7.3	3.1	0.4	0.4	0.05	1.7	0.2
8.5 m	6.8	3.3	0.5	0.2	0.03	0.8	0.1
9.5 m	7.8	5.6	0.7	0.3	0.04	1.0	0.1
10.5 m	4.1	2.0	0.55	0.5	0.1	1.5	0.4

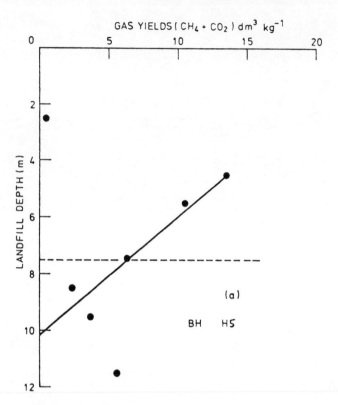

Figure 4.   Total gas yields (dm^{3}kg^{-1} dry refuse) from refuse
samples obtained from BH H5, Aveley Landfill.

How can these biochemical observations be accounted for
in terms of landfill engineering, and what lessons are there
for the future evolution of landfill practices? It is, of
course, assumed that an "Aveley type" leachate is environ-
mentally desirable. It is also recognized that although a
high landfill water content gives rapid gas production, it
also presents physical problems during gas extraction.

At Aveley, refuse is placed with a compactor vehicle,
pushing refuse over a 2 m face to give an initial density of
~0.65-0.7 tonnes m^{-3}. The entire 18 ha area of the site is
used in the operation and only a small amount of soil cover
material is used. This operational method encourages the
generation of large volumes of leachate from incident rainfall. In addition there is evidence of groundwater intrusion
into the site resulting in the leachate level rising in the
landfill at the rate of 2 m per year. This also results in
considerable lateral movement of water. This flow of water

into the base of the landfill, and the absence of major intermediate clay covers favors the establishment of a homogeneous
water saturated landfill at ~7.5 m below the refuse surface.
The refuse age in borehole H5 varied between 1 year at the top
and 5 years at the 12.5 m depth.

It now seems that the early initiation of methanogenesis
and the concomitant development of a good quality leachate at
Aveley can be accounted for in terms of landfill engineering.
Major factors favoring the establishment of this bio-reactive
landfill are the following:

1. Refuse placed with minimal intermediate soil cover.
2. Refuse placed in a large operational area so that
    it was exposed to aerobic conditions at the surface
    for 9 months to 1 year before being covered with
    more refuse.
3. The lateral movement of groundwater in the site.
4. Upward migration of groundwater/leachate from the
    base of the site.

These conclusions are based on various field, lysimeter, and
laboratory studies which showed the importance of partial
aerobic processes and water movement on the rapid initiation
of methanogenesis (Rees, 1981; Stegmann and Ehrig, 1982;
Klink and Ham, 1982).

Where it might be difficult or inappropriate to implement
some of these strategies on some landfills there remains the
potential for the direct introduction of appropriate landfill
microbes early on in the history of a landfill. This strategy
will achieve the same end product and in overall management
terms may well be easier to implement. BioTechnica Limited
is actively pursuing strategies of controlled introduction of
microbes to landfills to achieve a wide range of objectives.
Clearly the previous work on understanding landfill processes
has not been most important in providing a firm foundation for
this emerging technology.

REFERENCES

Buivid, M.G., D.L. Wise, M.J. Blanchet, E.C. Remedios, B.M.
    Jenkins, W.F. Boyd, and J.G. Pacey, *Resources and Con-
    servation, 6*, 3-20 (1981).
Klink, R.E. and R.K. Ham, *Resources and Conservation, 8*,
    29-41 (1982).

Pohland, F.G. and R.S. Engelbrecht, "Impact of Sanitary Land-fills," *Report for American Paper Institute,* NY (1976).

Rees, J.F., *J. Chem. Tech. Biotechnol., 30,* 458-465 (1980).

Rees, J.F., "Major Factors Affecting Methane Production in Landfills," *Landfill Gas Symposium,* Harwell Laboratory, Oxon, UK (1981).

Rees, J.F. and I. Viney, *AERE Report No. R - 10328,* Her Majes-ty's Stationery Office, London (1982).

Stegmann, R., "Design and Construction of Leachate Treatment Plants in W. Germany," *Landfill Leachate Symposium,* Harwell Laboratory, Oxon, UK (1982).

Stegmann, R. and H-J. Ehrig, "Enhancement of Gas Production in Sanitary Landfill Sites," *Proc. Resource Recovery from Municipal and Hazardous and Coal Solid Wastes,* Miami, FL (1982).

INDEX

HAZARDOUS CHEMICALS DATA BOOK
Second Edition

Edited by

G. Weiss

This Second Edition of the *Hazardous Chemicals Data Book* supplies instant information on more than 1000 of the most important hazardous chemicals. The data given will provide rapid assistance to personnel involved with the handling of hazardous chemical materials and related accidents.

The compilation of hazardous chemicals is presented in a clear, concise, easy-to-locate format. It should be an indispensable source book for any library or laboratory. It is intended for use by scientists, engineers, managers, transportation personnel, and anyone else who might have contact with, or require safety data on, a particular hazardous chemical.

A large quantity of pertinent data is given for each of the chemicals, which are arranged alphabetically. **Examples** of types of properties and safe-handling-information provided are listed below.

COMMON SYNONYMS

FIRE, EXPOSURE, AND WATER POLLUTION EFFECTS

RESPONSE TO DISCHARGE
Containment
Warnings to Be Issued

LABELING
Classification
Code

CHEMICAL DESIGNATIONS
Compatibility Class
Formula
IMO/UN Designation
DOT ID No.
CAS Registry No.

OBSERVABLE CHARACTERISTICS
Physical State, Color, Odor

HEALTH HAZARDS AND TREATMENT
Personal Protective Equipment
Symptoms Following Exposure
Treatment of Exposure
Threshold Limit Value (TLV)
Short Term Inhalation Limits

Toxicity by Ingestion
Late Toxicity
Vapor (Gas) Irritant Characteristics
Liquid or Solid Irritant Characteristics
Odor Threshold
IDLH Value

FIRE HAZARDS
Flash Point
Flammable Limits in Air
Fire Extinguishing Agents
Special Hazards of Combustion
Behavior in Fire
Ignition Temperature
Electrical Hazards
Burning Rate
Adiabatic Flame Temperature
Stoichiometric Air to Fuel Ratio
Flame Temperature

CHEMICAL REACTIVITY
With Water
With Common Materials
Stability During Transport
Neutralizing Agents
Polymerization
Polymerization Inhibitor
Molar Ratio
Reactivity Group

WATER POLLUTION
Aquatic Toxicity
Waterfowl Toxicity
Biological Oxygen Demand (BOD)
Food Chain Concentration Potential

SHIPPING INFORMATION
Grades of Purity
Storage Temperature
Inert Atmosphere
Venting

HAZARD ASSESSMENT CODE

HAZARD CLASSIFICATIONS
Code of Federal Regulations
NAS Hazard Rating for Bulk Water
  Transportation
NFPA Hazard Classification

PHYSICAL AND CHEMICAL PROPERTIES

ISBN 0-8155-1072-1 (1986)   8½″ x 11″   1068 pages